EXPERIMENTAL FLUVIAL GEOMORPHOLOGY

EXPERIMENTAL FLUVIAL GEOMORPHOLOGY

Stanley A. Schumm
Colorado State University
Fort Collins, Colorado

M. Paul Mosley
New Zealand Ministry of Works and Development
Wellington, New Zealand

William E. Weaver
National Park Service
Arcata, California

A Wiley-Interscience Publication
JOHN WILEY & SONS
New York Chichester Brisbane
Toronto Singapore

Library of Congress Cataloging in Publication Data:

Schumm, Stanley Alfred, 1927–
Experimental fluvial geomorphology.

"A Wiley-Interscience publication."
Bibliography: p.
Includes index.
1. Watersheds—Experiments. 2. Alluvium—Experiments. 3. Erosion—Experiments. 4. Geomorphology—Experiments. I. Mosley, M. Paul. II. Weaver, William E. III. Title.
GB 561.S34 1986 551.4 86-22383
ISBN 0-471-83077-1

Printed in the United States of America

10 9 8 7 6 5 4 3 2 1

Preface

Beginning in 1969 a series of experimental studies of drainage basin evolution and hydrology, river-channel and valley morphology and behavior, and alluvial-fan morphology and sedimentology were performed at the Engineering Research Center of Colorado State University (CSU). Some experiments were performed to generate and test hypotheses, and others simply to document the morphology and evolution of model landforms. Numerous papers have presented the results of these experiments, which were undertaken by students working toward advanced degrees in geology, watershed science, and civil engineering, but much of the work remains unpublished, and some results have appeared only in a very abbreviated form. Therefore, a major objective of this book is to make the results of these experiments more generally available. We have attempted to assemble and integrate the experimental results, both published and unpublished, and to relate the experimental findings to the prototype ''real world.'' The emphasis is, of course, on landforms but landforms cannot be isolated from hydrologic and sedimentary processes; therefore, in addition to the chapters that deal with drainage basins, rivers, and alluvial fans, there are chapters on drainage basin hydrology, alluvial fan sedimentology, and placers. The book is organized in three parts, which deal with the morphology and dynamics of (1) drainage basins, (2) rivers and valleys, and (3) alluvial fans and sedimentary deposits.

Experimental work on rivers has been international and extensive, but much less has been accomplished regarding drainage basins, alluvial fans, valleys, and sedimentary deposits, and it is in these areas that the CSU contribution may be most significant. We have attempted to relate the CSU experiments to other work, although our review of the literature dealing with experimental studies is not exhaustive. In particular, an apparently considerable Russian and Chinese literature was not accessible to us.

STANLEY A. SCHUMM
M. PAUL MOSLEY
WILLIAM E. WEAVER

Fort Collins, Colorado
Wellington, New Zealand
Arcata, California
January, 1987

Acknowledgments

In addition to former students and colleagues who permitted us to use the product of their research efforts, we acknowledge the support of the program directors of the National Science Foundation, Richard Ray, John Lance, and John Maccini, and the U.S. Army Research Office, Finn Bronner and Steven Mock, who, depending on one's evaluation of this effort, were farsighted enough or deluded enough to support the research. We thank the National Science Foundation, the U.S. Army Research Office, Colorado State University, and the New Zealand Ministry of Works and Development for support in preparing this document. Water Engineering and Technology, Inc., of Fort Collins generously provided free word-processor time. We are grateful to Stuart Savig and William J. Spitz for drafting and to Nancy L. Hettinger for typing numerous versions of the text. Chapters relating to research that they performed as graduate students have been reviewed by R.S. Parker, C.D. Lidstone, Z.B Begin, T.W. Gardner, R.G. Shepherd, D.L. Macke, and M.L.W. Jackson. In addition, Dru Germanoski generously reviewed the penultimate draft, and John Costa reviewed the final manuscript. We thank them for this significant contribution to our effort.

This book is based on work supported by the U.S. Army Research Office and the National Science Foundation under Grant No. EAR-8211592. Any opinions, findings, and conclusions or recommendations expressed in this publication are those of the authors and do not necessarily reflect the views of the U.S. Army Research Office or the National Science Foundation.

S.A.S.
M.P.M.
W.E.W

Contents

EXPERIMENTAL FLUVIAL GEOMORPHOLOGY

1 Introduction

Throughout the history of geomorphology, the changing form of the landscape with time has been a topic of primary consideration. However, due to the short time available to the investigator, models of landscape evolution have depended largely on deductive reasoning (Cotton, 1941; Davis, 1909; King and Schumm, 1980), on measurements of erosion in restricted areas of rapid erosion, on the extrapolation of empirical relations (Schumm, 1984, 1985), and on ergodic assumptions (substitution of location for time) (Paine, 1985). The evolution of hillslope profiles in two dimensions has been extensively considered using mathematical models (Ahnert, 1970; Carson and Kirkby, 1972; Scheidegger, 1970) and, more recently, data on rates of erosion and sediment transport by different processes have become sufficiently reliable and voluminous (Saunders and Young, 1983) that computer simulation of landscape evolution in three dimensions has been attempted (Ahnert, 1976; Bridge, 1977). Nevertheless, all approaches, including the apparently most rigorous, make many simplifying assumptions, and divergent opinions on the origin and evolution of a landform are common.

In order to understand a large, complex, and slowly evolving geomorphic system the earth scientist may resort to study of a smaller and simpler analog. For example, Tanner (1960) and Davies and Tinker (1984) noted the similarity between a thread of water winding down a sheet of glass and a meandering river; Romey (1982) saw glaciers in his oatmeal; and the senior author once refused to let his daughters eat a dish of chocolate pudding for two days because, as it desiccated, fractures that resembled lunar features were forming on its surface. Another strategy is to compress time by studying very rapidly eroding natural landforms such as badlands (Brown, 1983; Howard and Kerby, 1983; Schumm, 1956). A logical combination of these two strategies is to construct an analog of the landform of interest that is sufficiently small, simplified, and rapidly evolving to permit close study. Such analogs, termed physical models by Ward (1971, p. 119) and hardware models by Chorley (1967) and Mosley and Zimpfer (1978), provide the opportunity to observe and measure the form and evolution of small landforms that are analogous to features of the earth's surface. Stream tables and similar devices have long been used as teaching aids (Ellis, 1912; Tarr and von Engeln, 1908), and even the simplest experiment will generate questions and hypotheses that may lead to a new understanding of the form and evolution of the earth's surface.

Although this book deals with fluvial geomorphology, earth scientists have used experimental techniques to study a variety of geologic phenomena (Woldenberg, 1985), such as igneous intrusion (Howe, 1901), diatreme mechanics (Woolsey, McCallum, and Schumm, 1975), tectonics (Ramberg, 1981), landslides (Seed and Wilson, 1967), volcanism (Wohletz and McQueen, 1984), coastal erosion (Sunamura, 1982), moon craters (Gilbert, 1893; Mills, 1969; Schumm, 1970; Wegener, 1921), Mars channels (Schumm, 1974), eolian processes on Mars (Greeley et al., 1974), rainsplash transport of sediment (Mosley, 1973), and impact cratering (Fink et al., 1981; Piekotowski, 1980).

Daubrée (1879) was one of the first to perform extensive geomorphologic experiments, and his book describes experiments involving a wide range of geologic features. He realized that there were major problems associated with experimental work, but he suggested that the primary value of his experiments was to generate hypotheses, which could then be tested by reference to natural landforms. Several other geologists working at the turn of the century used laboratory models to reproduce natural landforms or to examine theories of landform development. For example, Hubbard (1907) modeled drainage networks by placing a mud slurry on a glass plate, which was tilted to develop small rill systems; Howe (1901) experimented with the erosional evolution of a model laccolith, a domed surface of sand and marble dust that was sprayed with water; and Jaggar (1908) reproduced meanders, stream piracy, deltas, and drainage networks in the laboratory. Both the value of such work and the problems associated with experimentation are indicated by Jaggar's remark (1908, p. 300) that "the foregoing experiments suggest many questions and answer few."

Davison (1888a, b), on the other hand, employed experimental techniques to test hypotheses regarding the movement of scree material that were first generated by field observation. Observing the movement of slate fragments on a quarry waste heap, he assumed that movement was caused by expansion of the stones due to heating by the sun, and he set up experiments to measure this effect. He verified his hypothesis and then made estimates of the rate of denudation associated with rock creep on typical scree slopes.

A famous example of a laboratory study is Gilbert's (1917) work on the transportation of debris by running water. The data collected by Gilbert during a series of experiments have been a source of information and have been used repeatedly to test models of sediment transport.

In Germany Würm (1935, 1936), aware of this earlier work in the United States, applied water with a vineyard sprayer to a small (70 cm^2) surface in order to investigate the erosional development of "stepped land" (*schichtstüfenlandschaft*) resulting from the erosion of horizontal beds of differing resistance. Würm (1935) and Gavrilovic (1972) also studied slope evolution and the development of river profiles experimentally.

The early studies point to the many ways that experimental studies are of use to the geomorphologist. First, they provide a physical respresentation of the form and evolution of a landform that, because of its small size and rapid development, is comprehended more readily than the prototype, and, in addition, relationships or features may be noticed that would otherwise have remained hidden. Second, the model is a source of hypotheses about the behavior of the natural landscape, which may then be tested by fieldwork or further experiment. Third, hypotheses developed as a result of field observations may be tested experimentally under controlled conditions. Fourth, measurements made in the laboratory of geomorphic processes may be used to estimate the manner and rate of change of natural landscapes. Finally, physical processes that are obscure and sometimes dangerous in the field may be observed in the laboratory.

EXPERIMENTAL GEOMORPHOLOGY

Experimental geomorphology may be defined as the study, under closely monitored or controlled experimental conditions, of a physical representation or model of a selected geomorphic feature (Mosley and Zimpfer, 1978). According to Flueck (1978), there are two types of experiments, exploratory and confirmatory. The exploratory experiment is an attempt to determine the existence of a relation between events, and the confirmatory experiment is designed to test a hypothesis.

Chorley (1967) identified three broad classes of physical models, namely, segments of unscaled reality, scale models, and analog models. The 3.3-km-long reach of the East Fork River, Wyoming, studied by the U.S. Geological Survey (Andrews, 1979), is an example of a closely monitored segment of unscaled reality that is taken as representative of a wide range of physical objects and that is particularly suited to study. Alternatively, the model may be a replica of a natural landform, scaled in such a way that ratios of significant dimensions and forces are equal to those in nature, although their absolute magnitudes may be greatly different. Scale models need not necessarily reproduce a specific prototype, but they may be a scaled-down version of a general class of geomorphic features. Finally, a physical analog model may reproduce some significant aspect of the form and function of a natural phenomenon, but the forces, materials, and processes may be quite dissimilar to those in nature. An example is the use of a thread of water winding down a glass plate to simulate a meandering river with erodible bed and banks (Davies and Tinker, 1984; Gorycki, 1973; Tanner, 1960).

The monitoring of segments of unscaled reality is a widely used approach, and it has a long history. A particularly fruitful use has been by agricultural engineers who, by collecting runoff and sediment yield data from field plots having a variety of soils, cover types, and management regimes, have arrived at a good understanding of surficial erosion processes on hillslopes. This has led to the development of models such as the Universal Soil Loss Equation (Wischmeier and Smith, 1965). The scale at which this type of study is carried out has varied widely about the (21–27-m) plot size commonly used by North American agricultural engineers. At one end of the scale, Kirkby (1967) removed an 80 × 30 × 27-cm block of soil to the laboratory for minutely detailed observation; at the other end, the paired or multiple watershed approach to the study of drainage basin hydrology, erosion, and sediment dynamics may involve watersheds that are several hectares in area, which must perforce be studied at a coarser scale (Ward, 1971).

Although many watershed studies are referred to as experimental, they are not unless manipulation of the watershed characteristics (e.g., vegetation cover or managemnt) is involved (Church, 1984; Slaymaker et al., 1980). Thus there is a gradation from true experimental work to standard observation and measurement of geomorphic processes.

Scale modeling is also widely used and has a long history (Henderson, 1966; Johnson, 1970). Its fundamental characteristic is that the spatial and temporal dimensions of the model are manipulated according to a stringent set of requirements to maintain a known quantitative relationship—or similarity—between the model and the prototype it represents. Three types of similarity should be maintained for the model to precisely reproduce the prototype: (1) geometric similarity, in which the ratios of homologous dimensions are equal and equivalent angles are the same; (2) kinematic similarity, in which paths and patterns of motion are geometrically similar to those of homologous occurrences in the prototype; and (3) dynamic similarity, in which the ratios of homologous forces and masses affecting motion are equal at all times (Thornton and Romer, 1975). Procedures for scale modeling have been progressively refined, particularly in the field of hydraulics and river engineering, but experience, skill, and artistry are necessary for success. This is particularly the case where the shape of the model is free to evolve with time (i.e., in "movable-bed" models), and Henderson (1966) even suggested that "good fortune" played a part in the success of Reynolds's (1887) pioneering work.

As Albertson et al. (1960) pointed out, "complete similarity in the model is sometimes difficult or impossible to obtain. In these cases it is frequently possible to obtain only a partial (qualitative) or approximate answer by means of a model." Moreover, Warnock (quoted in Rouse, 1950, p. 165) commented that

> during cut-and-try verification tests it may be found necessary to manipulate the time scale, discharge scale, rate and manner of bed-load feeding, perhaps the slope scale . . . , and perhaps the gradation of the bed material. Often it is necessary to use time and discharge scales which vary with stage in order to make each model stage produce its proportional share of bed movement.

One problem related to the maintenance of scaling relationships between model and prototype is the strength of the analogy between the material of which the model is constructed and the composition of the prototype (Hubbert, 1944). Formal scale modeling principles can be used to define the size and specific gravity of the sediment required in a movable-bed hydraulic model (Henderson, 1966); coal dust or plastic granules of various sizes are commonly used to model bed sediment in sand-bed rivers. However, most landforms are at least partially composed of material that is cohesive, and its behavior is not controlled simply by grain size and specific gravity. There are no formal rules for selecting model material, the erodibility and transportability of which has a known quantitative relationship to that of the prototype material, whether it be the cohesive bank of a meandering river or the type of rock underlying a watershed.

In addition to the difficulty of maintaining scaling relationships between model and prototype, experimental work using scale and analog models suffers from uncertainties introduced by initial and boundary conditions. For example, in an experimental drainage basin rigid sidewalls form an unyielding boundary to the watershed, whereas in nature neighboring drainage basins are free to compete, and their divides may shift.

Barr (1968) considered that "once true dynamic similarity is lost, a hydraulic model becomes a hydraulic analogue." In this case, extension of quantitative relationships from model to prototype may not be strictly possible. Because formal scale-modeling procedures may be difficult to follow, too restrictive, and unsuited to geomorphologic studies when information on general principles rather than specific prototype is required, Hooke (1968) proposed a fourth, informal, similarity criterion, "similarity of process" or "similar performance" (Chery, 1969). Basic requirements for his process-similar study are that (1) gross scaling relationships be met, (2) the model reproduce some morphologic characteristics of the prototype and (3) the process which produced this characteristic in the experiment can logically be assumed to have the same effect on the prototype (Hooke, 1968, p. 392). In effect, Hooke recommended that the experimental model be regarded as a small system in its own right. Amorocho and Hart (1965) designated such a model a prediction-analysis prototype.

There is increasing evidence that at least some geomorphic processes are scale dependent (Schumm, 1985; Wolman and Gerson, 1978), so that extrapolation may lead to errors in interpretation if gross scaling relationships are not met (Eagleson, 1969). Clearly, scale modeling of landforms is too inexact a science at present to provide reliable quantitative results that can be applied directly to natural systems (Amorocho and Hart, 1965; Chery, 1969; Hooke, 1968; Woolhiser, 1973), but Hooke's (1968) similarity-of-process procedure offers geomorphologists a straightforward approach to laboratory studies. Therefore, the experimental drainage basins, stream channels, and alluvial fans, as discussed later, are simply considered to be small landforms. This prevents the direct extrapolation of results from the analog model to the prototype. However, similarity-of-process studies can provide new information regarding processes, reveal trends not obvious during field studies, and provide a basis for hypothesis generation. The similarity-of-process approach allows application of experimental results to natural systems, although such comparisons should be made with caution.

Mosley and Zimpfer (1978) reviewed the advantages and disadvantages of such experimental research. They considered the major advantages of experiments to be as follows:

1. They permit the identification, isolation, manipulation and precise measurement under controlled conditions of processes and variables that, for one reason or another, cannot be investigated in the field.

2. They permit the study of evolving geomorphic systems, of the differences between equilibrium and non-equilibrium systems, and of the implications of stage of evolution on the distribution of energy and matter within a system.

3. They allow several processes or aspects of the landscape to be examined in a single study.

4. They permit the study of various boundary and initial conditions.

5. Careful observation of hardware models may reveal hitherto unsuspected phenomena and open new lines of enquiry.

6. They provide easy visualization of geomorphic phenomena and thereby aid understanding and education.

They consider the major disadvantages to be as follows:

1. Initial and boundary conditions in the model may not be analogous to those in nature, or may influence model behavior to an indeterminate or undesirable extent.

2. Materials and processes in the model may be dissimilar to those in nature, and there may be no obvious way of relating model behavior (e.g., rate of evolution) to that of the prototype. In particular, it is difficult to relate model behavior under constant rates of operation of processes (constant energy and material input) to prototype behavior, where highly variable rates of operation of processes are usual.

3. Study of only one or two processes or independent variables may mask interactions that occur in nature.

4. As model size decreases, there is a trade-off between precision of measurement and observation, and accuracy of representation of the prototype (realism). Confidence in the model results may therefore decline.

5. Hardware models can be time-consuming and tedious to operate, expensive to build, and demand much physical labour.

6. They do not have the fashionable aura of some other techniques.

7. They cannot be easily stored, and

8. They cannot be the final step in the development of a theory.

Minshull (1975, p. 122) pointed out that there is a danger of oversimplification in model work, and Beveridge (1957, p. 65) suggests that experiments may be misleading:

> There is an interesting saying that no one believes an hypothesis except its originator but everyone believes an experiment except the experimenter. Most people are ready to believe something based on experiment but the experimenter knows the many little things that could have gone wrong in the experiment.

If geomorphic experiments are viewed simply as studies of small landforms, then it is information about small landform evolution that is obtained. The major difficulty is to transfer the experimental results by analogy to larger landforms. Confirmation of the experimental results must come from field investigations, and frequently this can be done by considering existing field data from the new perspective that has been provided by the observation of the small landforms as they evolve and react.

In conclusion, physical models can be used in the generation and testing of hypotheses and for predictive and descriptive purposes. However, the inherent uncertainties in the relationships between the models and their prototypes always require field validation before the results can be utilized with confidence for prediction or postdiction (Schumm, 1985). In spite of the significant problems associated with the design and prosecution of experimental studies of landforms, they can provide an insight into landform evolution and dynamics that can be obtained in no other way.

I DRAINAGE BASIN MORPHOLOGY AND DYNAMICS

The drainage basin has been described by Chorley (1967) as the fundamental geomorphic unit. Within this unit the channel network is of premier importance because it evacuates the water and sediment produced within the drainage basin. Understanding the morphology and dynamics of drainage basins and channel networks is elusive because of their complexity, but experimental studies provide a glimpse of the system in operation that aids the fuller understanding of the prototype.

In Part I, the evolution of drainage networks (Chapter 2), some aspects of their hydrology, sediment yields, and placer formation (Chapter 3), and drainage basin behavior (Chapter 4) will be considered.

2 The Drainage Network

The drainage network occupies only a small part of a drainage basin, but it has been the subject of great geomorphic and hydrologic interest, especially since the publication of Robert Horton's (1945) seminal paper. The techniques developed by Horton for quantitative description of a drainage network opened the way for the study and understanding of this complex geomorphic system.

Horton's approach to network description was employed and expanded by the "Columbia School" of A. N. Strahler and his students, and many studies of drainage basins and stream networks have followed the lead of Horton and Strahler. Reviews of this work have been published by Abrahams (1984), Dunne (1980), Gardiner and Park (1978), Jarvis (1977), and Smart (1972).

MODELS OF DRAINAGE NETWORK GROWTH

Horton (1945) not only pioneered the quantitative description and analysis of channel networks and established "laws" of network composition, but he also proposed a model of network growth by overland flow. Horton suggested that on a steep, newly exposed surface a series of parallel rills develop and that, with time, cross-grading and micropiracy among these rills produce an integrated dendritic network (Fig. 2.1*A*). A second model of network growth is the headward growth type (Howard, 1971a; Smart and Moruzzi, 1971b) in which the drainage network develops fully at the edge of an undissected area (Fig. 2.1*B*). That is, as the channels grow headward and bifurcate, they fill the space available and form a fully developed dendritic network. In a third model, suggested by Glock (1931), the drainage area is rapidly subdivided by channels and the addition of tributaries then fills the available space (Fig. 2.1*C*).

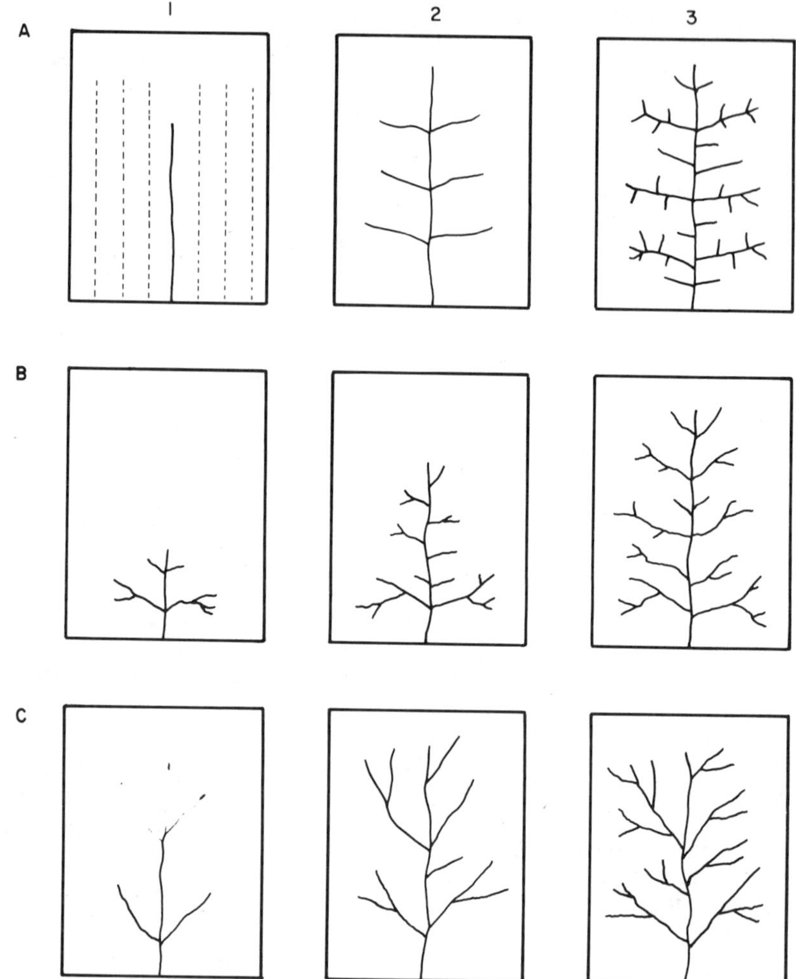

FIGURE 2.1. Models of drainage network growth. (*A*) Replacement of parallel rills by an angular dendritic pattern (see Fig. 2.5). (*B*) Expansion by slow headward growth into available area. (*C*) Extension of rapidly growing, long channels that block out drainage areas. Tributaries are added subsequently to fill in drainage networks.

Thus three very different models of network growth which may represent end members of a continuum of different growth types, have been proposed. At one extreme is the Horton model (Fig. 2.1*A*), in which parallel channels develop almost instanteously over the surface, and the final pattern of the network progressively emerges by internal changes (capture) and replacement of the initial pattern of rills. At the other extreme is the headward-growth model in which a "wave of dissection" (Howard, 1971a; Schumm, 1956) can be envisioned at the tips of first-order channels, which is the active zone of network headward growth (Fig. 2.1*B*). As this wave progresses into the undissected basin, the fingertip channels lengthen and bifurcate, leaving behind a channel system that is almost fully developed. In this developed portion of the network few additions or losses of channels occur during the continued extension of the network. The significant feature of this model is that the network is almost fully developed as the wave of dissection passes a particular point. The model proposed by Glock (Fig. 2.1*C*) lies between the two extremes and is probably the most common mode of network growth.

Random Growth Models

Shreve (1966) built on Horton's work by introducing to the study of networks a branch of topology known as graph theory. He demonstrated that Horton's law of stream numbers is consistent with the hypothesis that networks that have developed in the absence of geologic controls are topologically random, and he suggested that the law is largely a consequence of random development of channel networks.

Additional work by Shreve (1969), Smart (1969), and others has examined the laws of stream lengths (Horton, 1945) and basin areas (Schumm, 1956), and they have demonstrated that the model of random stream development gives satisfactory results. A particularly convincing demonstration of the validity of the model of random stream development is provided by simulations of stream networks that are created by random walk processes. Although many of the earlier simulations were generated by a downslope-stepping process, which bears no relation to natural processes of headward channel extension, they produced networks (Figs. 2.2, 2.3) that conformed to the laws of network composition, although their geometric (as opposed to topologic) characteristics were sometimes very unnatural (Hack, 1965; Leopold and Langbein, 1962; Mosley, 1972a; Smart et al., 1967). An improvement was the use of headward-growth simulation (Howard, 1971a; Smart and Moruzzi, 1971a, b) and the capture model of Howard (1971b), which simulated the effect of stream capture on a preexisting network. Such networks conform to the laws of drainage network composition, resemble natural networks, and may also simulate the stages of network growth.

The simulations frequently use a matrix or grid coordinate scheme within which the network grows. Each grid element represents a unit length of possible growth. In the Leopold–Langbein model (Fig. 2.2) the length of a stream from initiation to its joining with another stream has a minimum and maximum limit, which is dependent on separation or cell size, and the actual length of an individual link is therefore a function of the cell size in the matrix. The unit size of a cell in the computer model is clearly related to the concept of a constant of channel maintenance (Schumm, 1956, p. 607), the unit area required to support a unit length of channel. Thus the simulation models have a built-in deterministic component.

In addition, the topologically random model is based upon the assumption that channel links combine randomly to produce a topologically random population of channel networks. Therefore there is an equal likelihood of all topologically distinct channel networks of equal magnitude. This hypothesis is based on the assumption

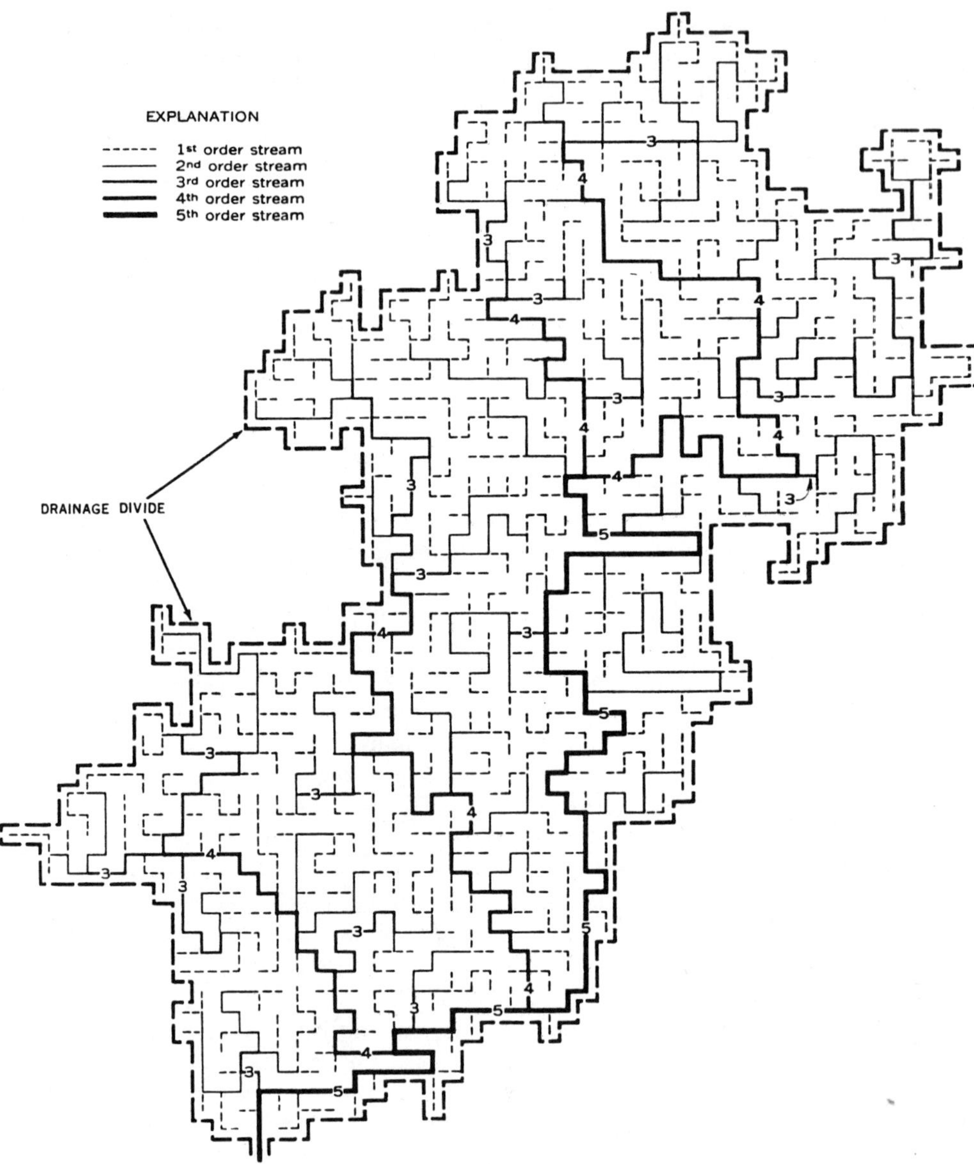

FIGURE 2.2. Random-walk drainage network (from Leopold and Langbein, 1962).

that in the absence of geologic controls, no topological patterns should be favored over any other in nature's selection process. Nevertheless, many topologically random patterns are similar to those which can be explained by reference to geologic or geomorphic controls. For example, patterns 1, 2, 5, 9, 13, and 14 of Fig. 2.4 suggest strong structural control by faults or joints. Patterns 4 and 10 represent

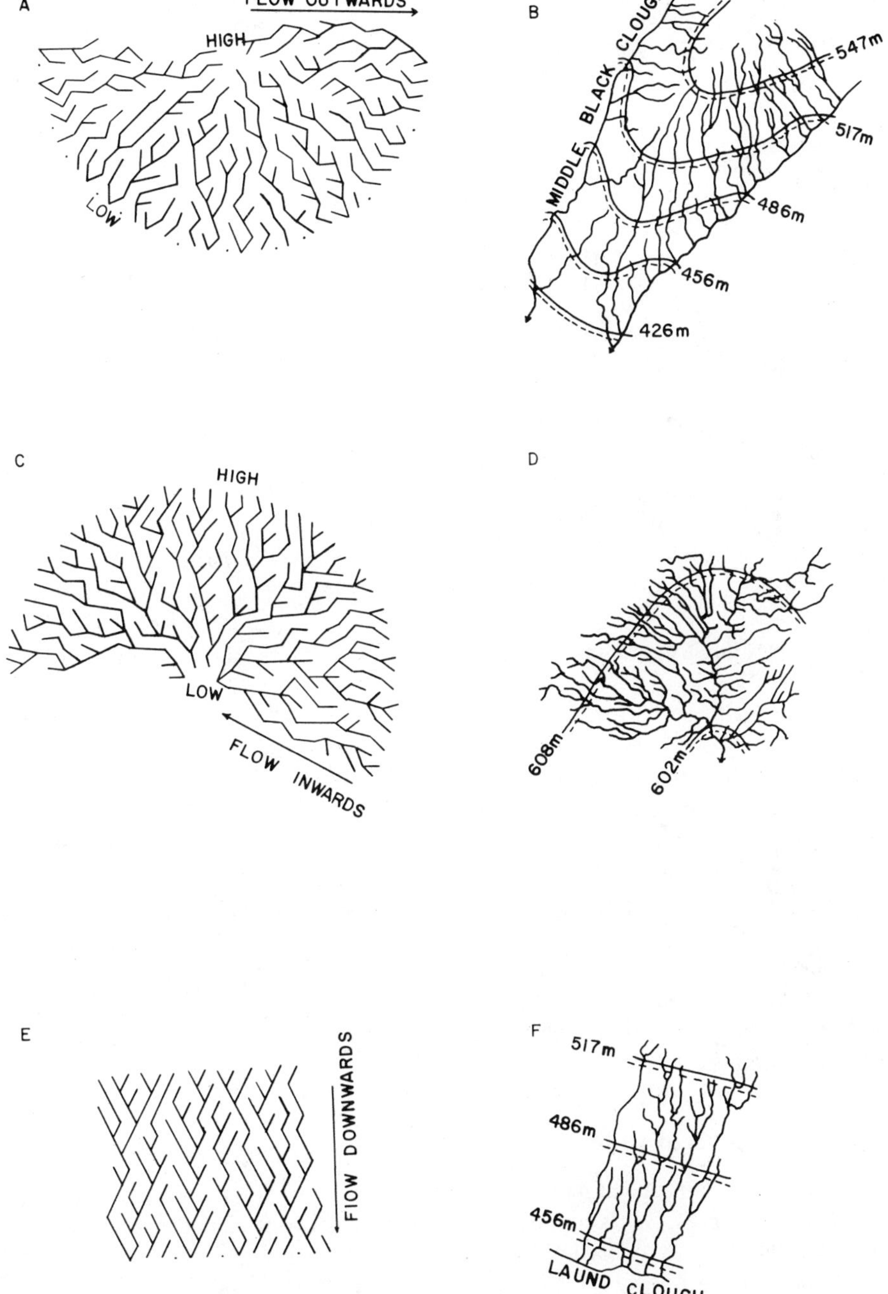

FIGURE 2.3. Actual and simulated drainage networks. (*A*) and (*B*) on convex surface; (*C*) and (*D*) on concave surface; (*E*) and (*F*) on plane surface. Actual networks developed on Bleaklow Hill, U.K. (from Mosley, 1972b).

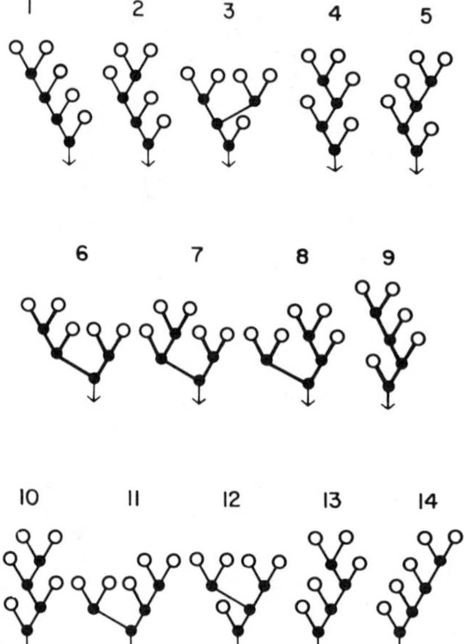

FIGURE 2.4. Topologically random networks with five source links (open circles).

development on steeper slopes than patterns 3, 6, 7, 8, 11, and 12, which appear to be dendritic.

Despite all these efforts at simulation, Howard (1971a) commented that "the areal random growth models simulate many of the characteristics of natural stream systems, but it remains uncertain to what extent inferences may be drawn about the processes responsible for natural stream systems." He also suggested that although natural stream systems appear to exhibit numerical relationships nearly identical to those of topologically random channel networks, the relevant geomorphic processes are deterministic and the apparent randomness arises from independent variation of a large number of factors, such as microclimate and lithology (Abrahams, 1984). It has also been demonstrated that networks developed by strict rules of growth are similar to those generated by random processes, which caused Stevens (1974, p. 130) to conclude that "randomness can appear regular and regularity random."

Deterministic Growth Models

Horton's (1945) conceptual model of network growth by rilling, cross-grading, and micropiracy (Fig. 2.1*A*) is an attempt to describe network characteristics created by a specific process, overland flow on a plane. He suggested that downslope of a critical distance from the divide necessary for overland flow to occur, a steep, newly exposed surface is quickly covered with a series of parallel rills. With time the rills with greater drainage areas become more deeply incised, and their neighbors are captured and then eliminated to create the final integrated dendritic network.

Horton's model (Fig. 2.5) seems most appropriate to planar areas on which overland flow is the dominant geomorphic process, and it is difficult to envisage its application to large drainage basins. Horton (1945, p. 341–346) did, however, propose an extension of his model of rill development to the more useful case of network development on newly exposed coastal lands. This model reflects two major factors: (1) streams develop successively at points where the length of overland flow exceeds the critical length (x_c); (2) competition results in the survival of those streams which have the earliest start and/or the greatest length of overland flow, and which are therefore able to absorb their competitors by cross-grading. The end point of stream development is considered to be reached when the greatest remaining length of overland flow is less than the critical length.

In another model of network evolution, Glock (1931) proposed four stages of network growth (Fig. 2.6):

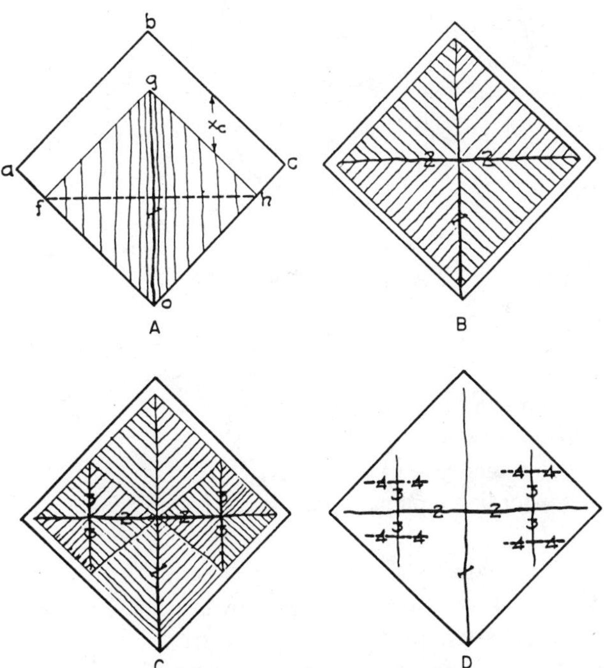

FIGURE 2.5. Development of a drainage network according to Horton (1945). (*A*) On a uniform surface sloping toward o, the maximum length of overland flow is ob. Rilling begins at a length greater than X_c. Because the most water is conveyed along distance bo, a channel forms there, and this incision forms a valley. (*B*) The development of a valley obliterates the channels parallel to og, and the second generation of channels forms (2). They in turn form valleys, and third- and fourth-generation tributaries (3, 4) form, as shown on (*C*) and (*D*).

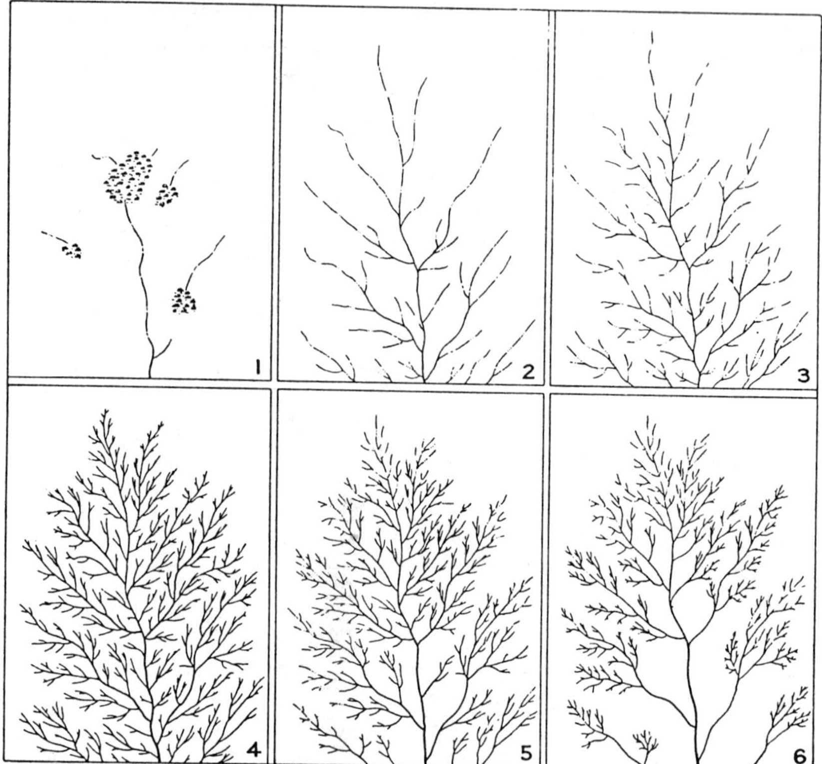

FIGURE 2.6. Drainage network from initiation (1) through extension (2) and (3), to maximum extension (4), followed by reduction (5) and (6) (from Glock, 1931).

1. Initiation.
2. Extension (growth of network).
   a. Elongation (headward growth)
   b. Elaboration (addition of tributaries)
3. Maximum extension (complete elaboration and maximum growth of the network).
4. Integration (reduction of the network).
   a. Abstraction (loss of identity suffered by a secondary stream by encroachment of a primary stream)
   b. Absorption (disappearance of a stream except immediately after rainfall).

Glock (1931, p. 479) characterized initiation by (1) a lack of streams over a large percentage of the surface, (2) indefinite termination of many streams without junction with a main stream, (3) the failure of many streams to have commenced active conquest and extension into the available drainage area.

The stage of extension is a period of growth for the initial drainage system.

Glock envisaged elongation as the dominant process at first, as major streams are established and subdivide the undissected drainage area. This is followed by elaboration of the skeletal form of the initial stream system by addition and growth of tributaries. At the stage of maximum extension, the network has grown into all the available drainage area.

The stage of integration is marked by the reappearance of the skeletal form of the network, as channels are eliminated by abstraction and absorption during the erosional reduction of the drainage basin (Glock, 1931, p. 481). The continued lowering of divides does not provide sufficient internal relief to maintain all channels, and some are lost by absorption. In addition, lateral migration of major streams eliminates small tributaries by abstraction. The net result is a loss of streams and a reduction in total channel length and drainage density.

Unfortunately, Glock's terminology is not clear. *Integration* means "to incorporate or to bring together." In fact, his process involves a subtraction of channels. The network does not shrink or contract but rather is reduced by channel loss; therefore the late stage of drainage network evolution should be referred to as *reduction*.

FIELD OBSERVATIONS OF NETWORK GROWTH

It is difficult to test models of network growth, whether they be Horton's, Howard's, or Glock's, because generally we cannot observe the evolution of natural networks. However, Ruhe (1952; see also Kashiwaya, 1983, and Wells et al., 1985) compared drainage networks and densities on glacial till sheets of different age in Iowa (Fig. 2.7), and he was able to show how drainage density changed over a period of about 40,000 years. His maps suggest a Glockian style of network evolution with elongation (Cary and Mankato), followed by elaboration (Tazewell and Iowan) and perhaps maximum extension (Pre-Iowan).

A few opportunities have arisen actually to monitor network growth. After the Hebgen Lake earthquake exposed a strip of the lakeshore to subaerial erosion, Morisawa (1964) mapped the small drainage systems that developed on the initial sand and silt surface one and two years later. The manner and rate of evolution of the networks depended largely on the slope of the initial surface and type of material. Flat, silty areas with a low infiltration capacity and high runoff became covered with an intricate network of many short, steep tributaries flowing into a large, low gradient main valley. Networks developed on sand or on steep surfaces were slower to form, more permanent, and simpler. The networks very rapidly achieved a degree of stability and subsequent pattern changes were minor. She noted that network expansion in the headwaters of a growing system is simultaneously compensated by loss of tributaries in the lower part and that a major cause of changes in the drainage network was lateral migration of the main channel.

Morisawa concluded that network development followed the outline established by initial drainage traces and swales, and changes in the pattern were conservative.

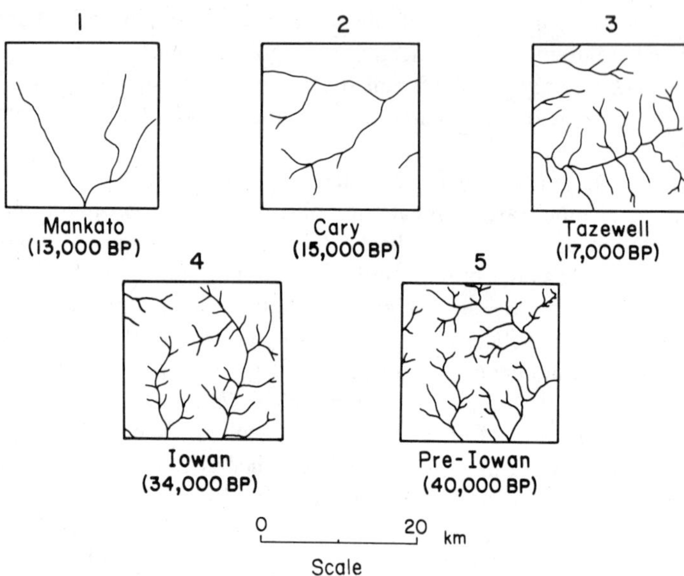

FIGURE 2.7. Drainage networks developed on till sheets of different ages in Iowa (from Ruhe, 1952).

Reduction of a network occurs simultaneously with extension, but an overall state of dynamic equilibrium is attained at a very early stage, so that network form is related less to stage in the erosion cycle than to lithology and initial surface configuration.

Schumm (1956) had a similar opportunity to study network growth on a clay–sand fill at Perth Amboy, New Jersey, although the area had been eroding for 20 years before the first survey. He selected 11 second-order basins having similar areas but for which between 2.4 and 77% of total erosion had occurred as estimated by hypsometric analysis (Fig. 2.8). The upper horizontal line (100% elevation) represents the original surface into which the basins have been eroded and the lower line (0% elevation) is baselevel. Each curve, which represent a separate basin, suggest how one basin will change with time as it is lowered by erosion from the original surface to baselevel. The bulk of early erosion is in the lower part of each basin (curves 1–6). Then maximum erosion is in the central part of the basin (curve 7). Curve 8, which is almost straight, represents a stage when 50% of the erosion is complete, and then curves 8–11 show that maximum erosion has shifted to the upper part of the basin. Such an erosional progression will occur when drainage basins are rejuvenated.

Schumm (1956, p. 619–620) examined network change by comparing maps of selected drainage networks made four years apart. During the four-year period, 12 new tributaries were added to the fifth-order drainage system, but 12 were eliminated. Of those lost, 6 were by abstraction, which involved lateral expansion of a more competent neighbor, 2 by angle reduction to the minimum with collapse of

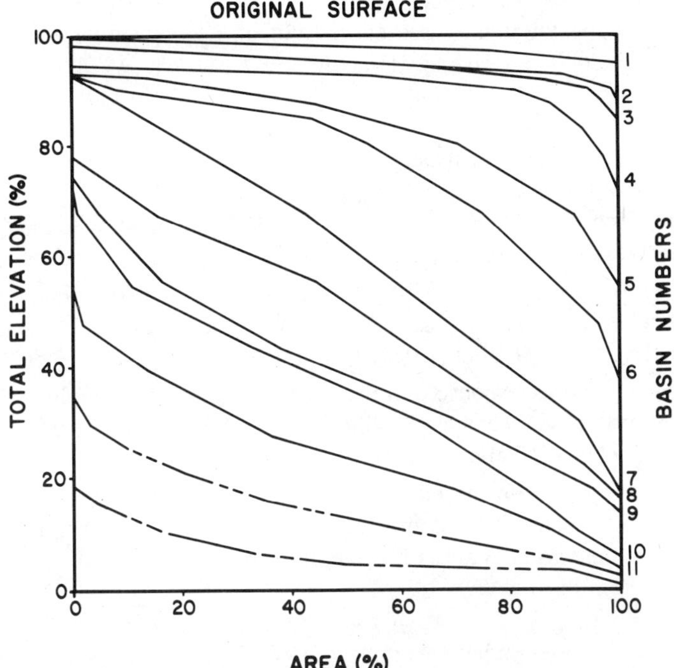

FIGURE 2.8. Sequence of hypsometric curves for 11 second-order drainage basins. Percentage of drainage area above a certain elevation in each basin is plotted against percentage of total relief above that elevation of the Perth Amboy badlands. Dashed lines show possible future evolution of the basins (from Schumm, 1956).

the divide and union of the streams, and 4 were in small shrinking basins surrounded by headward-growing channels. Hence Glock's (1931) stages of extension and reduction (primarily by abstraction) were simultaneous, as at Hebgen Lake, and capture was also occurring.

The three growth models may each be valid under different initial conditions. Horton's model, which requires replacement of parallel channels by a dendritic network, will probably occur on a plane surface, where one of a series of parallel channels becomes dominant. The Glock extension model requires a surface that tends to concentrate flow, which may also be the case for the expansion by headward-growth model (Fig. 2.1 *C*) except that some factor (gentler slope?) prevents rapid extension of the channels.

EXPERIMENTAL STUDIES OF NETWORK GROWTH

The information provided by studies of network development in rapidly eroding materials in the field (Morisawa, 1964; Schumm, 1956) suggests the potential of taking a further step and creating in the laboratory a small-scale landscape that is susceptible to precise measurement under controlled conditions. For this reason, a series of experimental studies was carried out in a large rainfall-erosion facility (REF) at Colorado State University (Harvey, 1980; McLane, 1978; Mosley, 1972a; O'Brien, 1984; Parker 1977; Zimpfer, 1982), and a study in badlands (Bergstrom, 1980) provided additional information on small landforms.

The general objectives of these studies were

1. to describe the evolution of drainage systems from initiation to the final stages of network development (McLane, 1978; Mosley, 1972; Parker, 1977);
2. to compare the morphology of networks generated with different initial slopes and relief but with the same homogeneous materials and watershed area (Mosley, 1972; O'Brien, 1984; Parker, 1977);
3. to document the response of a drainage system to baselevel lowering and to variations in sediment availability (Bergstrom, 1980; Harvey, 1980; McLane, 1978; and Parker, 1977);
4. to relate changes in the intensity of operation of geomorphic processes, as revealed in sediment yield and runoff patterns, to the morphology of a drainage basin at various stages of evolution (Mosley, 1972; Parker, 1977; Zimpfer, 1982).

Other experimental studies of drainage network development have been more limited in scope (Bones and Ford, 1971; Flint, 1973), but they demonstrated that Horton's relations prevail for small experimental networks.

Experimental Facilities and Procedure

The rainfall-erosion facility is a container, constructed of 2-cm plywood, that is 9.1 m wide, 15.2 m long, and 1.8 m deep (Fig. 2.9, 2.11). A walkway around the sides provides access and an observation platform. A carriage mounted on rails spans the width of the REF. A point guage, mounted on the carriage, can be used to give *X–Y–Z* coordinates of any point within the container. A flume attached to the front of the REF provides an exit for water and sediment. Baselevel is controlled by raising or lowering the flume.

Sprinkler System

The sprinkler system, constructed of 5.1-cm aluminum irrigation pipes, supplies water to lines of sprinklers mounted 3.0 m apart on both sides of the REF (Fig. 2.10). Each sprinkler is attached to the supply line by a 3.0-m-long galvanized steel pipe. The sprinkler heads are commercial irrigation sprinklers (Rainjet brand with nozzle No. 78). Sprinklers were grouped into four sets, so that by activating the different sets of sprinklers four average rainfall intensities of 20, 30, 45, and 60 mm/hr could be produced. The four sets of sprinklers were controlled by a 24-V solenoid valve, so that they could be activated nearly instantaneously. After the

FIGURE 2.9. Rainfall-erosion facility (REF) Colorado State University. (*A*) Start of first experiment. (*B*) Drainage network development.

first season of operation, the REF was enclosed in a building to eliminate problems caused by wind deflection of the rainfall. This necessitated removal of the four sprinklers at the extremities of the supply lines.

The water pressure at each sprinkler was controlled by a Watts Low Pressure Regulator; varying the pressure controlled both the water drop size and areal distribution of precipitation. The most even areal distribution was achieved with the regulators set at 23 psi. At this pressure the nozzles produced a drop size distribution with a mode of 1.52 mm and a range from near zero to 3.71 mm (Holland, 1969). Natural rainfall at an intensity of 50 mm/hr produces drops with a mode of 2.75 mm and a range between approximately zero and 7 mm (Holland, 1969). Because different intensities were produced by shutting off groups of sprinklers

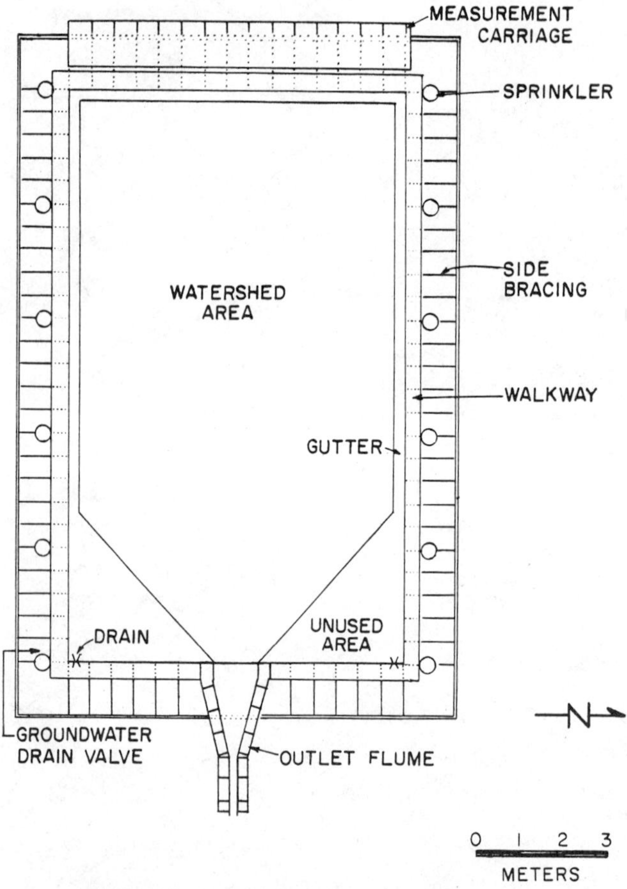

FIGURE 2.10. Plan of rainfall-erosion facility (REF).

rather than by changing pressure, the drop size distribution remained constant at the various precipitation intensities, whereas the drop size of natural rainfall increases with increasing intensity.

The sprinkler system was calibrated by placing large cans in a grid over the watershed surface. The highest precipitation intensity ocurred near the center of the REF during all calibration runs (Fig. 2.11). Replication was excellent, with an average difference of 2.5% between replications.

Experimental Material

Suitable sediment was difficult to find, primarily because of the quantity required to fill the REF (250 m^3). Preliminary tests indicated that a 50–50 mixture of commercial-grade plaster sand and floodplain sediment (Table 2.1) provided the cohesiveness needed to establish channels with stable side slopes but also permitted rapid erosion. X-ray diffraction of the clay-size material showed quantities of quartz,

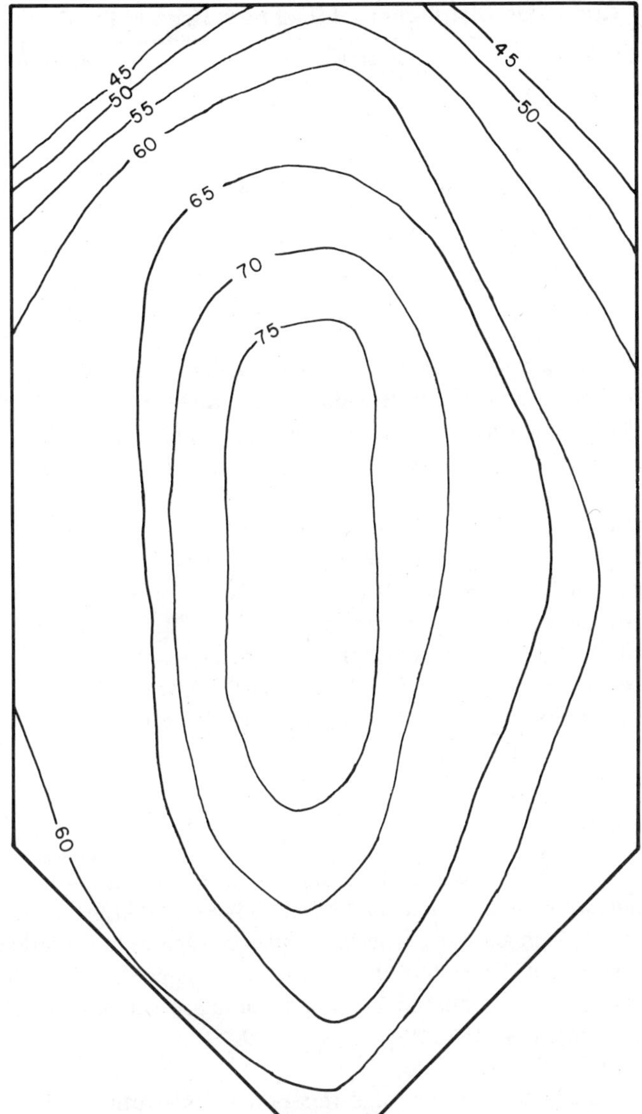

FIGURE 2.11. Areal distribution of simulated precipitation over surface of REF during a 61-mm/hr storm. Contours are in millimeters (from Parker, 1977).

plagioclase feldspar, and biotite, with approximately half being disordered kaolinite. Hence about 5% of the experimental materials was kaolinite, and there was also 0.41% magnetite in the sediment.

A 1.5-m depth of the experimental material was placed in the REF in 25-cm layers. The layers were compacted by rolling, and the final upper surface was graded toward the outlet. During experiments, sediment samples were collected at

TABLE 2.1 Grain-Size Distribution of Sediment Used in the REF

Size (mm)	Percentage	Cumulative Percentage
>2	1.17	1.17
1–2	11.55	12.72
0.5–1	15.96	28.68
0.25–0.5	17.04	45.72
0.125–0.25	15.38	61.10
0.062–0.125	12.08	73.18
<0.062	26.82	100.00

the outlet by catching total runoff in a large container. The volume of the total water and sediment sample collected during any time period (e.g., 10 sec) was measured. The sample was set aside until the sediment had settled, when the excess water was decanted. The sample was then oven dried before it was weighed.

The sediment used in the REF is classified as a loamy sand, which typically has infiltration capacities of 25–50 mm/hr (Kohnke and Bertrand, 1959). However, a comparison of rainfall and runoff during 1-min storms indicated that the average infiltration rate in the REF was 19 mm/hr. This was probably the result of the compaction of the sediment by the roller.

One-minute hydrographs were generated on several occasions during evolution of two drainage networks in the REF. An analysis of variance showed no significant changes of infiltration rate, so it may be assumed to be independent of the stage of drainage basin evolution.

Erosion Processes

The REF was intended to be a model drainage basin, although bearing no defined scaling relation with any known natural basin. However, it is also a drainage basin in its own right and, in addition, is in many respects similar to an agricultural field or a road embankment. Erosion of the experimental materials and evolution of the drainage basin was accomplished by several processes commonly observed on natural surfaces (Mosley, 1972a).

Rainsplash. Rainsplash detached and transported soil particles. Fluorescent orange beads the size of large sand grains, which were scattered over the surface to facilitate observation, were constantly displaced, occasionally by several centimeters. After 1 hr of precipitation, beads scattered at a given location were no longer visible, having been buried, displaced some distance, or, mainly, moved into rills and transported downstream.

Sheetflow. When precipitation was applied, water soaked directly into the soil, but after 2 or 3 min, runoff commenced. Instead of flowing downslope as a thin sheet, the water collected almost immediately into more or less distinct flow lines.

Although sheet erosion has been indicated to be an untenable concept (Schumm, 1956; Schwab et al., 1966) sheetflow did occur between flow lines and rills, but for only very short distances. Close examination demonstrated that water flowing in thin sheets and at low velocities rolled grains towards the nearest channel. When the soil surface was sheltered from the impact of falling water drops, transport virtually ceased. Ellison (1947) and others have noted that the main erosive function of waterdrop impact is to facilitate soil transport by overland flow.

Concentrated runoff. As runoff concentrated into distinct flow lines, its erosive power was enhanced. Flow lines thus suffered greater erosion than neighboring areas upon which sheetflow occurred and channels incised below the level of the initial soil surface. During runs in which baselevel was lowered at the outlet, nickpoints moved rapidly upslope along flow lines, cutting channels. Where flow lines bifurcated, nickpoints moved up both branches. Both incision and nickpoint retreat were restricted to flow lines, which could be identified at a very early stage in the experimental runs.

Mass movement. Incised channels tended to cut laterally and meander. Bank undercutting was frequent, leading to bank caving and progressive widening of the main valleys. Once mass movement had widened the valley sufficiently to give the channels room to meander freely, surface erosion continued to reduce the slope angles of the steep valley sides.

These processes are all observed in natural watersheds, particularly in arid, semiarid, or cultivated watersheds, where vegetation cover is sparse and does not protect the soil surface against water erosion, or in youthful terrain, where deeply incised channels remove support from the valley sides and induce side slope retreat by mass movement. Some processes that are important in natural landscapes, notably (creep) and solution, were, however, insignificant in the REF.

NETWORK DEVELOPMENT

Effect of Initial Topography

Recent work on computer modeling of network growth, except for that by Seginer (1969), has tended to ignore the slope and initial topography of the surface upon which networks are initiated. McLane's (1978) observations in the REF confirm the influence of the microtopography of the experimental surface upon network growth, even though every effort had been made to create initially smooth plane surfaces.

For this experiment the surface of the REF was graded into two intersecting planes with an overall slope toward the outlet of 0.0075 (Fig. 2.12*A*). Baselevel was lowered 20 cm and precipitation was applied at a rate of 58 mm/hr for a total of 12 hr, when maximum extension was attained. During this time the network was mapped every 2 hr.

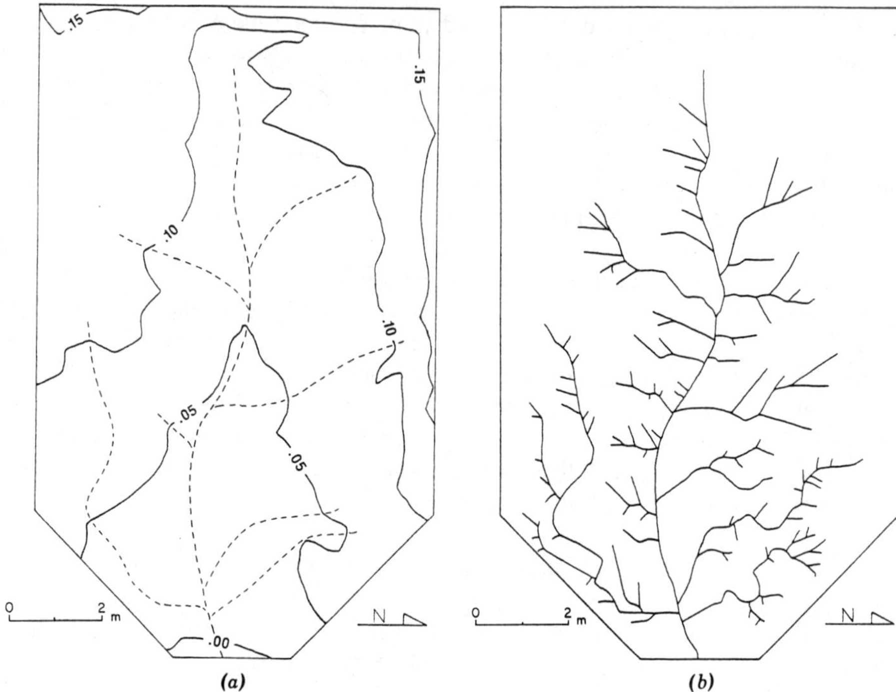

FIGURE 2.12. McLane's (1978) drainage patterns. (*A*) Contour maps of initial surface. Dashed lines show initial drainage network. Contour interval is 0.05 m. (*B*) Fully developed drainage network after 1022-mm total precipitation on baselevel 1.

Precipitation was completely absorbed by the smooth, dry surface of the basin during the first 8 min of application. Then the formation of shallow ponds (depression storage) indicated that infiltration rates were decreasing, and during the next 10 min, small streams of surface runoff connected the ponds to form a broad, shallow drainage network. The initial experimental surface directed runoff toward the centerline of the REF (Fig. 2.12*A*), where it concentrated and flowed to the basin outlet. Thus the longest and deepest link in the pond-and-stream drainage system was located along the centerline of the basin, and channel erosion progressed rapidly headward along this line to form a wide main channel with vertical banks. Points at which the interconnected pond systems drained over the banks became sites of tributary initiation, and nickpoints migrated headward into the ponds and along the rudimentary drainage system to form major tributary channels (Fig. 2.12*A*). Channels formed and advanced along the lines of greatest water supply (Figs. 2.12*B*). The initial topography shaped the rudimentary drainage system, which in turn provided the paths of maximum water supply that were followed by the headward-developing channels. Thus initial topography controlled the network pattern in the REF (Fig. 2.12*B*).

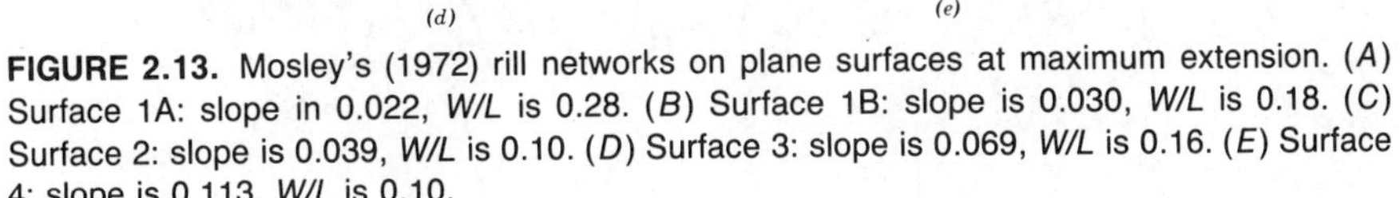

FIGURE 2.13. Mosley's (1972) rill networks on plane surfaces at maximum extension. (*A*) Surface 1A: slope in 0.022, *W/L* is 0.28. (*B*) Surface 1B: slope is 0.030, *W/L* is 0.18. (*C*) Surface 2: slope is 0.039, *W/L* is 0.10. (*D*) Surface 3: slope is 0.069, *W/L* is 0.16. (*E*) Surface 4: slope is 0.113, *W/L* is 0.10.

Effect of Slope Angle and Shape

The persistent effect of initial conditions on final network form is also demonstrated by an experiment conducted by Mosley (1972). The initial experimental surfaces were shaped into five planar (Fig. 2.13), two convex-up (Fig. 2.14), and two concave-up (Fig. 2.15) surfaces with inclinations between 0.022 and 0.121. At the downslope end of each surface a 10-cm-high vertical face of sediment was exposed. Precipitation was applied at an intensity of 60 mm/hr for periods for 4–10 hr until the rill networks had reached maximum extension. As already described, the final networks were strongly controlled by flow lines established at an early stage on the initial surfaces, and final network patterns clearly reflected the initial conditons.

For the planar surfaces, the relative width of the small watersheds that developed on the surface of the REF, given by the ratio of maximum width (W) to length (L), decreased with slopes (S) from 0.3 to 0.1 according to the equation

$$W/L = 0.04 - 0.174S \qquad (r = 0.37) \tag{2.1}$$

That is, watersheds were more elongated on steeper initial slopes. The correlation coefficient in Eq. (2.1) is surprisingly poor. This is because the greatest change takes place between slopes of 0.02 and 0.03, with the drainage patterns changing from subdendritic on a 0.022 slope to parallel on a 0.03 slope (Fig. 2.13*A*, *B*). Nevertheless, Eq. (2.1) reflects the increased likelihood that on steeper slopes a channel will flow directly downslope for longer distances before joining with a neighbor, so that there is a tendency for a larger number of elongated, infrequently branching networks on steep slopes. The networks were also markedly more elon-

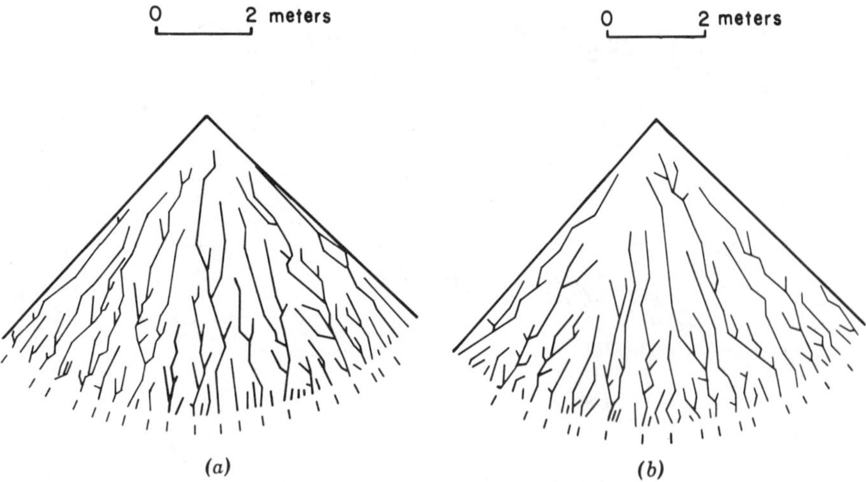

FIGURE 2.14. Mosley's (1972) rill networks on convex surface at maximum extension. (*a*) Surface 5: slope is 0.109. (*b*) Surface 6: slope is 0.080.

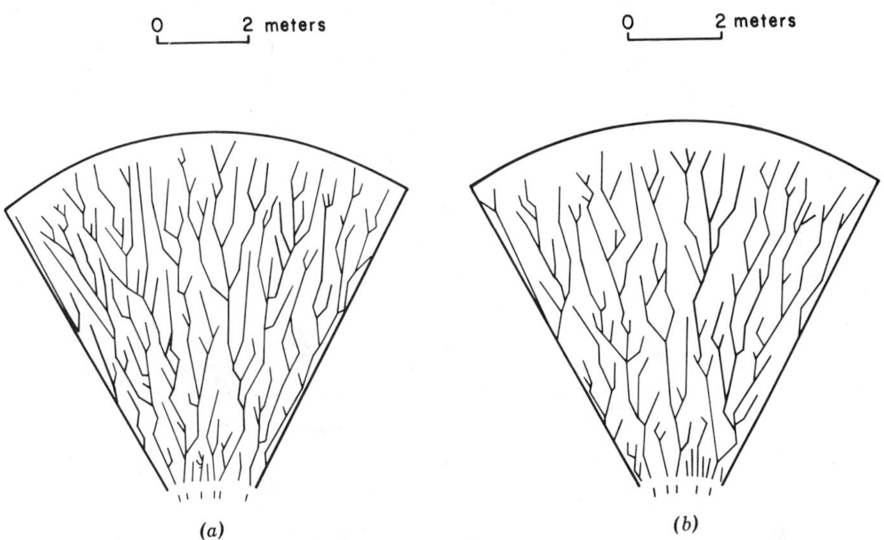

FIGURE 2.15. Mosley's (1972) rill networks on concave surface at maximum extension. (*a*) Surface 7: slope is 0.121. (*b*) Surface 8: slope is 0.076.

gated on the convex surfaces and relatively wider on the concave surfaces, as might be expected. This conclusion is similar to Mosley's (1972) from a study of natural gully networks incised on slopes of different shapes and of simulated networks (Fig. 2.3).

Drainage density (D in m/m^2) of the entire surface increased with initial surface slope (Fig. 2.16) according to the equation

$$D = 0.91 + 22.4S \qquad (r = 0.974) \tag{2.2}$$

which is very good considering that the three different surface shapes are involved.

Associated with an increase in drainage density is a reduction in the distance that sheetflow must travel before reaching a channel (Fig. 2.13). For example, the channels are closer to the top of the REF on the steeper slopes; therefore a larger proportion of the surface is under the direct influence of channelized flow. Nevertheless, it is noteworthy that the convex and concave surfaces (Figs. 2.14, 2.15) do not produce significantly different values of drainage density. At a slope of about 0.08 the converging surface has a slightly higher drainage density; but on the steepest, 0.10 slope there is little difference between straight, convex, and concave surfaces, which suggests that a maximum value for drainage density has been achieved for the experimental material, and this obscures any effect of the shape of the experimental surface. This is an important point that may partly explain why some geomorphic and hydrologic relations are weak when drainage basins are very steep.

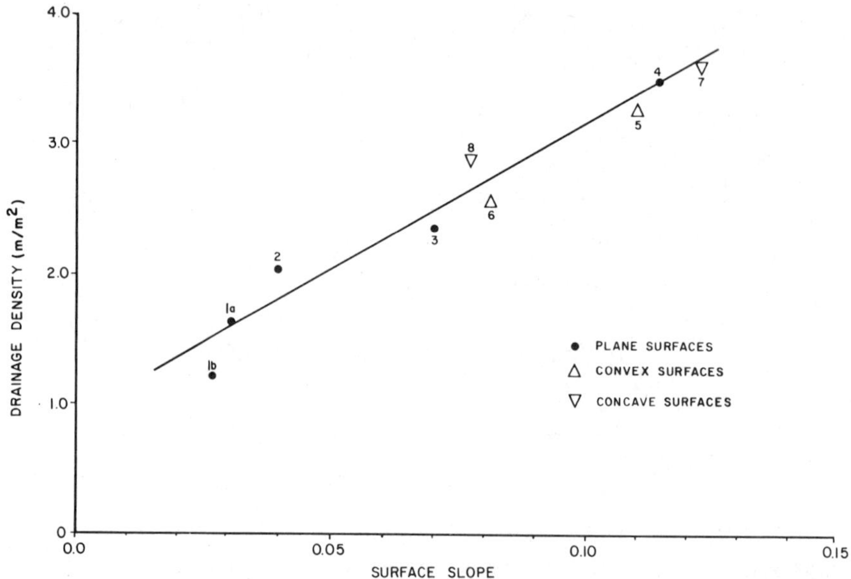

FIGURE 2.16. Drainage density as a function of slope and surface shape (from Mosley, 1972).

Effect of Contributing Area

Channel extension is the result of progressive erosion at a channel head. The erosive power of running water is controlled primarily by discharge, depth, and slope. The only two factors that varied significantly during periods of channel extension in the REF were length of overland flow, an index of the area contributing runoff to each channel head, and slope. Since an actively growing channel head receives runoff from the surrounding area, it is more appropriate to express runoff in terms of the basin area that is contributing flow to the channel tip than the length of overland flow (Fig. 2.5; Horton, 1945).

There is a critical value of shear stress ($\tau_c = \gamma dS$, in which γ is unit weight of water, d is water depth, and S is slope) necessary to initiate erosion. This requires a certain critial contributing area A_c, which provides a sufficient discharge and depth for actual shear stress τ to exceed τ_c. This critical area, Schumm's (1956) constant of channel maintenance, is analogous on a nonplanar watershed to Horton's (1945) critical distance of overland flow (x_c) on a plane. Downslope, erosion increases in proportion to the excess of τ over the critical value τ_c, and therefore it increases in proportion to some power of excess contributing area ($A - A_c$). Conversely, as the stream head advances upslope, the rate of headward extension declines until the stream head has reached the location at which $A = A_c$.

As the head of the main channel migrates upslope, a tributary advances in its subbasin at a growth rate slower than that of the main channel because it is supplied by a smaller area. As it extends, it may trigger the growth of even smaller tributaries at the mouth of areas of sufficient size to produce and maintain a channel. The headward growth of all channels will slow and eventually cease as the areas supplying runoff to their heads decrease.

McLane's (1978) experiment provides information on the rate of network growth and network response to baselevel lowering. Drainage network development was most rapid initially, but its rate declined progressively with time (Fig. 2.17). Baselevel lowering then rejuvenated the basin and a nickpoint migrated up the main

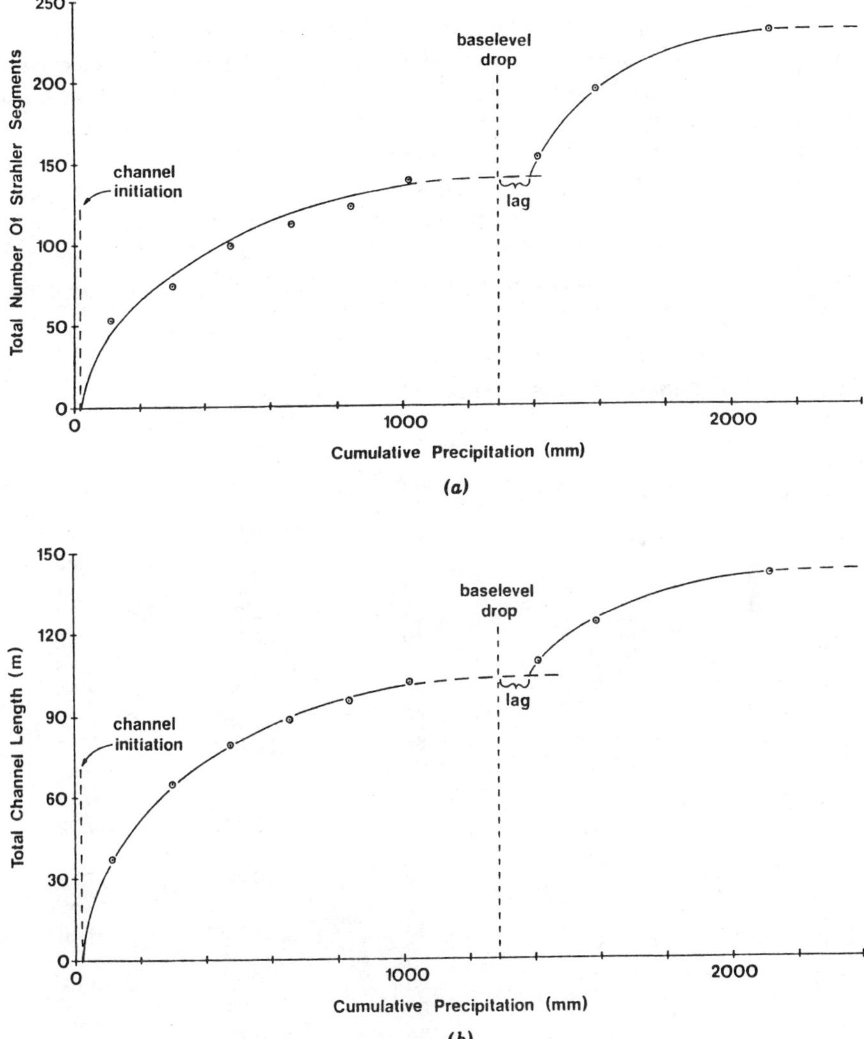

FIGURE 2.17. Drainage network extension as shown by change of (*A*) stream frequency and (*B*) total channel length (from McLane, 1978).

channel and into the tributaries. No significant network extension occurred until nickpoints had migrated through the system and rejuvenated the first-order streams, so that there was a time lag between baselevel lowering and the initiation of a new network growth curve (Fig. 2.17). Once the effect of the new baselevel reached the periphery of the network, the form of the rejunvenated network curve was very similar to that of the initial curve, with the rate of network extension being initially rapid and then slowly declining to zero.

Network Evolution

Although the ultimate form of the networks in the REF was effectively fixed by the microtopography of the initial surfaces (Fig. 2.12*A*), the growth of the channel network into the available drainage area permits evaluation of the models of drainage network evolution (Fig. 2.1). During the first two experimental seasons, an attempt was made to document the entire evolution of a drainage network and to relate hydrology to drainage basin morphology (Parker, 1977). In order to determine the effects of the initial slope and changes in baselevel on the drainage pattern, each season was begun with different initial conditions (Table 2.2). Data are identified by experiment and network numbers (e.g., El, Nl). These first experiments

TABLE 2.2 Summary of Experiments 1, 2, and 3

Experiment	Subset	Network	Elapsed time (hr)	Relief (m)	Drainage Density at Maximum Extension (m/m^2)
1	1	1	2	0.3	0.21
		2	4	0.3	0.32
		3	6	0.3	0.63
		4	8	0.3	0.79
		5	10	0.3	0.92
	2	6	14	0.3	1.16
		7	27	0.39	1.39
		8	48	0.48	1.78
		9	65	0.67	2.07
		10	86.5	0.94	2.47
2	1	1	2	0.48	0.44
		2	4	0.48	1.02
		3 }	10	0.48	1.33
2	2	3 }			
		4	33	0.53	1.50
2	3	5	96	0.78	1.58
		6	106	0.78	1.63
		7	150	0.78	1.51
		8	216	0.78	1.20
3	1	1	2	0.98	—

were carried out with caution because the facility and the experimental material were essentially untested and no guidelines from previous studies were available. Therefore, instead of lowering baselevel to a depth necessary for complete dissection of the basin at the beginning of the experiment, baselevel was lowered by small increments.

The plan to follow each network through its evolution was not entirely successful. Complete erosional reduction of the basin was not possible because the first experiment (El) was terminated by cold weather (Fig. 2.18, Networks 1–10). During the second experimental season (Fig. 2.19, E2, Networks 1–8) the procedure was modified to ensure that the later stages of development would be documented.

Experiment 1. For the first experiment, the initial surface was graded into two intersecting planes, with a total area of 115 m^2 and a slope towards the outlet of

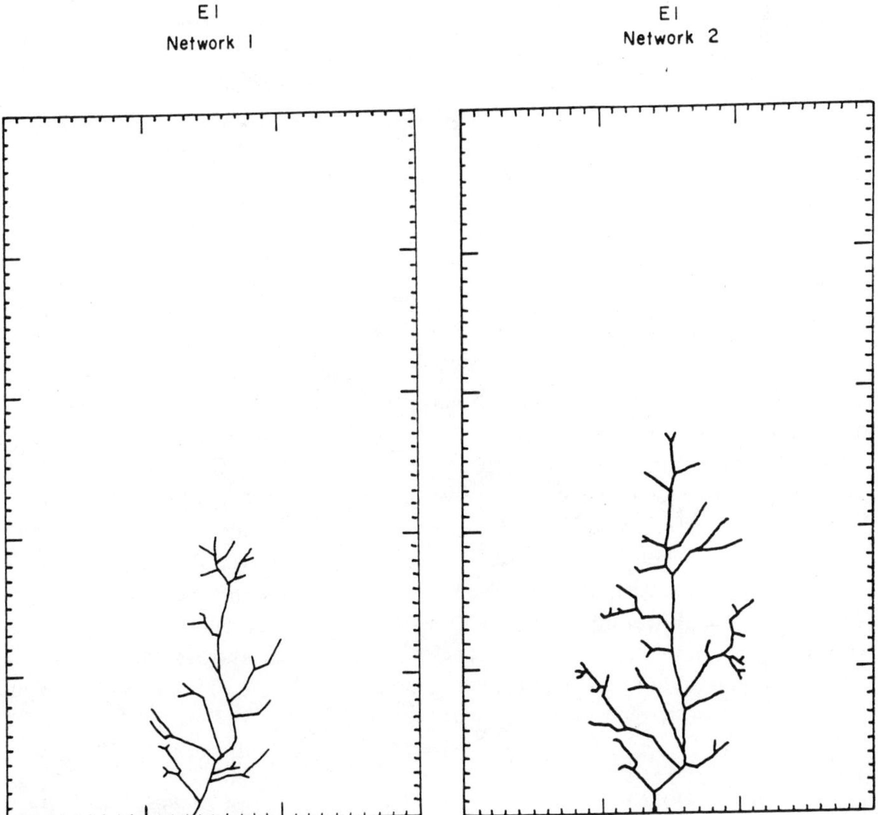

FIGURE 2.18. Drainage networks developed during Parker's (1977) Experiment 1 (Networks 1–10). Rectangle is 9.1 × 15.2 m.

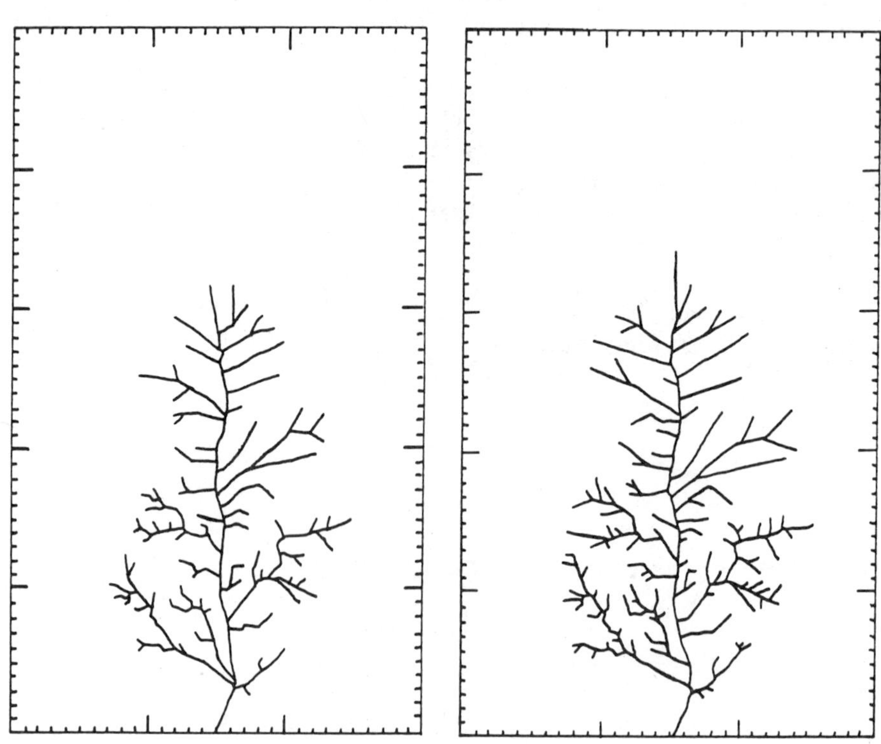

FIGURE 2.18. (continued)

0.0075. Baselevel was lowered 22 cm and precipitation was applied at an intensity of 66 mm/hr for five 2-hr periods. At the end of each period the drainage network was mapped and the longitudinal profile was measured. After an additional 4 hr, the network had essentially ceased growing and vertical stereo photographs were taken with a Wild RC-8 aerial camera suspended from a crane (Fig. 2.20, El, N6).

Baselevel was then lowered an additional 0.1 m; precipitation was applied for 13 hr more until network growth ceased, and another set of photographs was taken. This sequence of lowering of baselevel, application of precipitation, and taking of vertical aerial photographs at the point of maximum extension was repeated to give five sets of photographs taken at elapsed times of 14, 27, 48, 65, and 86.5 hr (Fig. 2.20, Networks 6–10).

Experiment 2. The surface was graded into two intersecting planes, with a maximum slope towards the outlet of 0.032. Precipitation was applied at an intensity

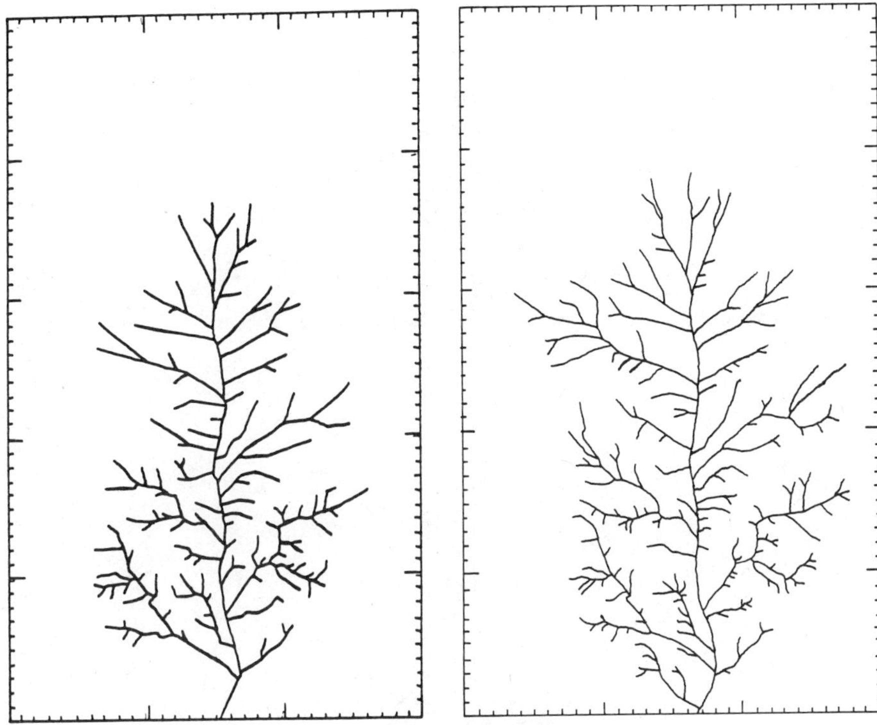

FIGURE 2.18. (continued)

of 61 mm/hr without baselevel lowering, and the growing network was mapped after 2 and 4 hr, and again after 10 hr, at maximum extension for this baselevel.

Baselevel was then lowered 0.05 m and precipitation was applied for 23 hr, at which time the network had reached maximum extension, and it was remapped. Baselevel was again lowered 0.28 m and the network was mapped at 96, 106, 150, and 216 hr elapsed time; maximum extension was achieved at 106 hr (Table 2.2).

Modes of Growth

The evolution of the drainage networks was markedly affected by the different initial conditions, and two growth modes were apparent (Fig. 2.21). In Experiment 1, baselevel was lowered before precipitation was applied, and an intricate network slowly extended headward on the initial surface. This is the *expansion mode* of growth. In Experiment 2, on a steeper slope and without an initial baselevel lowering, a skeletal network was established rapidly throughout the watershed, but dissection was less intricate. This *extension mode* of growth ultimately gave a

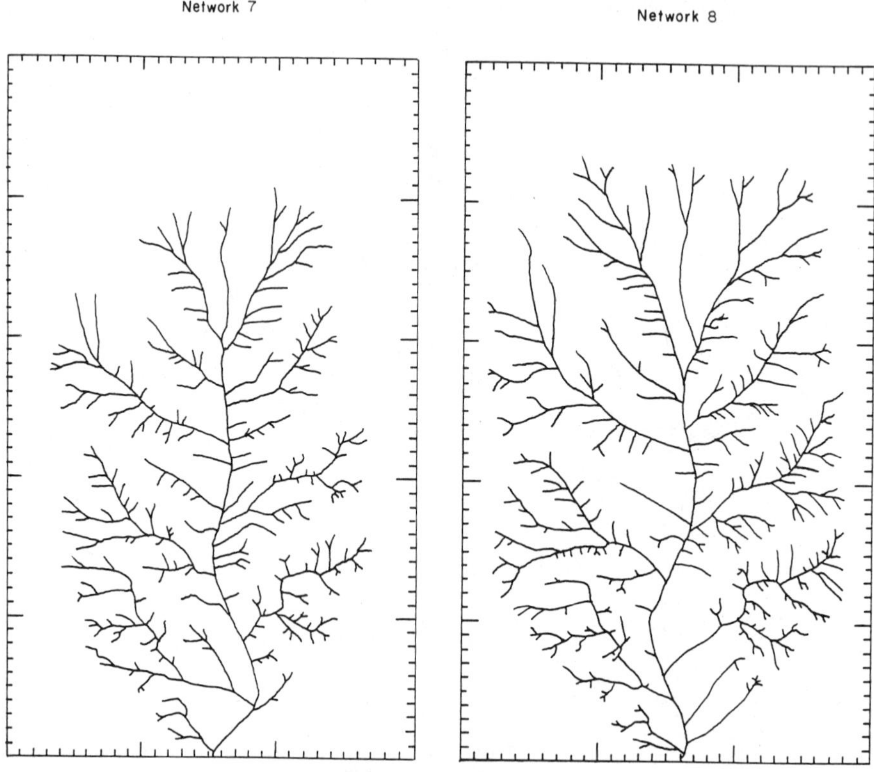

FIGURE 2.18. (continued)

maximum total channel length less than that produced by headward growth for approximately the same relief (Fig. 2.22).

Drainage density increased to a maximum during both Experiments 1 and 2 (Fig. 2.22). Experiment 1 was terminated at what was judged to be the point of maximum extension due to the onset of winter. During the following year, Experiment 2 was continued beyond the prolonged period of maximum extension and drainage density slowly declined. Glock's stages of network evolution could be identified, although initiation and extension could not be separated. In Experiment 1, initial baselevel lowering produced a network that was well developed behind a headward-migrating "wave of dissection." Elongation and elaboration (Glock, 1931) were simultaneous. However, in Experiment 2, without baselevel lowering, long tributary channels developed (elongation). The areas between these tributaries were filled by subsequent bifurcation and addition of new, shorter first-order tributaries (elaboration) and the mean length of first-order streams declined.

Initially, the network on the steeper surface (Experiment 2) achieved a higher

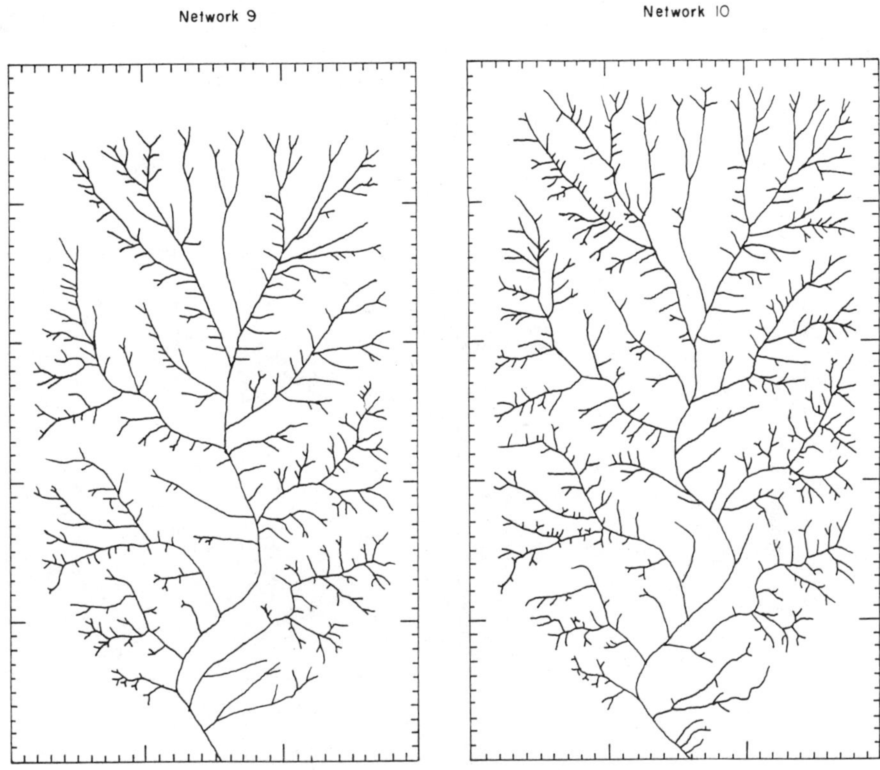

FIGURE 2.18. (continued)

drainage density because the low-order tributaries were longer and they extended further into the available area. However, the network had a lower drainage density at maximum extension (Fig. 2.22). It appears, then, that a greater initial slope produced more rapid network growth, but drainage density was lower at maximum extension. This possibility is difficult to assess by comparison between Experiments 1 and 2 because the effect of baselevel lowering is also present. However, another experiment was performed with a slope of 0.12 and with a baselevel lowering of 0.1 m. During this third experiment precipitation was applied for 2 hr and then the drainage network was mapped (Table 2.2). Experiments 1 and 3, with slopes of 0.0075 and 0.121, respectively, both had initial baselevel lowering. The frequency distributions of first-order stream lengths of Experiments 1 and 3 during network growth were similar, with a preponderance of short streams, and the geometric mean lengths were not significantly different (0.34 and 0.41 m, respectively) (Fig. 2.23*A*, *C*). The mode of network growth was the same, but the rate of extension was greater on the steeper surface of Experiment 3. The different mode of network

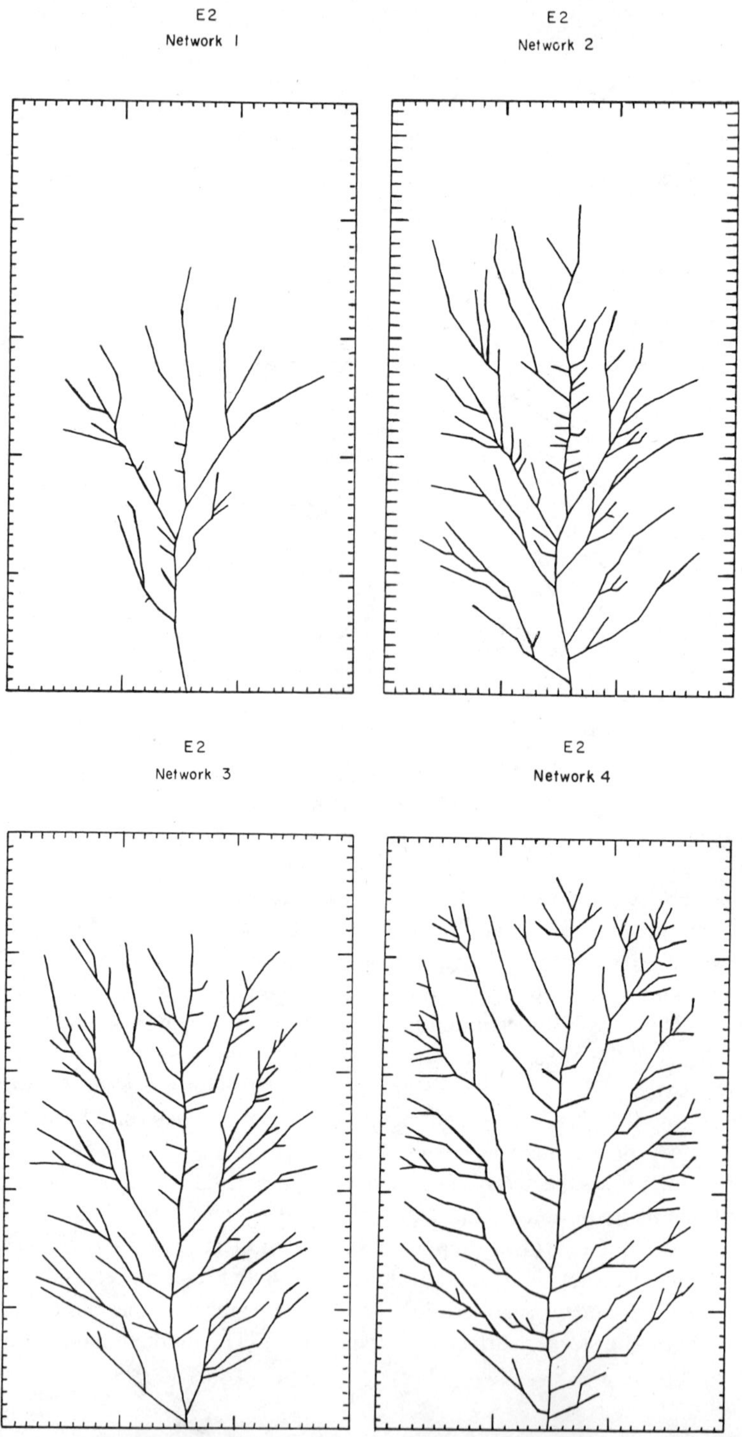

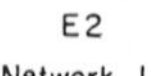

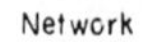

E2
Network 3

E2
Network 4

FIGURE 2.19. Drainage networks developed during Parker's Experiment 2 (Networks 1–8). Rectangle is 9.1 × 15.2 m.

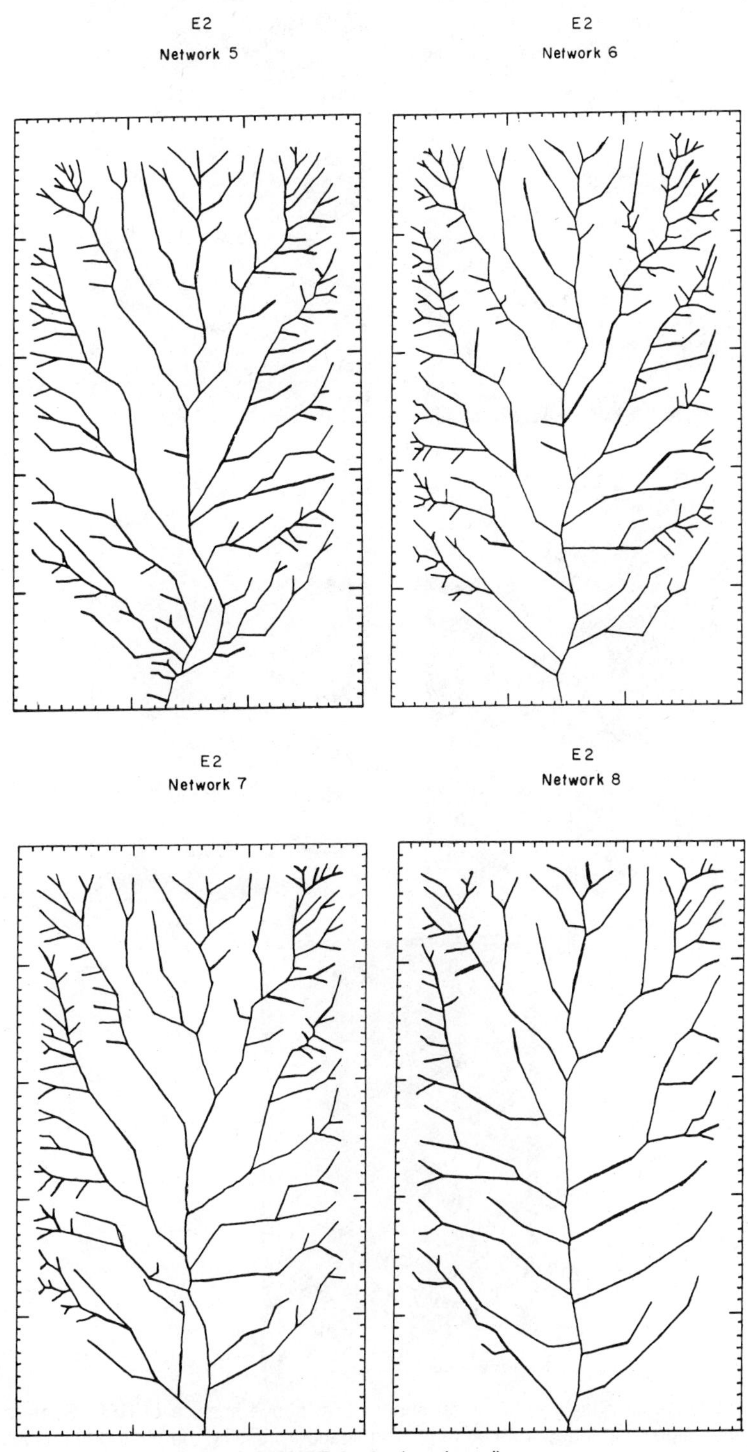

FIGURE 2.19. (continued)

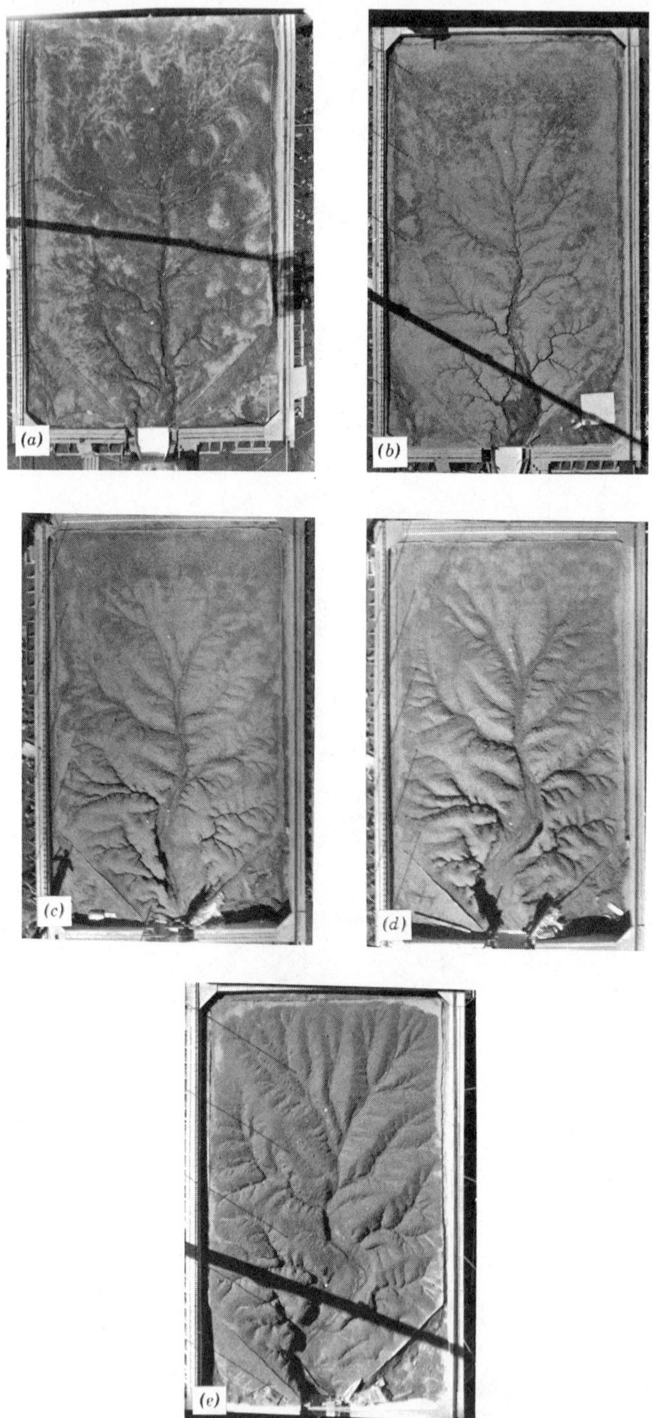

FIGURE 2.20. Aerial photographs taken about 16 m above surface of REF. Camera was mounted in a cage suspended from a crane. The shadow crossing the photograph is the arm of the crane. (*A*) Network 6 of Fig. 2.19; (*B*) Network 7 of Fig. 2.19; (*C*) Network 8 of Fig. 2.19; (*D*) Network 9 of Fig. 2.19; (*E*) Network 10 of Fig. 2.19.

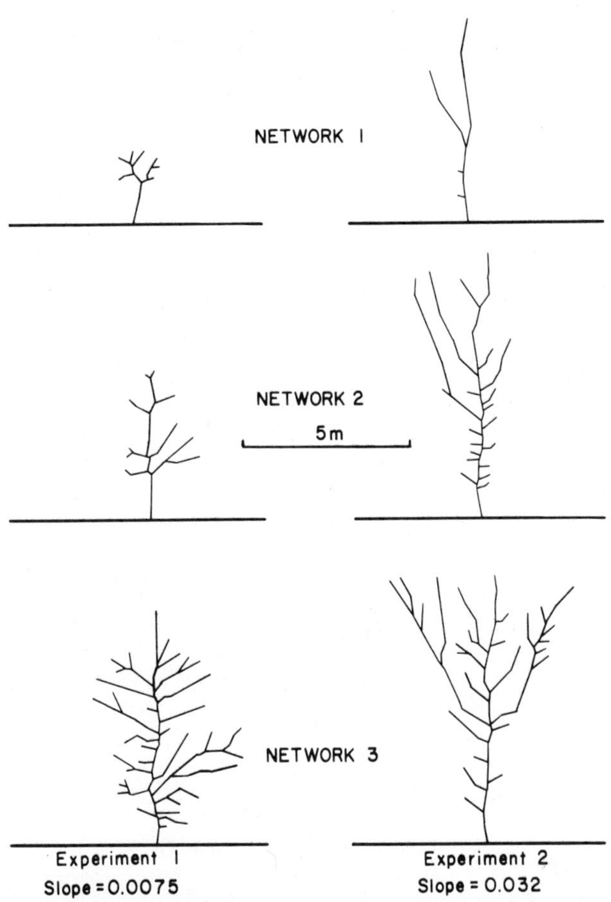

FIGURE 2.21. Drainage network growth under different initial conditions but at equivalent times during Experiments 1 and 2 (from Parker, 1977).

growth during Experiment 2 (Fig. 2.21) is also revealed by the uniform frequency distribution of first-order stream lengths (Fig. 2.23*B*) and the geometric mean of 0.93 m, which is significantly greater than that for Experiments 1 and 3.

Stream Length and Numbers. At maximum extension, the frequency distributions of first-order-stream lengths of Experiments 1 and 2 were generally similar in shape (Fig. 2.24), but the geometric mean lengths of 0.24 and 0.46 m, respectively, were significantly different. Hence, at maximum extension, first-order stream lengths had decreased by 30% in Experiment 1, and by 50% in Experiment 2, from that of early network growth (Fig. 2.23). For example, first-order geometric mean length declined significantly during Experiment 2 as a linear function of elapsed time (Fig. 2.25). However, the slope coefficient of Experiment 1 was not significantly different from zero (i.e., L did not decline significantly). This difference

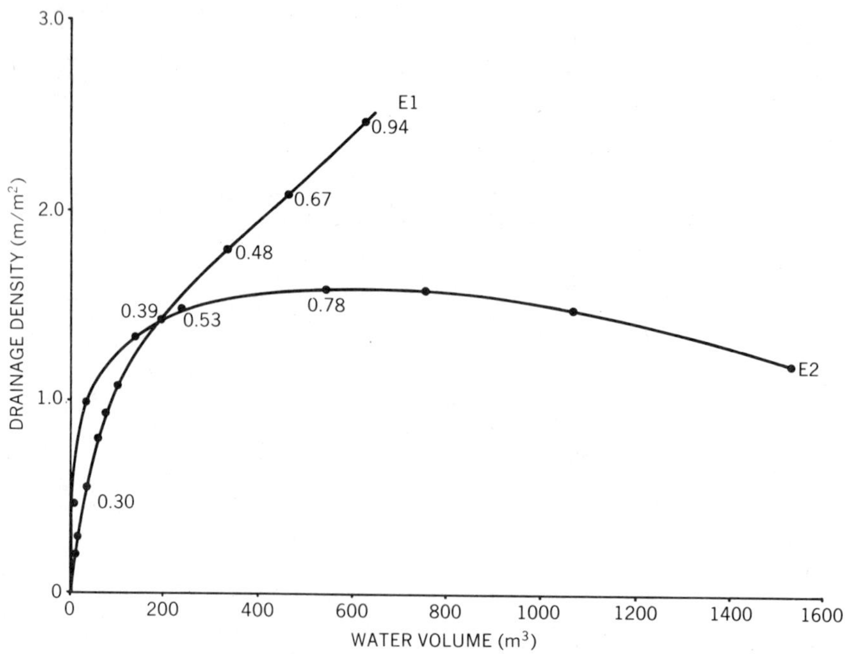

FIGURE 2.22. Change of drainage density during Experiments 1 and 2 (E1 and E2). Time is expressed as volume of water applied to surface of REF. Numbers refer to relief in meters after baselevel change.

reflects the greater significance of channel bifurcation and network elaboration during Experiment 2.

Following the stage of maximum extension in Experiment 2, the mean length of first-order channels again increased (Table 2.3, Networks 6, 7, and 8), which is to be expected as tributaries are lost. Of the first-order tributaries, approximately half were source (S-type) links and half were tributary–source (TS-type) links [an S-type first-order stream joins another S-type to create a second-order stream; a TS-type first-order stream joins a higher-order stream (Shreve, 1966)]. The TS-type tributaries were significantly longer than the S-type tributaries (Table 2.3); they drained large areas within the skeletal drainage network, whereas the shorter, S-type tributaries tended to be in the high relief area of the basin periphery. Interior links also increased significantly in length during network reduction. Horton's (1945) law of stream numbers—"the numbers of streams of different orders in a given drainage basin tend to approximate a geometric series in which the first term is unity and the ratio is the bifurcation ratio R_b"—was obeyed from an early stage of network development.

Conclusions about the law of stream numbers have been based on data collected

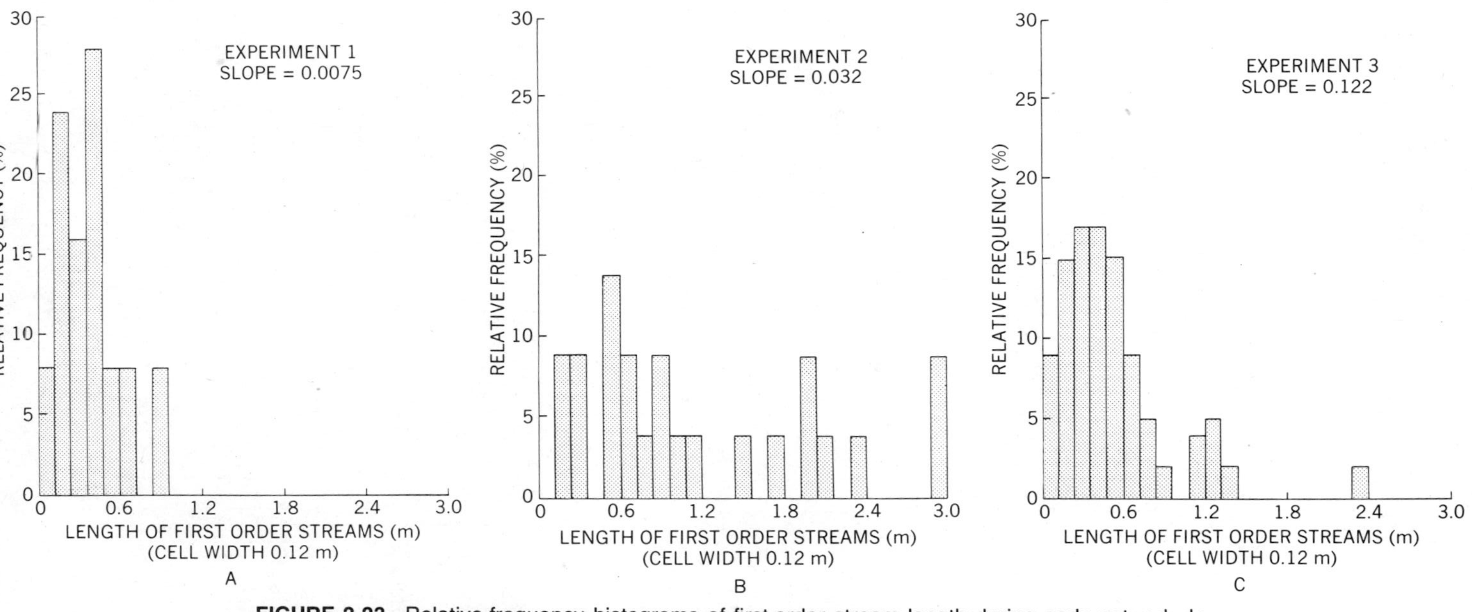

FIGURE 2.23. Relative-frequency histograms of first-order stream length during early network development in Experiments 1, 2, and 3 (from Parker, 1977).

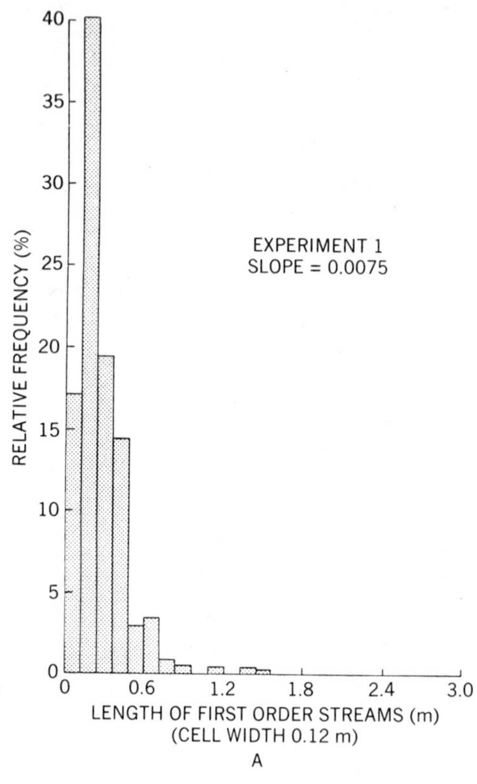

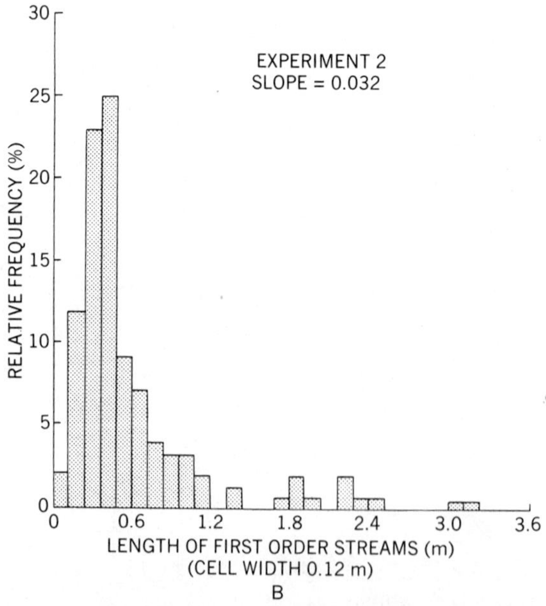

FIGURE 2.24. Relative-frequency histograms of first-order streams at maximum extension, Experiments 1 and 2 (from Parker, 1977).

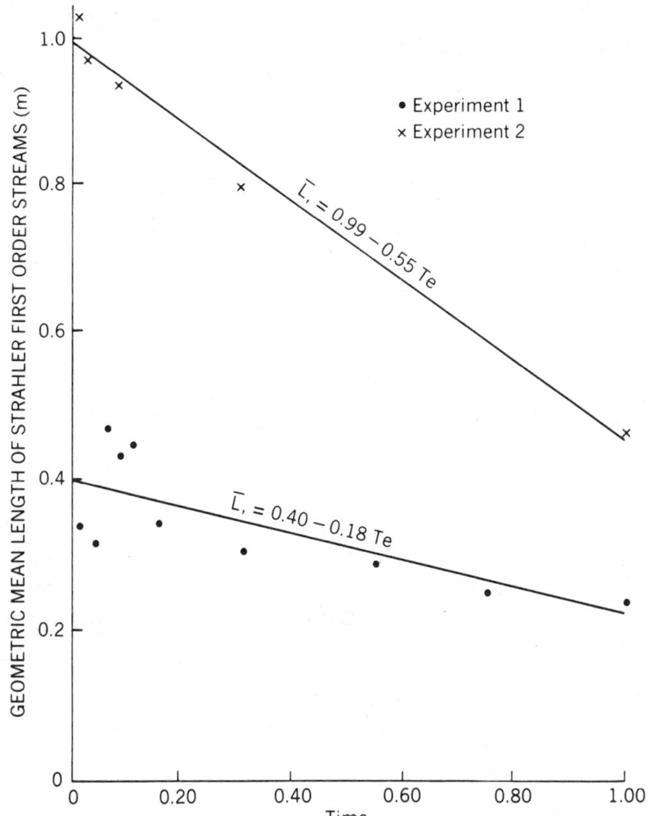

FIGURE 2.25. Change of geometric mean lengths of first-order streams during Experiments 1 and 2. Time is expressed as a ratio of the precipitation applied to form a given network to the total precipitaion applied up to maximum extension of the drainage network (after Parker, 1977).

from drainage basins which have evolved over a long period of time. Hence the networks are relatively stable. In this experimental study the networks were initiated on an undissected surface. Thus changes in stream numbers were observed as each network evolved from a very simple initial pattern. Such observations through time can be used to identify changes, if any, in the law of stream numbers during network growth, and four subbasins of Experiment 1 were selected for detailed study. The number of each Strahler-order stream was counted at each stage of growth as the networks extended into the initial surface (Fig. 2.27). As the number of streams increases, the Strahler order increases, so that the lines shift to the right; the Strahler-order abscissa is therefore an index of time.

The number of first-order streams increased with time until eventually, when there were 7 first-order streams, an additional second-order stream was formed. Similarly, 9 second-order streams were added (from 2 to 11) before an additional

TABLE 2.3 Statistical Summary of the Lengths of Exterior Links (First-Order Streams) for Experiment 2, Networks 5–8.

	Total Exterior Links			S-Type Links			TS-Type Links		
Network	*n*	Mean (cm)	Stand. Dev.	*n*	Mean (cm)	Stand. Dev.	*n*	Mean (cm)	Stand. Dev.
5	115	65.9	5.2	56	49.4	48.2	59	86.6	54.6
6	152	46.4	61.6	78	35.1	55.2	74	62.5	60.7
7	117	55.5	56.7	60	46.1	52.8	57	68.0	57.3
8	65	91.5	65.0	28	58.3	52.5	37	121.9	64.4

third-order stream was formed, which, in turn produced a fourth-order drainage network. This growth pattern produces concave lines on Figure 2.26, which explains the deviation from Horton's law of stream numbers discussed by Eyles (1968), Maxwell (1960, p. 12), and Schumm (1956, p. 603). It also produced marked variation in the bifurcation ratios (R_b, ratio of number of streams of one order to the number of streams of the next lower order) of first- to second-, second- to third-, and third- to fourth-order streams, particularly in the early stages of network growth. Addition of the lower-order tributaries repeatedly caused R_b to increase until an additional higher-order channel was created, which caused it to decrease again (Fig. 2.27). However, this oscillation of R_b stabilized as the networks grew (or became older), and the number-versus-order plots for fully developed basins were almost straight (Fig. 2.28). Therefore, the experiments support Eyles's (1968) view that a concavity in such plots is indicative of rejuvenation and an excess of first-order streams, rather than Smart's (1968, p. 25) appeal to geologic controls.

As the network of Experiment 2 continued to evolve after the stage of maximum extension, abstraction occurred. First-order streams declined from 150 to less than 70, and the network declined from order 5 to order 4 (Fig. 2.29). However, even during this period of abstraction, new first-order streams were initiated at the basin periphery, so that the number of first-order streams actually lost was even greater than appears from Fig. 2.29.

Network reduction began near the basin outlet and extended progressively headward with time. This indicates that drainage density was not constant within the total watershed at a given time. By dividing the watershed into two equal areas, internal and peripheral, the effect of network extension and reduction at different locations in the watershed may be noted (Fig. 2.30). While drainage density at the network periphery continued to increase by network extension, it was already declining by reduction in the network interior.

A description of network change was provided by Melton (1958), who demonstrated that stream frequency (F = number of Strahler segments per square mile) and drainage density (D-mi/mi^2) are related for mature drainage basins as follows:

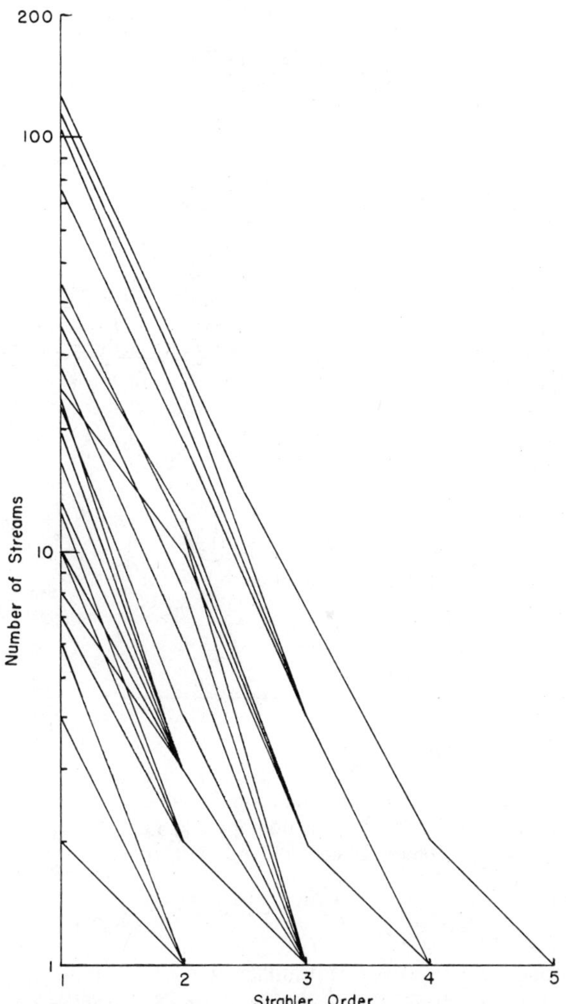

FIGURE 2.26. Number of streams for each Strahler order during network growth, Experiment 1, Networks 1–5 (from Parker, 1977).

$$F = 0.694D^2 \tag{2.3}$$

However, Abrahams (1972) showed that the exponent of similar equations relating F and D for five Australian study areas approached 1.0, which was the case for the experimental results (Fig. 2.31). The experimental data were, of course, obtained from an evolving network. Therefore, a network will not shift along Melton's line with time as it evolves. Rather, it will shift across Melton's line, with F

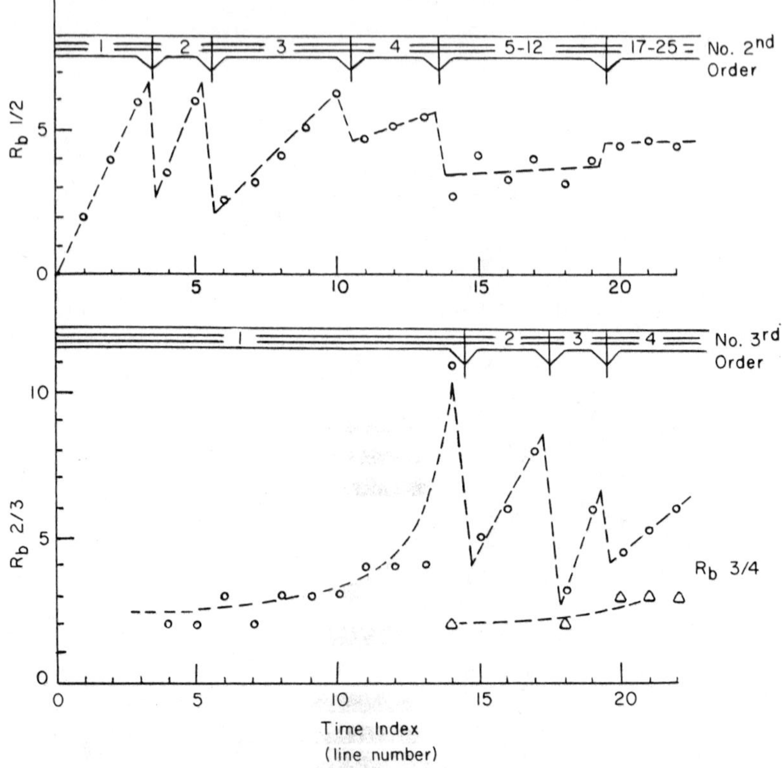

FIGURE 2.27. Changes in bifurcation ratio during early stages of network development. Lines from Fig. 2.26 were numbered consecutively from the origin and used as a time index. Numbers at top of plots give the number of second- and third-order streams present during that time period (from Parker, 1977).

increasing at about the 1.0 power of drainage density. This result emphasizes the need actually to follow a drainage basin through its evolution experimentally rather than to compare drainage basins of different apparent age from maps.

Network Change

The quantitative discussion above omits much qualitative information on network change, and a comparison of the photographs and network maps (Figs. 2.18–2.20) provides additional insight into network adjustments during the experiments. The maps (Fig. 2.18, 2.19) represent the channel thalweg within a valley, so that the sometimes spectacular changes of channel position are the result of thalweg shift on a wide alluvial valley floor.

The network in Experiment 1 was remarkably stable for the first 14 hr, as it extended into the undissected drainage area (Fig. 2.18). Lateral migration of the

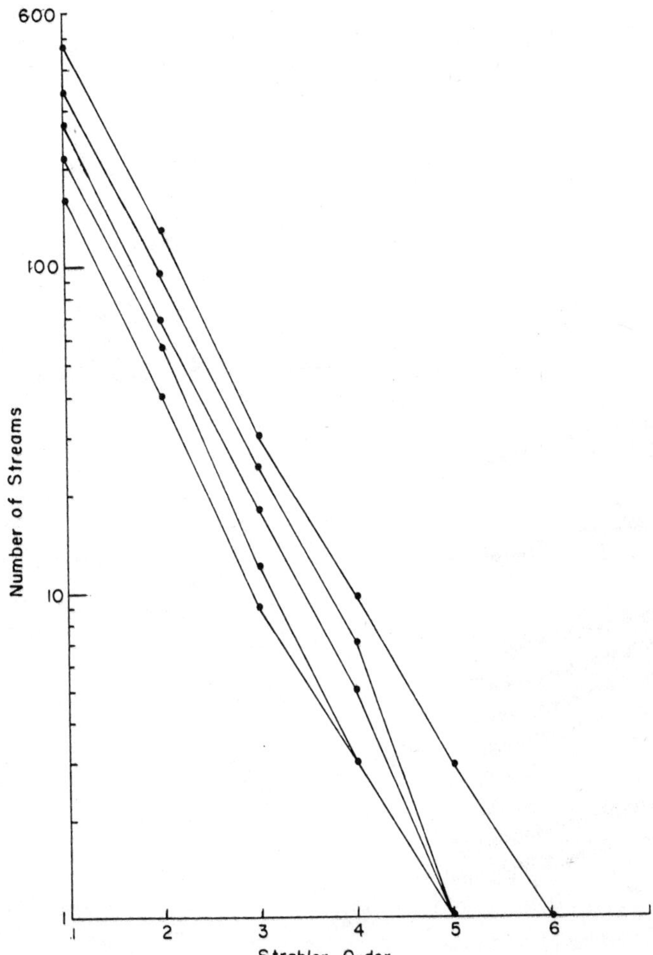

FIGURE 2.28. Number of streams for each Strahler order for fully developed Networks 6–10 during Experiment 1 (from Parker, 1977).

main channel widened the lower valley near the exit, rejuvenating or lengthening near-mouth tributaries. Network 6 (Figs. 2.18, 2.20*A*) had a relatively straight main channel with deeply incised tributaries in the lower basin and poorly developed tributaries in the upper basin. Baselevel lowering rejuvenated headwater tributaries and produced bank caving, lateral migration, and valley widening along the main channel. Incision of lower-basin tributaries produced sufficient sediment to build alluvial fans in the main valley and to initiate low-sinuosity valley bends (Network 7, Figs. 2.18, 2.20*B*). Additional baselevel lowering caused further

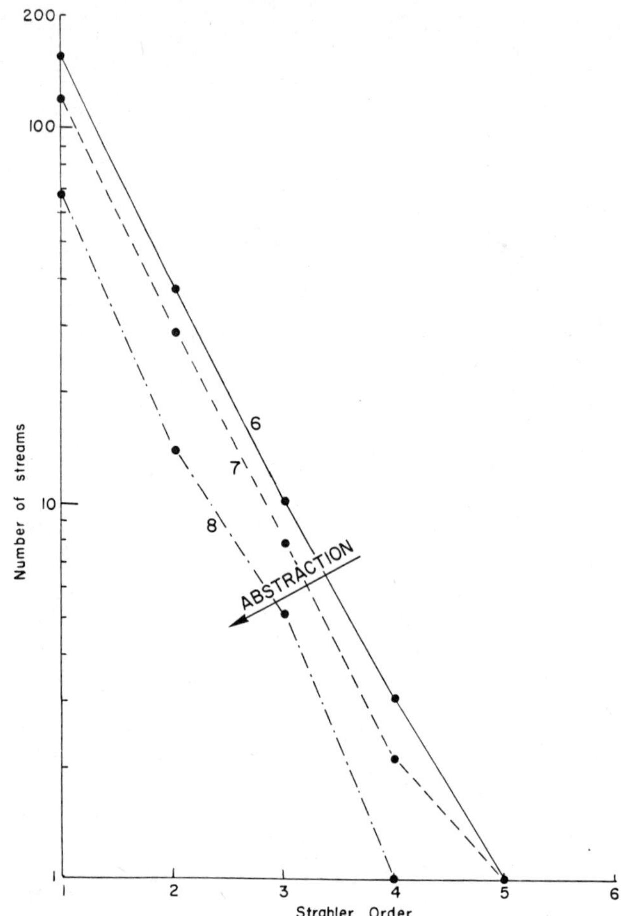

FIGURE 2.29. Number of streams in an abstracting network, Experiment 2, Networks 6 (maximum extension), 7, and 8 (from Parker, 1977).

network development, and lateral migration of the main channel influenced the behavior of the lower-basin tributaries and the position of tributary junctions in the lower right of Networks 9 and 10 (Figs. 2.18, 2.20*D, E*).

In Experiment 2, network growth was similarly orderly until baselevel lowering induced main channel migration and capture between Networks 4 and 5 (Fig. 2.19). Between Networks 5 and 6, a main channel shift was responsible for the migration of tributary junctions in the lower basin (Fig. 2.19), and some interior first-order streams were abstracted as the network expanded to the periphery. Between Networks 6 and 7, headward growth at the periphery was minor, abstraction of internal tributaries accelerated, and tributary junction shift continued in response to main

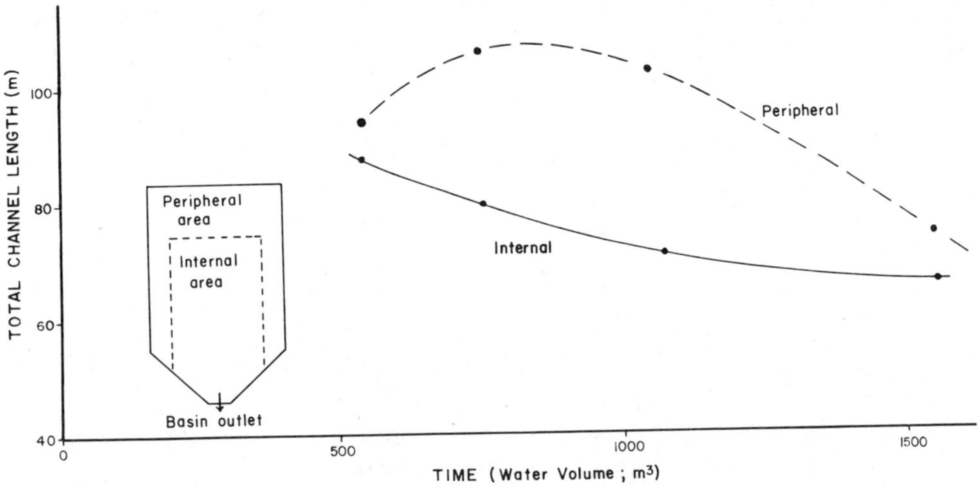

FIGURE 2.30. Change of channel length with time in peripheral and internal portions of REF (from Parker, 1977).

channel migration (Fig. 2.19). Between Networks 7 and 8, abstraction became dominant and the lowering of divides and valley alluviation reduced internal relief to the extent that streams could shift across divides, primarily near junctions of larger tributaries in the basin interior (Fig. 2.19). Lateral migration of both the main channel and major tributaries was a major process at this stage of maximum extension and early reduction that allowed tributary-junction shift in both the down-

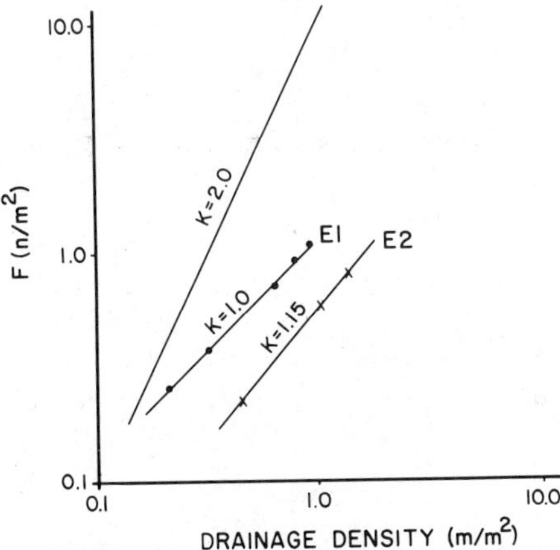

FIGURE 2.31. Comparison of relation between stream frequency and drainage density for Experiments 1 and 2 and Melton's regression line. *K* is exponent of Melton's equation.

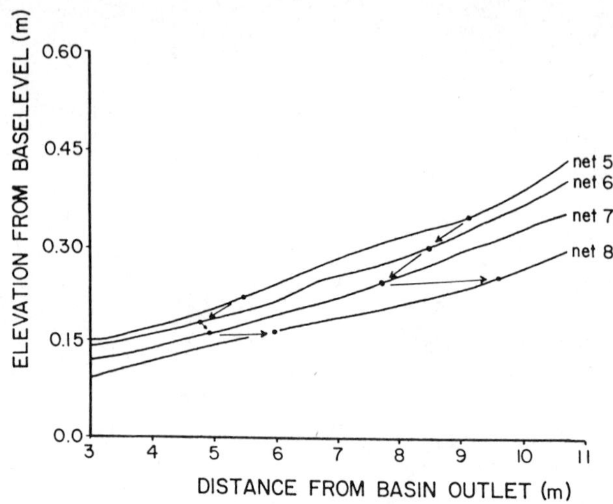

FIGURE 2.32. Successive longitudinal profiles of the main channel for Networks 5–8, Experiment 2. Shift of fourth-order tributary junctions are shown.

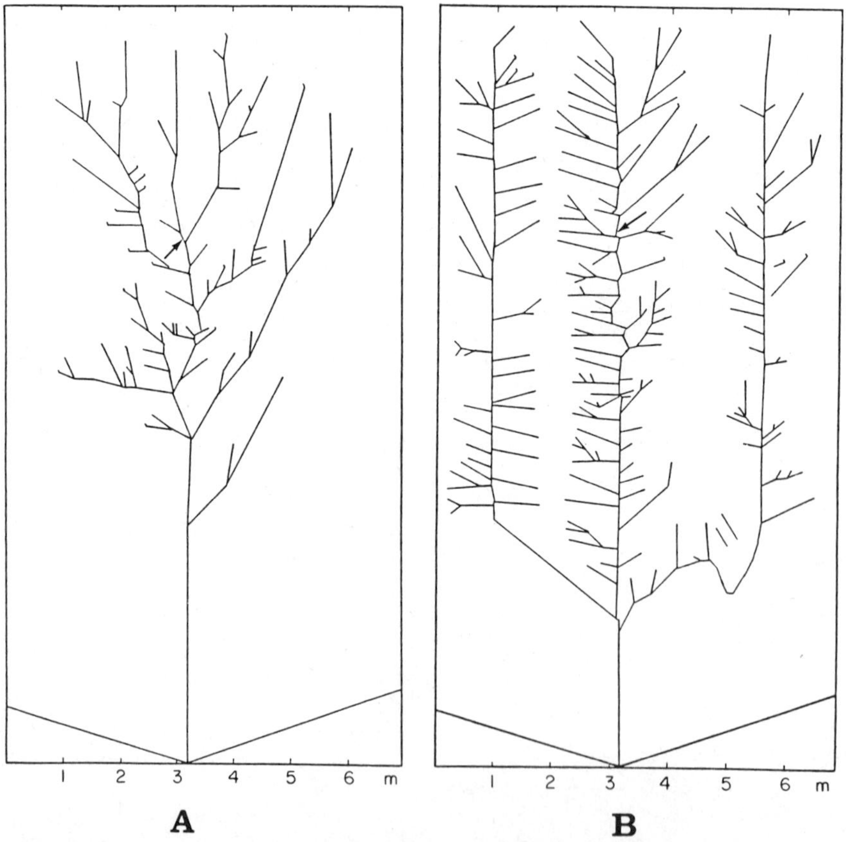

A

B

FIGURE 2.33. Drainage networks developed during O'Brien's (1984) experiments. (*A*) Drainage pattern developed on a plane with a 0.037 slope toward the outlet (O'Brien, 1984). (*B*) Drainage pattern developed on a surface with three precut valleys with sideslopes of 2.5% and an overall slope toward the outlet of 0.033 (O'Brien, 1984).

stream and upstream directions (Fig. 2.32). Shift appeared to be downstream during extension and upstream during reduction, causing reduction of tributary entrance angles during extension and an increase during the later stages of evolution, as suggested by Lubowe (1964) and Schumm (1956).

The different modes of network growth discussed above can also result in different generations of channel formation in one drainage network as well as different drainage patterns. For example, in Horton's model (Fig. 2.5) there are four generations, or ages, of channel development, and in Fig. 2.1*A* there are three, whereas during the expansion growth mode (Fig. 2.1*B*) channels progressively form as the network grows headward. Finally, the extension-growth mode (Fig. 2.1*C*) is probably intermediate in character, with the bulk of the channels forming early and then a second generation of channels filling in the voids.

A comparison of two extreme types of networks is provided by O'Brien (1984), who showed that a network that developed by extension on a 0.037 slope has a dendritic or subdendritic pattern. In this pattern only a few first-order channels developed at obtuse angles as the valley sides were formed (Fig. 2.33*A*). In a second experiment O'Brien cut a rudimentary network into a surface with about the same slope (0.033) and created valley sides by building 2.5% side slopes adjacent to the main channel. This is analogous to channels that have rapidly incised to form valley sides on which low-order tributaries could develop. The resulting two-generation pattern has large angles of junction between the first-generation channels and numerous second-generation tributaries (Fig. 2.33*B*). Therefore, and one- and two-generation drainage networks may display very dif-

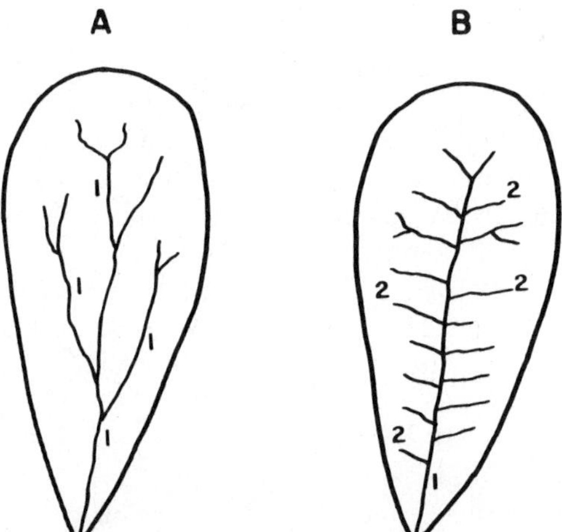

FIGURE 2.34. Effect of growth mode on drainage pattern and tributary generaton. Numbers refer to tributary generation. (*A*) Subparallel pattern expected with dominant elongation (extension mode). (*B*) Pinnate pattern expected when main channel incises and second-generation tributaries form on valley side slopes.

ferent drainage patterns ranging, for example, from subparallel to pinnate (Fig. 2.34*A*, *B*).

DISCUSSION

Elements of Glock's (1931) model could be identified during evolution of the experimental drainage networks, but they were less distinct than he envisaged. Although he noted that extension and reduction could coexist regionally or in different parts of a drainage basin, there was marked overlap between these growth stages in the experimental networks. The interior of the basin progressed through all of the stages to maximum extension and then into reduction while some parts of the basin had not progressed beyond initiation. Therefore, there was never a basin-wide state of maximum extension as Glock had envisaged (Fig. 2.6). The relative importance of the growth stages and erosion processes were also very much modified by different initial conditions in the experiments, although the same agents of erosion were acting. The most obvious difference was in the relative importance of the processes of nickpoint recession and channel incision; in Experiment 1, with initial baselevel lowering, the former was dominant, whereas in Experiment 2 it was, at least for 33 hr, of little importance. The effect of baselevel lowering probably decreases as the initial surface slope increases, because in steeper basins the amount of relief (and hence potential energy available for erosion) is already large. On low-angle, low-relief surfaces, baselevel lowering at the outlet concentrates potential energy and relief at one point so that there is intense localized erosion. On steeper surfaces, potential energy is more evenly distributed, and this causes erosion throughout the basin, whether or not baselevel lowering provides a local concentration of relief difference.

The evolution of the experimental drainage patterns from two different initial conditions (Parker, 1977) produced very different dendritic drainage patterns. Two growth modes were apparent (Figs. 2.1, 2.21). One can be referred to as the *expansion mode* (Fig. 2.1*B*) because the network expands slowly but with an essentially fully developed pattern into the available area. The other is the *extension mode* (Fig. 2.1*C*), so called because long low-order channels elongate rapidly and block out the network drainage areas. The first-order streams are longer and the pattern is more treelike (dendritic), so that tributaries join the main channel at a smaller angle of junction (Figs. 2.1, 2.33*A*, 2.34*A*).

Stream capture is a process that has fascinated geomorphologists, and it has a long history of discussion (Davis, 1909; Horton, 1945; Howard, 1971b). Mosley (1972a) observed many examples of capture during his rill experiments, particularly on the steeper surfaces, but only during the early stages of network development. Apparently, once the channels became incised, they were locked into place. Although a falling baselevel might, by increasing relief, induce landscape and network instability and allow capture, the experiments provide little support for Horton's model of network evolution (Fig. 2.5), which includes major changes in network geometry as a result of "micropiracy" and cross-grading. Instead, the experiments indicate that there is a close correspondence between the initial pattern of lines of water concentration and flow and the final incised channel network. This situation should be expected, as the formation of channels is dependent upon initial and continued flow concentration.

The correspondence between the characteristics of natural drainage networks and branching networks simulated by random growth processes has been used, in somewhat circular fashion, both to justify the computer simulations as analogs of the real world and to conclude that natural networks are topologically random (in the absence of geologic controls). Apart from the fact that all the simulation models have an implicit deterministic component because of their matrix or grid-cell basis (analogous to Schumm's (1956) constant of channel maintenance concept), such claims are unfortunate when they turn attention from deterministic, causal explanations of network form to the treatment of networks based on topology. For an alternative view, the reader is referred to Kirby (1976) and Shreve (1975). The experiments demonstrate that network growth is a deterministic process; indeed, the observations of McLane (1978) and Mosley (1972) indicate that final network form is largely determined by initial microtopography. Furthermore, variations in drainage density within the experimental watersheds both at one time and at one place during the course of an experiment (Fig. 2.30) demonstrate that simulations of network growth that are based on a single constant grid size (constant of channel maintenance) are inadequate.

During reduction the drainage network loses channels, starting in the core of the basin. Internal reduction and peripheral expansion can occur simultaneously, thus prolonging the period of maximum drainage density (Fig. 2.30). Again, it is clear that any simple model of drainage network evolution will be inadequate to describe the complex development of a drainage system on surfaces of different configuration and inclination. The variability of drainage density within the basin also casts doubt that a constant of channel maintenance can meaningfully be identified, except perhaps at the beginning of the period of maximum extension (Fig. 2.18, N10; Fig. 2.19, N5).

In spite of the complexity of network growth, the overall regularity of network geometry at all stages of growth, as demonstrated by conformance to Horton's laws, is consistent with the observations of regularity made by Carter and Chorley (1961) and Morisawa (1964) in natural expanding stream systems. The networks in all experiments bore the marks at maximum extension of the imposed initial conditions, although no part of the initial surface had escaped surface lowering. Therefore, despite the topologic regularity associated with observance of Horton's laws, the experiments demonstrated that network geometry is not time independent (Hack, 1960). Hence the landscape is not explicable simply in terms of dynamic equilibrium concepts in which landforms are seen as a function of the interaction between landscape materials and landforming processes without reference to landscape history.

From a practical point of view the studies show that the development of channels into an uneroded area such as a mine-tailings pile can be predicted if the configuration of the surface is known. The rate of erosion and growth of the network will depend both on the gradient and shape of the surface, with a convex surface from which flow diverges being the least affected by network development. O'Brien (1984) also demonstrated the stability of a drainage network. After full network development he obliterated all first-order channels, and then he resumed the application of precipitation. Almost all first-order tributaries redeveloped, although not in identical locations, which indicates that all were required for the export of water and sediment. Such observations of evolving drainage networks have obvious relevance to the management of highway cut-and-fill slopes, mine tailings, and other artificial landforms, as well as to the understanding of natural drainage networks.

3 Runoff, Sediment Yield, and Placers

One of the objectives of the experimental investigations was to establish the effect of drainage basin morphology on runoff and sediment yield. Parker (1977) and Zimpfer (1982) collected considerable data on runoff and hydrograph characteristics, and Mosley (1972), Parker (1977), Harvey (1980), Weaver (1984) and Zimpfer (1982) obtained data on sediment yield, and Adams et al. (1978) and Lidstone (1981) studied heavy-mineral movement and placer formation.

The landscape in the REF was generally steep and rugged, and so the results are most closely related to the hydrology of high-energy drainage systems with large sediment yields. A major disadvantage arising from this was the rapid response of the REF. Hydrograph changes were measured in seconds, and slight changes of main channel sinuosity altered flow duration, rise time, and start of flow. For this reason only a few general statements can be made concerning runoff. However, one advantage of the experiments is that unlike the prototypes, where comparison among drainage basins may involve some differences of soil and sediment type, the materials in the REF did not change.

RUNOFF

Despite numerous investigations, many details of drainage basin hydrology and geomorphology remain poorly understood. For example, since the 1930s, the Hortonian theory of overland flow has been one of the underlying concepts in hydrology; however, the assumption that the entire watershed contributes overland flow is not applicable in many areas, and it is being replaced by the variable-source-area and partial-source-area concepts (e.g., Dunne and Leopold, 1978, pp. 259, 271).

Experimental Studies

Black (1970, 1972) conducted a series of experiments with small laboratory models to investigate the relation between watershed hydrology and geomorphology. In the first of these studies, the effects of slope, drainage density, and soil depth upon the outflow hydrograph were investigated. When the storms were of sufficiently long duration for the basin to reach equilibrium, the resulting hydrographs were the same regardless of soil depth, slope, or drainage pattern. However, for shorter-duration storms and as slope increased, the peak discharge increased and decay time decreased. Surprisingly, hydrograph changes could not be systematically related to drainage density.

In later studies, Black (1972) broadened the investigation to include the effects of rainfall intensity, antecedent moisture, drainage pattern, channel storage, basin shape, and direction of storm movement. When rainfall intensity was increased, peak discharge increased, decay time decreased, and time of concentration and lag time remained constant. As the antecedent moisture of the drainage basin was increased, peak discharge increased, but time of concentration and lag time decreased. In this second study, peak discharge was inversely related to drainage density. In the final study in this series (Black and Cronn, 1975), peak discharge, time of concentration, decay time, and flow duration were found to increase as the basin area was increased. Lag time and the rate of recession were not related to basin size.

Another model study (Roberts and Klingman, 1970), using three different rainfall intensities, provided information on the effect of various storm and basin variables upon the runoff hydrograph. The high two intensities produced similar hydrographs, one proportional to the other. For the low-intensity rainfall, the runoff peak was rounded and delayed. The higher the rainfall intensity, the more quickly equilibrium discharge was achieved. Storms moving downbasin had a delayed rising limb, a high peak discharge, and a rapid recession limb. As permeability increased, the time required to each equilibrium increased. Peak discharge increased as antecedent moisture was increased, and the rising limb of the hydrograph was considerably delayed under dry antecedent conditions.

For the most part, none of the results of the experimental studies is surprising except for the lack of clear relations with drainage density. One would assume that drainage density strongly influences the hydrologic character of a drainage basin because it reflects the efficiency with which water can be drained from the drainage basin.

Hydrology and Network Evolution

Data were collected in the REF that provide information on the effect of drainage network evolution on the hydrologic character of a drainage basin. In order to examine changes, if any, in the hydrologic response of the basin to changes in its geomorphic configuration, a series of hydrographs was generated at the available intensities. Discharges were measured at a 60° V-notch weir at the mouth of the REF. Zimpfer (1982) collected hydrologic data at 2-hr intervals during drainage network development. When the hydrographs that were generated during the first 6 hr of Zimpfer's experiment are examined, the most noticeable change is hydrograph shape (Fig. 3.1). At the end of the first 2-hr period, the drainage network was poorly developed (D = 0.3 m/m^2) and overland flow was more important than channel flow in conveying water to the basin outlet (Fig. 3.1*A*). After an additional 2-hr of network development drainage density had doubled (Fig. 3.1*B*) and hydrograph shape reflected a more efficiently drained system. The center of mass of the runoff shifted to the left, and the duration of the rising limb of the hydrograph was shorter than the duration of the falling limb. During the next 2 hr of network development, the hydrograph continued to change shape (Fig. 3.1*C*) and peak discharge increased. A better-defined peak with steeper rising and falling limbs was associated with a tripling of drainage density between 2 and 6 hr (Fig. 3.1). Although baselevel was lowered after 6 hr, the general shape of the hydrograph remained unchanged during the remainder of the experiment. Slight variations in hydrograph characteristics did occur, but this constancy illustrates that there was equilibrium between drainage network morphology and the hydrologic character of the drainage basin.

Parker (1977) studied the effects of changing basin morphology on hydrology for Networks 6, 8, 9, and 10 of El (Fig. 2.18). Unlike Zimpfer, he began his investigation after the basin was well developed, but drainage density increased as a result of baselevel lowering (Table 2.2). Parker concentrated on the effects of drainage density and relief, which he combined by multiplication ($D \times R$) to obtain a ruggedness number (Strahler, 1958).

To evaluate changes in the hydrographs with changes of both relief and drainage density, Parker used the ratio between discharge at the hydrograph peak from the 1-min-duration events (Q_p) and the equilibrium discharge rate (Q_e) when precipitation was applied for a long time at the same intensity. An increase in ruggedness number produced an increase in this Q_p/Q_e ratio up to a ruggedness number of about 1.0 (Fig. 3.2) for all intensities. The hydrographs thereafter appeared to maintain a nearly constant Q_p/Q_e ratio as the ruggedness ratio continued to increase.

One of the factors that affected the large changes observed in the Q_p/Q_e ratio at lower ruggedness numbers was the change in the amount of overland flow. In networks with low ruggedness numbers the channel system was fully developed only in the lower two-thirds of the basin (see Fig. 2.18, N6) when the first hydrograph set was generated. Thus nearly one-third of the basin was characterized by overland flow. With time the upper one-third of the basin was dissected, which produced higher ruggedness numbers and higher peaks. After this dissection, further channel growth produced little change in the peak flows (Q_p) of the hydrographs. Clearly, channelized flow is a more efficient way to move water, and therefore Q_p will increase as overland flow decreases.

The ruggedness numbers of the REF basin (from 0.348 to 2.314) are comparable to natural basins. Strahler (1958, p. 296) reports ruggedness number values of 0.022 for the Gulf Coastal Plain of Louisiana, 0.35 of the Ozark Plateau of Missouri, 1.0 for the Verdugo Hills in California, and 1.10 for the Perth Amboy Badlands of New Jersey. Patton and Baker (1976) report values as high as 1.2 in

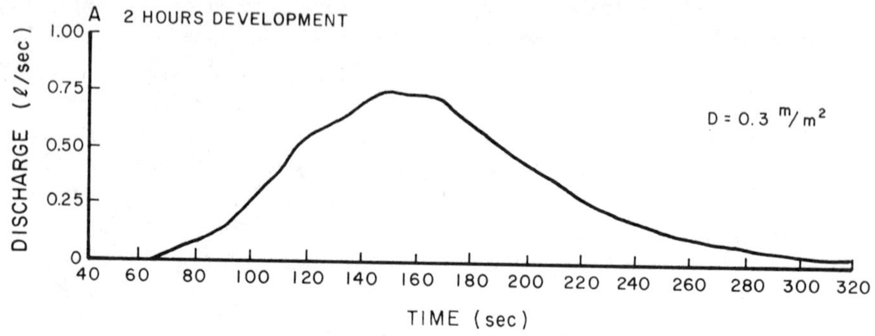

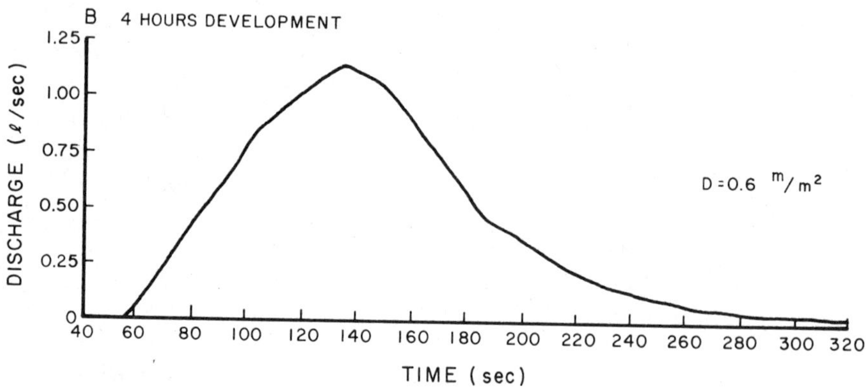

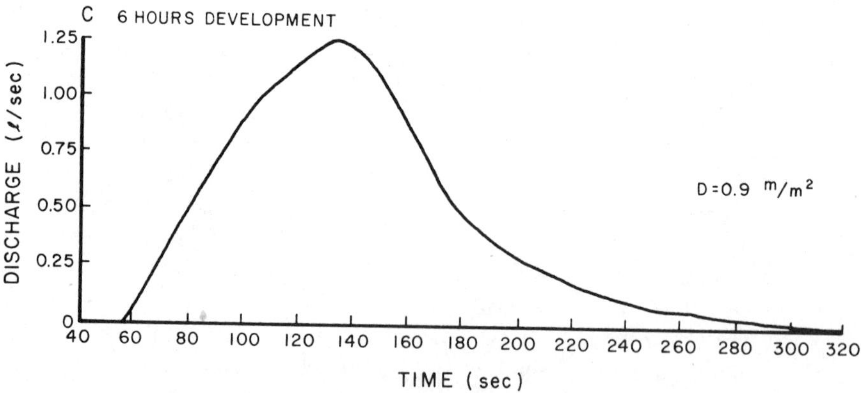

FIGURE 3.1. Hydrographs showing the change of hydrograph shape after 2, 4, and 6 hr of drainage network development (2-min, 5.8-cm/hr storms) (from Zimpfer, 1982).

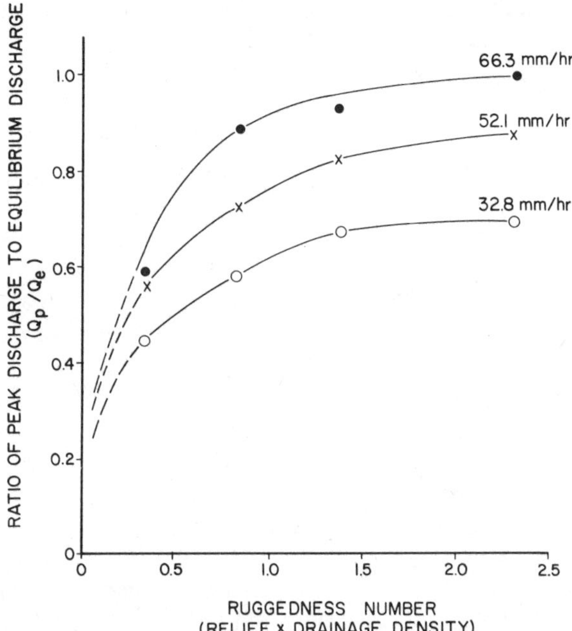

FIGURE 3.2. Changes in the ratio of peak discharge to equilibrium discharge (Q_p/Q_e) with changes in the ruggedness number (EI). Each precipitation intensity is shown (from Parker, 1977).

central Texas, 10 in the San Dimas Experimental Forest, southern California and 11.6 in north central Utah. Patton and Baker (1976) conclude that high ruggedness numbers are associated with flash flooding. Figure 3.2, however, suggests that, beyond a certain value of ruggedness, increasing relief and drainage density should not significantly affect flood peaks, because at a ruggedness number of about 1.0 to 1.5 the water leaves the basin as rapidly as possible.

Effects of Source Area Location

The location of runoff source areas within a drainage basin should also affect the shape of the outflow hydrograph (McLane, 1978; Zimpfer, 1982). To study the effect of runoff source area locations on hydrographic characteristics, McLane (1978) and Zimpfer (1982) created source areas of runoff by placing 20 × 30-cm rectangular plastic sheets on the surface of the REF basin. Rainfall falling upon the plastic sheets immediately produced runoff.

Ten percent of the drainage basin was covered with the plastic runoff-source-area sheets. Four different patterns were used by placing sheets upon divides, hillslopes, the floodplain, and randomly throughout the basin. For a random pattern, the locations were chosen using a random number table to select grid coordinates.

After the plastic source areas were placed in the basin in the desired pattern, a 10-min high-intensity "saturation" storm was applied. Then storms of different intensity and duration were applied to the basin. For each runoff event, the hydrograph was characterized by recording the start of flow, arrival of the flood peak, and end of flow.

The placing of plastic to create source areas of runoff may simulate geologic variations within the basins as well as the effects of urbanization fire, and timber harvesting. Figure 3.3 illustrates the effects of source area location on time to start of flow. The source area patterns had the most dramatic effect at the lowest intensity (Fig. 3.3*A*). Time to the start of flow decreased as the source areas were placed closer to the floodplain. Flow from 1-min storms took longer to reach the outlet than flows from 5-min storms. In addition, the effects of pattern were greater for 1-min storms than for 2-min storms. This may have been because the 1-min storm event had ceased before the channel flow reached the basin outlet. The difference between the 1- and 2-min storms decreased as the source areas were moved from the divides to the floodplain (Fig. 3.3*A*). The response to intermediate-intensity storms was less clear (Fig. 3.3*B*) and the effects of pattern for high-intensity storms was negligible (Fig. 3.3*C*).

Time of cessation of flow in each case remained essentially unchanged, so that the duration of runoff was strongly influenced by the rapidity with which runoff started. Thus flow duration increased as source areas were moved from the divides toward the floodplain for the lowest intensity, although there were no clear relationships at intermediate- and high-intensity precipitation.

Storm intensity, storm duration, and the pattern of runoff source areas had little clear influence on peak discharge. For the lowest intensity, peak discharge increased slightly as the source areas were moved toward the floodplain. This reflected a decreased opportunity for water to infiltrate before leaving the basin. Moderate- and high-intensity storms showed no relation between peak discharge and source area location.

The response of basin water yield was similar to that of peak discharge. Water yield for both 1- and 2-min storms at the lowest intensity increased as the runoff source areas were moved closer to the stream channels. Untreated basins and basins with only the divides covered provided increased opportunity for infiltration to occur and a larger portion of the total precipitation infiltrated during a 1-min storm than during a 2-min storm.

To summarize, the results (Fig. 3–3) reveal an interaction among source area pattern, storm duration, and storm intensity with start of flow and flow duration. Peak discharge and total runoff, although affected by precipitation intensity and duration, showed little effect of source area pattern. At the lowest intensity, both storm duration and source area pattern had an effect on the hydrograph characteristics. However, at intermediate and high intensity, the effect was not clear. Similar effects have been observed in studies on the effect of urbanization on storm runoff (e.g., Waananen, 1969).

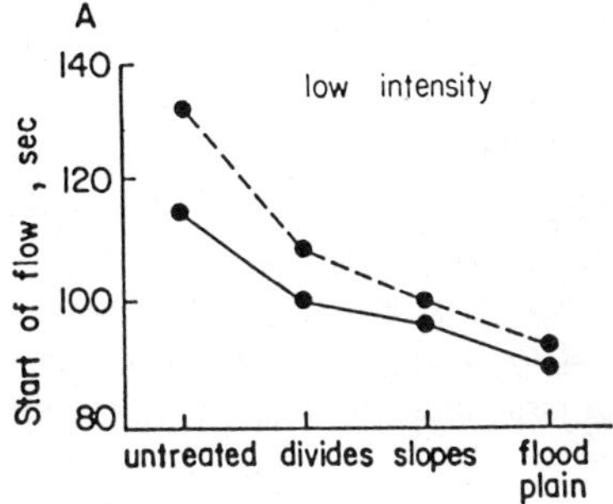

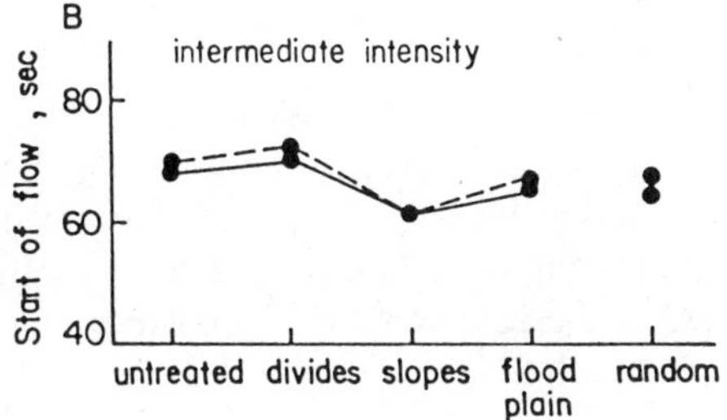

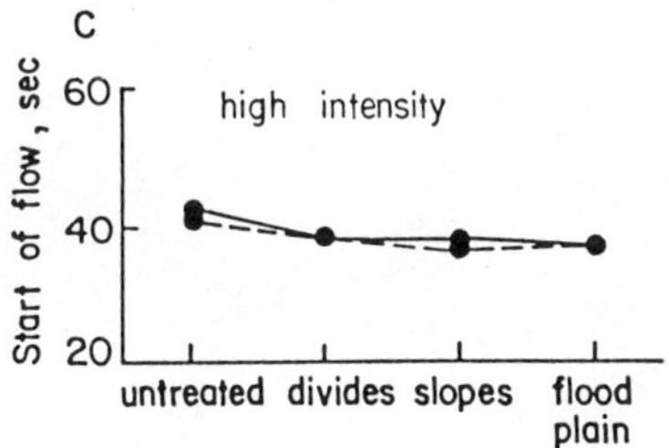

FIGURE 3.3. Plots showing the relationship between the time to the start of flow and the pattern of runoff source areas for 1-min (dashed line) and 2-min (solid line) storms. (*A*) 2.0-cm/hr rainfall intensity; (*B*) 2.9-cm/hr rainfall intensity; (*C*) 5.8-cm/hr rainfall intensity (from Zimpfer, 1982).

SEDIMENT YIELD

Sediment yield is closely related to drainage basin morphology, both because its magnitude is controlled by the intensity of erosion processes, which are themselves influenced by morphology, and because erosion and sediment transport are the mechanisms whereby drainage basins are molded. Accordingly, sediment samples were taken during the experimental studies in the REF to elucidate the relationships between sediment yield and drainage basin morphology.

Controls on Sediment Yield

The rate of drainage basin evolution was shown in Chapter 2 to vary during the experimental runs. Following the start of the experiment or any change in experimental conditions such as baselevel lowering, rates of change were initially high, but then they rapidly declined. Sediment yields reflected these trends at constant precipitation intensity. For example, in Experiment 1 (E1) there was an exponential decline of sediment yield, with the erosion rate at maximum extension being less than 10% of the initial rate (Fig. 3.4). There is a large amount of scatter about the best-fit relationship

$$Y_1 = 850\ V^{-0.86} \tag{3.1}$$

in which Y_1 is sediment yield (g/sec) and V is volume of water applied to the basin (m^3), but Eq. (3.1) explains 76% of the variability of sediment yield (Parker, 1976).

Each time that basin relief was increased by baselevel lowering during E1, nickpoint retreat and lateral shift of the incising channels increased the rate of surface erosion and channel downcutting and sediment yield increased dramatically (Fig. 3.5). As the drainage basin adjusted to the new conditions, an exponential decline in sediment yields was observed; but each time baselevel was lowered, the

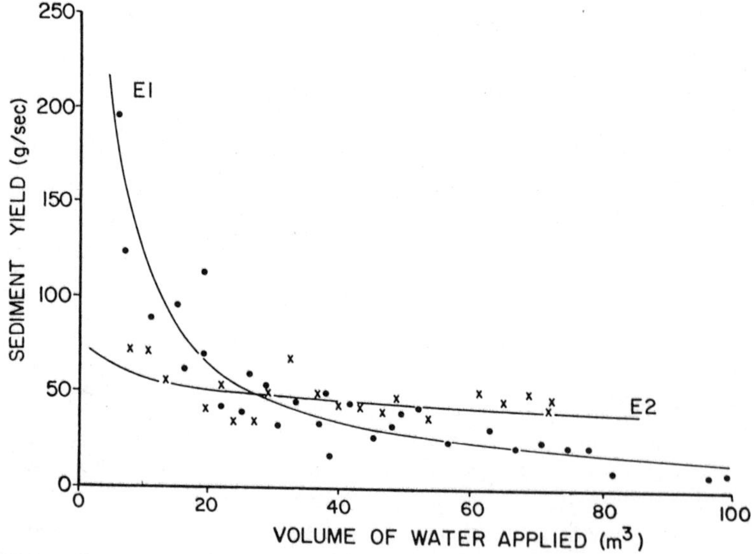

FIGURE 3.4. Change of sediment yield with time (expressed as volume of water applied to REF) during early stages of Experiments 1 and 2 (EI started with baselevel lowering) (from Parker, 1977).

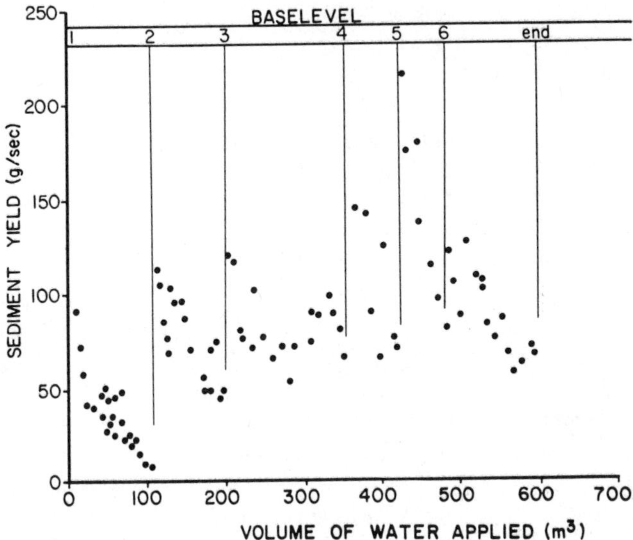

FIGURE 3.5. Change of sediment yield during Experiment 1 with six baselevel changes (from Parker, 1977).

initial peak sediment yield and the final "background" yield was higher than before. This occurs because the volume of material above baselevel and potential energy was increased with each baselevel lowering, and, of course, increased relief and slope promoted higher erosion rates (Schumm, 1956). The increased basin relief induced deeper valley incision, particularly near the outlet, leading to an increase in valley wall instability and slumping. Since slumps tended to occur randomly, thereby adding large quantities of sediment to the drainage system, sediment yields fluctuated dramatically with each baselevel lowering.

During E2 there was no initial baselevel lowering, the network developed by direct channel incision and minor nickpoint development, and sediment yields were lower, despite the greater initial slope (Fig. 3.4). The best-fit relation

$$Y_2 = 78\ V^{-0.15} \tag{3.2}$$

has a much smaller exponent than Eq. (3.1) reflecting the slow decline of sediment yields as the network developed, but the final equilibrium yield, produced by surface erosion processes such as rainsplash, was higher because of the greater initial slope. As the E2 channel network extended throughout the basin, sediment transport became more efficient, and an essentially constant sediment yield was maintained. During E1, on the other hand, deposition in and widening of the lower main channel tended to reduce transport efficiency and to increase sediment storage and decrease sediment yield. The tendency in E1 for an increase of initial sediment yields, a higher final background yield, and greater variability of sediment yield after each baselevel lowering (Fig. 3.5) was repeated during E2 (Fig. 3.6). Increased variability was more pronounced for the later baselevel changes, when the drainage network was well established and valleys were deeper before baselevel lowering. This combination produced more frequent mass failure of valley walls.

In summary, following baselevel lowering (time zero), there is an initial exponential decay of sediment yield values. This is followed by an increase in sediment yield (or at least a higher variance) and then a further decrease (Fig. 3.6). This entire sequence takes place before maximum extension or while the network is still growing. After maximum extension, when network reduction is the dominant process, sediment yield continues to decrease until it apparently stabilizes at a minimum value, which probably reflects the effect of raindrop impact on the experimental surface.

The overall trend of sediment yield through time is an exponential decay (Fig. 3.7), but variability is so great that trends and relations between basin morphology and sediment yield are difficult to establish. Such variability is characteristic of all sediment yield data, and it points to the importance of evaluating not only mean values of sediment yield but also the patterns of variability. In the field, considerable sediment yield variability is assumed to result from climatic fluctuations and

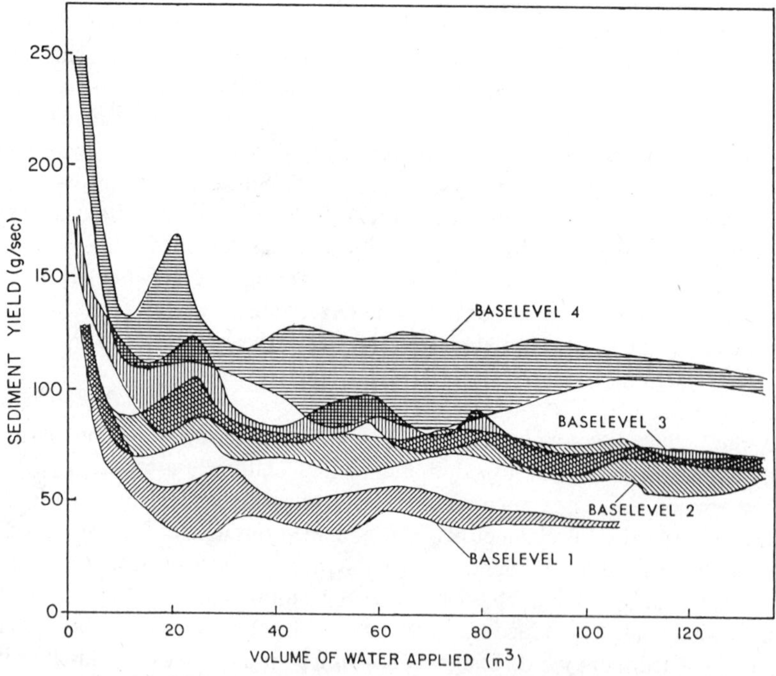

FIGURE 3.6. Change of sediment yield with time at baselevels 1–4 during Experiment 2. Patterned zones indicate the variability of individual observations (from Parker, 1977).

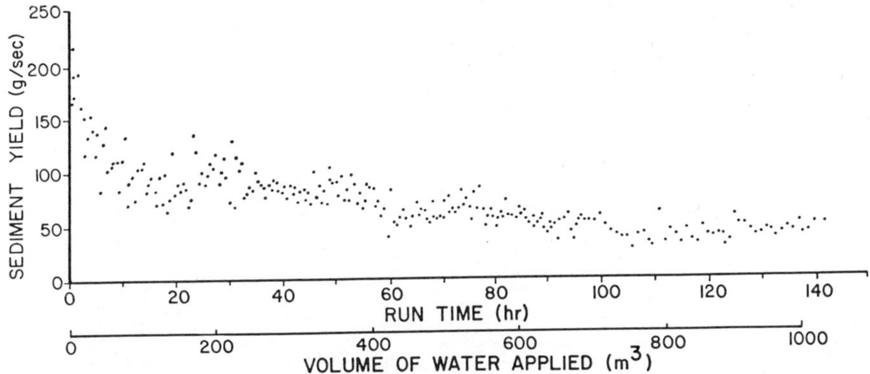

FIGURE 3.7. Sediment yield variations with time following baselevel lowering, Network 5, E2.

land use changes. However, the experimental results show that sediment yields naturally maintain a high variance. Discussion concerning the cause of this variability is presented in Chapter 4.

Weaver (1984) also measured sediment yields from the REF. His experimental design was very different from that of Parker (see Chapter 9). Weaver was studying alluvial fan development, and he used only the upper half of the REF as a sediment source area while the fan grew in the lower half. He measured sediment yield both at the outlet of the sediment source area and by mapping total deposition on the experimental alluvial fan as it grew.

Weaver documented the same type of sediment yield changes and the great variability of sediment yield as described by Parker (Fig. 3.8). For example, his data show a well-defined secondary peak in sediment yield at roughly 105 m^3 followed by two broad peaks at 145 and 180 m^3, respectively. Data beyond 190 m^3 reflect the addition of impervious plastic squares on interfluves within the sediment source area, which caused a decrease of sediment yield (see discussion, Chapter 9).

Much of the inherent scatter in Weaver's sediment yield data can be smoothed by plotting a moving average of the measured values (Fig. 3.8). Once again, the sharp secondary peak at 110 m^3 as well as the two broad crests at 145 and 180 m^3 remain as anomalies from the overall exponential decline in sediment yield with time.

A second procedure was employed to measure sediment yield and to verify these longer-term trends. A 4.9 × 6.4-mm grid of 279 pins was established in the area of the accumulating alluvial fan, each pin representing 0.09 m^2 of depositional surface. Pin exposure was repeatedly remeasured during the experiment, and successive accumulations at each pin were summed to derive the total sediment deposited (sediment yield) during the 44 separate runs of the experiment (Fig. 3.9). By plotting a moving mean of these data, the resulting curve can be compared to

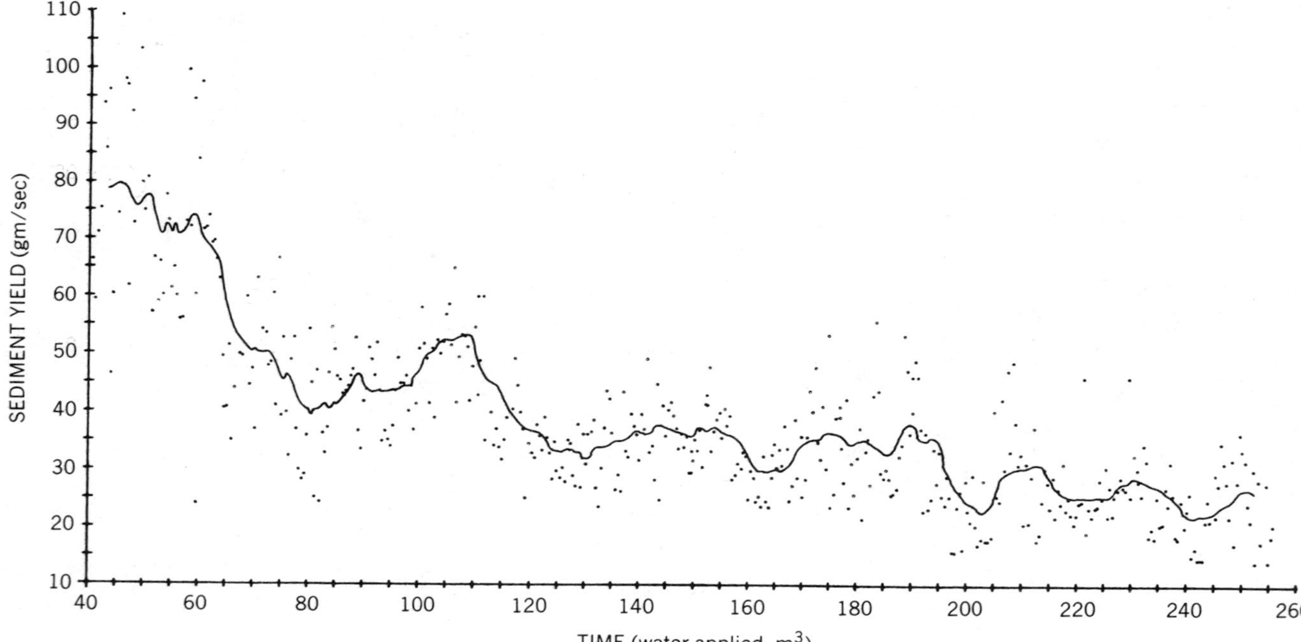

FIGURE 3.8. Plot of 450 sediment samples from the fluvial fan experiment. Samples were taken every 10 min at the watershed outlet. Note the large variability in sediment discharge over short periods of time and the long-term, regular fluctuations of sediment yield. The line represents a moving mean of every 20 samples (from Weaver, 1984).

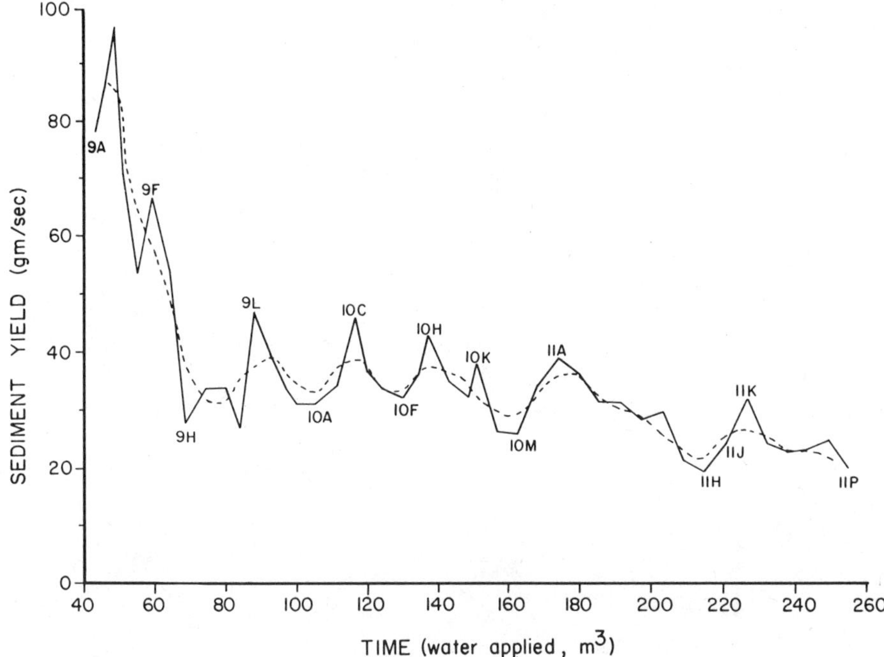

FIGURE 3.9. Plot of sediment yield as derived from the repeated measurement of erosion pins on the alluvial fan surface. The dashed line represents a moving mean of every 30 points and shows (1) the initial rapid decline in sediment yield, (2) a generally constant sediment discharge through 175 m^3 time (interrupted by regular perturbations), and (3) a declining yield following the artificial reduction in drainage basin erosion rates (from Weaver, 1984).

Fig. 3.8. In general, the trends are similar. Absolute values of sediment yield derived from the two methodologies appear comparable, as do the time and magnitude of the various pertubations about the trend, although the single peak at 105 m^3 on Fig. 3.8 is represented by two smaller peaks at 90 and 115 m^3 on Fig. 3.9. The initial rapidly declining curve segment is followed by a gradually declining midportion, but a more rapid decline occurs after 180 m^3 following placement of impermeable plastic squares on the source area.

Effect of Slope

The influence of initial surface slope upon sediment yield, indicated by the comparison between sediment yields during Parker's experiments E1 and E2 (Fig. 3.4), is clarified by Mosley's (1972a) comparison of channel network morphology on surfaces of different inclination and shape (Chapter 2). As would be expected there was a clear relationship between sediment yield at the stage of maximum extension and surface inclination and shape (Fig. 3.10). Steep slopes and concave slopes, which concentrated runoff, produced the greatest quantities of sediment. As slope increased, the concave-up slopes increased erosion rates more than did planar

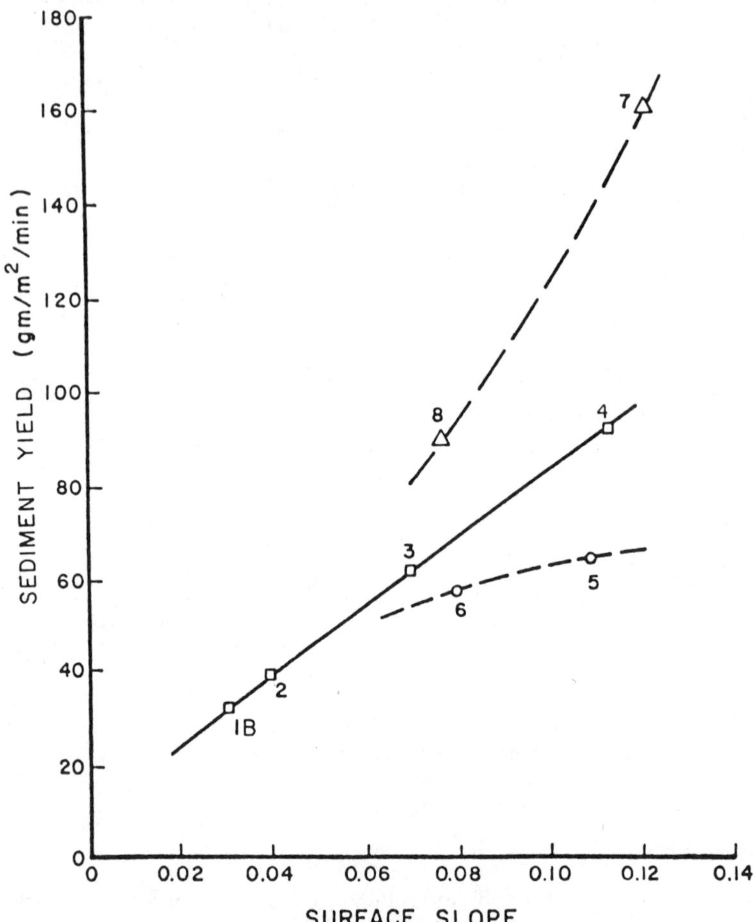

FIGURE 3.10. Relationship between sediment yield from rill networks and surface slope for straight (□), concave (△), and convex (o) initial surfaces (after Mosley, 1974).

slopes. Convex-up slopes caused flow to be divergent, less concentrated, and less erosive, and therefore sediment yields from these slopes were lower (Fig. 3.10). The effect of concentrating flow in a channel system on sediment yield was further indicated by the strong positive relation between sediment yield and drainage density for the four planar surfaces (Fig. 3.11).

Parker (1977) also collected sediment samples during the generation of 1-min-duration precipitation events, and a series of sediment hydrographs were prepared for Networks 8–10 of El. As ruggedness number ($D \times R$) increased, the peak sediment yield for a 1-min duration storm increased (Fig. 3.12). It is difficult to define the relationship accurately with only three data points, but the rate of increase of sediment yield declined at the higher ruggedness numbers, as did the ratio of

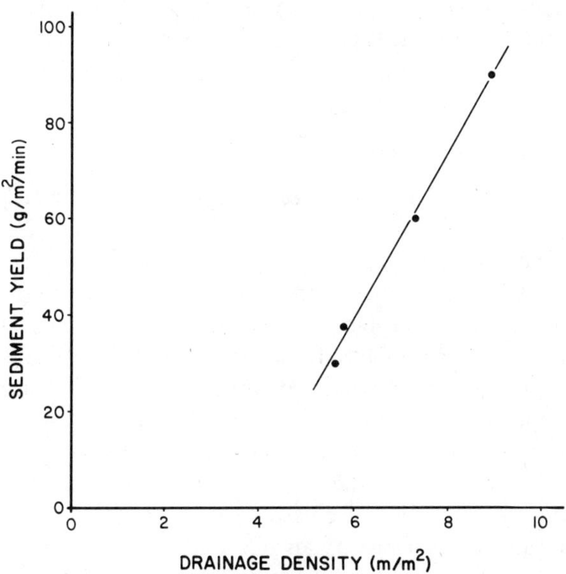

FIGURE 3.11. Relation between sediment yield and drainage density for plane surfaces (Fig. 2.13, from Mosley, 1974).

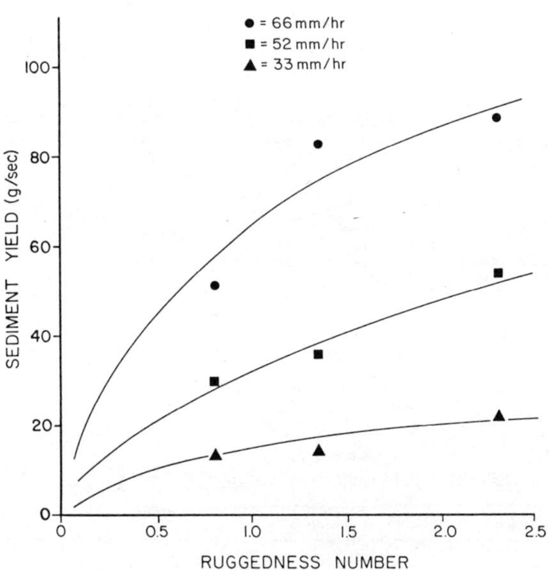

FIGURE 3.12. Effect of ruggedness (drainage density times relief [$D \times R$]) on peak sediment yields for different intensity precipitation events (EI, Networks 8–10) (from Parker, 1977).

peak of equilibrium discharge (Fig. 3.2), which again suggests that the maximum effect of relief and drainage density occurs at ruggedness numbers smaller than 1.0.

Components of Sediment Yield

The erosion rate at any time is a result of the operation of several processes, the relative importance of which varies during an experiment (Chapter 2). This suggests an analogy with the technique of hydrograph separation (Linsley et al., 1949), in which streamflow at any time may be separated into baseflow, interflow, and storm flow, which in idealized circumstances are equivalent to groundwater flow, flow through the soil, and surface runoff. In a "sedigraph" there is a baseflow component provided by processes, the rates of which are relatively constant during an experiment—rainsplash, sheet erosion, and slow channel incision.

In the REF, baselevel lowering superimposes a second component as nickpoints advance through the basin and channels incise. This component declines with time, because the advancing nickpoints decrease in height from 20 cm at the outlet to zero in the headwaters as their rate of advance declines (Chapter 4). A third component, analogous to storm flow in a hydrograph, is provided by mass failure of the valley walls. Initially, such collapses are frequent and involve large blocks of sediment, but as the drainage network adjusts to a new baselevel, the quantities of sediment they introduce to the channel become smaller. These components may be combined into an idealized sedigraph analogous to a hydrograph (Fig. 3.13).

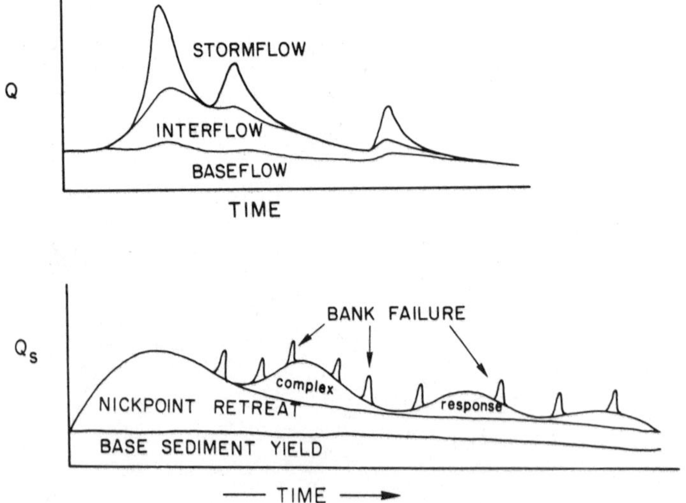

FIGURE 3.13. Idealized diagrams showing components of flow (Q) hydrograph (top) and sediment (Q_s) sedigraph (bottom).

In addition, Figs. 3.7 and 3.8 contain several major peaks of sediment yield that are unrelated to valley wall collapse. These are presumed to be related to flushing of sediment stored in the main channel, which is the "complex response" (Fig. 3.13) of the drainage network to rejuvenation (Schumm and Parker, 1973; see Chapter 4).

ALLUVIAL PLACERS

Experimental studies of drainage network development and basin sediment yield have shown that sediment produced by hillslope and channel erosion does not move through the system to the depositional zone continuously or at a constant rate. Sediment storage, erosion and flushing of storage sites, and selective sorting are important processes within the drainage network that can produce concentrations of heavy minerals, or placers, in channel, valley fill, and terrace alluvium (Fig. 3.14).

Formation of placers requires the production, transport, deposition, reworking, and concentration of heavy minerals. Sites of sediment deposition and reworking should therefore be ideal locations for placers, not only in valleys but on alluvial fans. Morphological traps significantly influence the loci of placers within the fluvial system. Such traps include (1) scour holes at channel confluences (Mosley and Schumm, 1977), (2) pools and riffles in meandering channels, and (3) topographic elements such as point bars, transverse bars, longitudinal bars, and breaks in channel slope. These traps, though not permanent and in some cases migrating downstream (Leopold et al., 1964), generally survive seasonal changes of flow conditions.

During experiments in the REF, heavy minerals were segregated into streaks and patches on the watershed surface and in the channels. Adams et al. (1978) and Lidstone (1981) studied the movement of heavy minerals and processes associated with their concentration into placer deposits within the experimental drainage net-

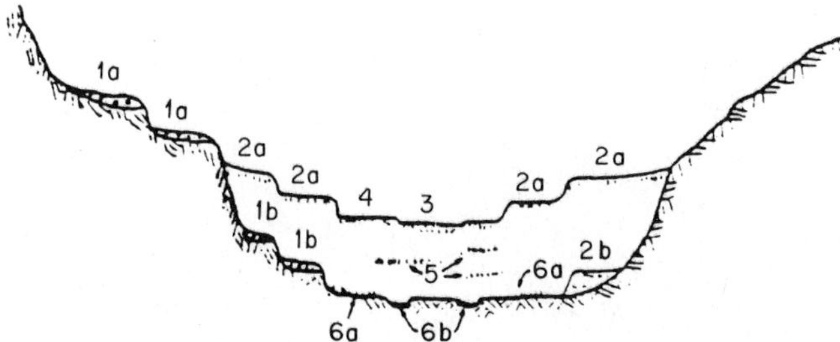

FIGURE 3.14. Classification of fluvial placers (after Schumm, 1977). (1) Bedrock terrace: (a) high level, (b) buried; (2) alluvial terrace: (a) high level; (b) buried; (3) channel lag; (4) floodplain, point bar; (5) lag in valley fill; (6) bedrock: (a) surface, (b) inner channel.

works. Magnetite was the predominant heavy mineral found in the sediment, averaging 0.41% by weight.

Heavy-Mineral Transport and Storage

During Zimpfer's experiments, sediment samples were taken regularly and the magnetite contents were determined (Adams et al., 1978). When the drainage network had reached maximum extension after 12 hr, baselevel was lowered 10 cm by removing a board at the mouth of the REF, and total sediment and magnetite discharges were periodically measured (Fig. 3.15). Sediment yield increased to a peak of nearly 300 g/sec immediately after baselevel lowering, and then it slowly declined to a "background" level of about 40 g/sec. Magnetite, however, showed no similar pattern, but was removed from the basin at a roughly constant rate of about 0.25 g/sec, with peaks up to 2 g/sec. The discharge rates of total sediment and magnetite were highly variable and pulselike, but the magnetite pulses were more marked and magnetite concentrations in the sediment samples varied from 0.068 to 2.01%. The average discharge concentration was 0.37%, which implies that there was progressive storage in the watershed (Fig. 3.16). At three times after baselevel lowering, magnetite was removed at least in proportion to its occurrence in the experimental material (samples 11, 15, 18), but after sample 18 was collected, there was always net storage (Fig. 3.16). At the end of the experiment, 2.5 kg of magnetite had been concentrated and stored in alluviated drainage basin channels.

The fluctuations of magnetite yield may be explained by observations made during the experiments. The first 10 samples were collected before baselevel was lowered, when the basin was near equilibrium and sediment transport rates were low. Samples 11, 15, and 18 were collected during periods of vertical channel

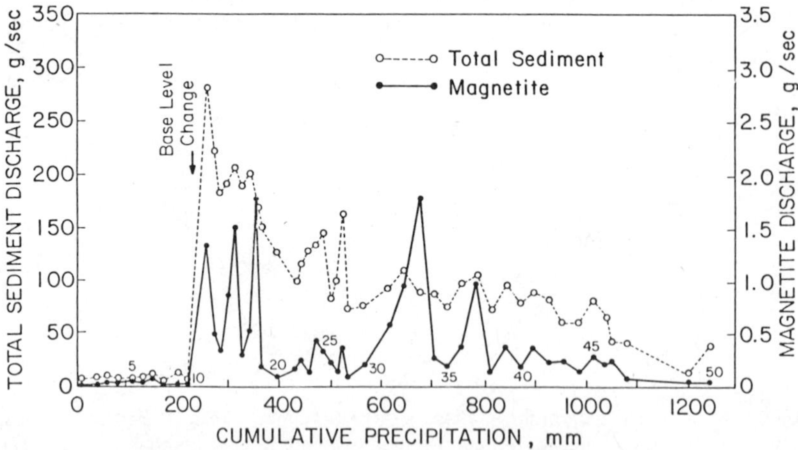

FIGURE 3.15. Discharge of sediment and magnetite from REF (from Adams et al., 1978).

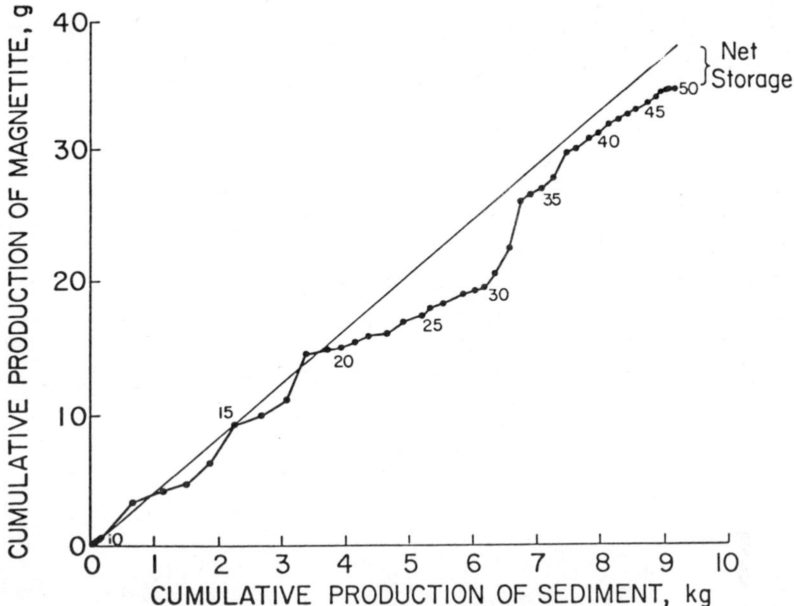

FIGURE 3.16. Cumulative discharge of magnetite as a function of cumulative total sediment discharge. Points that plot above the straight line ($P_{mag} = 0.0041\ P_{sed}$) indicate a period of net yield of magnetite, whereas points plotting below the line indicate net magnetite storage. Numbers are sample identifiers (from Adams et al., 1978).

incision, when nickpoint migration through valley fill deposits was responsible for entraining much magnetite. At the time that samples 12, 13, 14, 16, and 17 were collected, sediment from valley-side slumping was temporarily overloading the channel and causing braiding and storage of concentrated magnetite in the valley fill. Samples 19–29 were taken as magnetite was removed from the headwaters by nickpoint recession and incision of valley fill. However, at the same time, in the lower main channel there was aggradation and magnetite storage (Fig. 3.16).

No observations were made when samples 30–33 were taken, but following this period the alluvium in most tributaries had been eroded and the magnetite transported out of the REF. During the final period (Samples 38–50), the main channel was braided and lateral erosion was dominant. Local overloading of the channels was responsible for aggradation and magnetite storage.

Samples of the surface material were taken at 24 grid locations at the end of the experiment, and they were analyzed for magnetite concentration, which ranged from 0.91 to 4.88% and averaged 1.58%. Floodplain samples averaged 2.72%. Several valley fills had magnetite concentrations either at the ''bedrock'' surface or in the bottom of old channels (Fig. 3.17).

Storage and remobilization of magnetite are consistent with the concept of a complex response to baselevel lowering (Schumm and Parker, 1973). Also, during

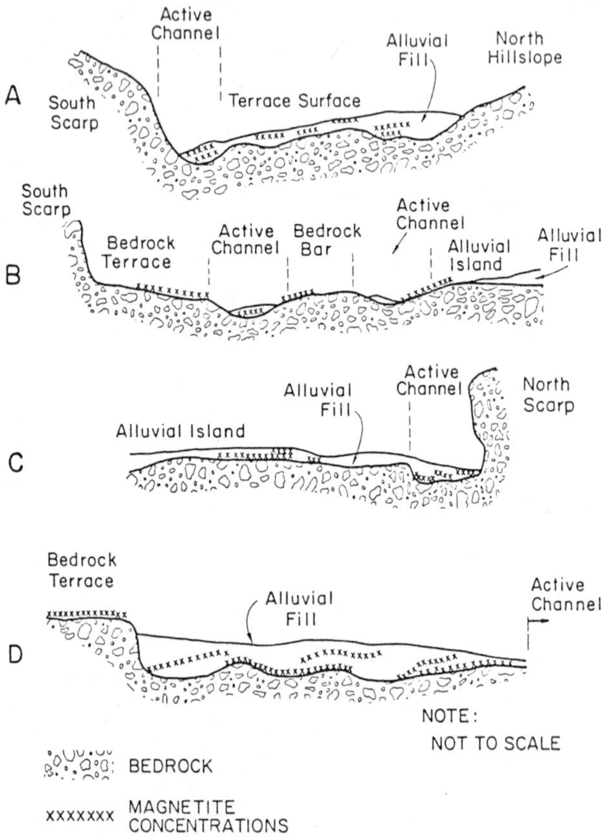

FIGURE 3.17. Sketches of cross sections of alluvial fills in the REF showing heavy-mineral concentrations (from Adams et al., 1978). Distances above outlet are as follows: (*A*) 7.7 m, (*B*) 5.0 m left side, (*C*) 5.0 m right side, (*D*) 2.4 m.

periods of low sediment supply from the headwaters, channels scour and winnow the valley fill to bedrock, which concentrates heavy minerals at the bedrock surface. Shallower scour of the fill results in "false bottom" deposits within the alluvium (Fig. 3.17). During this time, rates of heavy-mineral transport from the watershed are relatively high. However, during periods of excessive sediment supply, aggradation and braiding occur, channel competence is reduced, and heavy minerals tend to be deposited and stored within the basin. This model was recently used by Adams et al. (1978) to explain multistory placer deposits such as those of the Ross Goldfield, New Zealand. It is important to realize that a single uplift will induce the multiple reworking of valley fills that is necessary for placer formation. Although rejuvenation may destroy existing placers, it also creates a suitable environment for the formation of new ones.

Downstream Dispersion of Heavy Minerals

Hawkes (1976) proposed a formula to describe the dispersion and dilution of minerals downstream from an ore body by assuming, among other things, a curvilinear downstream dilution. However, Rozelle (1978) evaluated lead and molybdenum dispersion in a small, arid drainage basin in the Sonora Desert, Mexico, and found that heavy-mineral grains of vanadinite and wulfenite were segregated at preferred sites rather than declining steadily in concentration downstream. [A *preferred site* is a location where heavy minerals are preferentially concentrated over extended periods of time (Schumm, 1977).]

Wertz (1949) noted that mineral concentration zones occur in a logarithmic sequence coinciding with areas of erosion. In a series of 10 experiments in a wooden flume he found that areas of deposition and erosion alternate, the concentration zones coincide with erosion areas, and barren zones coincide with deposition areas.

Mosley and Schumm (1977) also used results from an experimental study to demonstrate that there is a preferred site immediately downstream from the confluence of two tributaries. The streambed is scoured where two or more streams join, and scour depth is greatest when the tributary discharge is at least one-half that of the main channel and when the two channels meet at angles between 60 and 90°. Hester's (1970) analysis of gold concentrations at the junction of Dominion and Sulfur creeks in the Klondike region substantiates Mosley and Schumm's (1977) experimental work. Hester also found that, although the overall gradient of Dominion Creek averaged 0.02, there was substantial gold enrichment below breaks in the slope along the stream profile, where the stream gradient suddenly decreased to about 0.01. This is consistent with Wertz's (1949) conclusion that the first zone of concentration coincides with a decrease of river slope.

Work by Theobald et al. (personal communication, 1980) in the Ajo Mining District in northern Mexico demonstrated that concentration zones within an ephemeral stream tend to reestablish at the same location. During a major flood the heavy minerals at one site are mobilized and redeposited downstream, but the original site is reenriched from an upstream source at either the same or a higher concentration.

SPATIAL AND TEMPORAL DISTRIBUTION OF PLACERS

Lidstone (1981) conducted an experimental study in the Rainfall Erosion Facility to examine the nature, formation, and spatial distribution of fluvial placers. The REF was filled with the same material used in earlier studies, and a watershed with four tributary valleys and an average initial relief of 0.45 m was constructed (Fig. 3.18). Two valleys (A and D) were graded to a 6% slope; two (B and C), to a 4% slope; and the main channel, to a 2% slope. The upper basin at cross

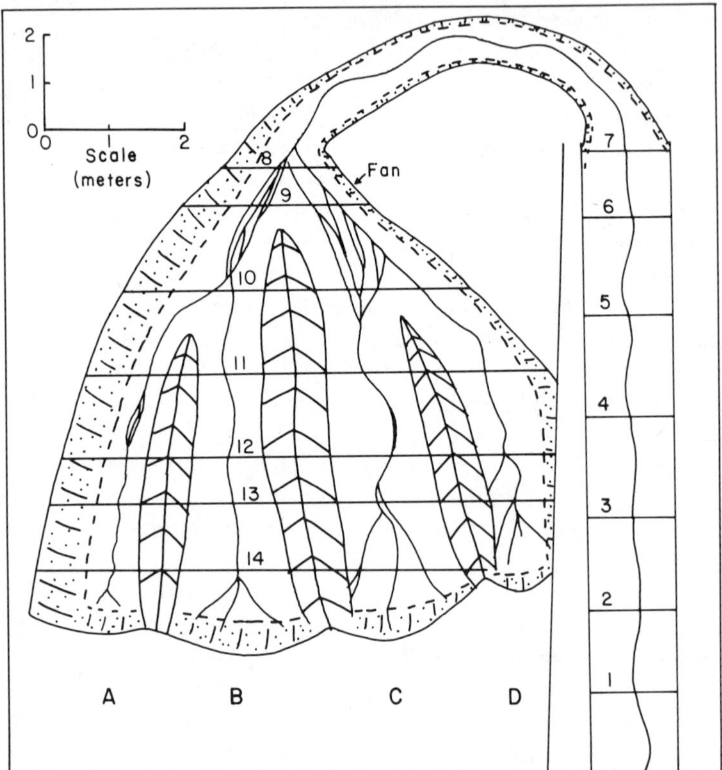

FIGURE 3.18. Lidstone's design of a drainage basin in the rainfall-erosion facility (REF).

section 13 was "salted" with 34 kg of a mixture of iron ore and taconite ore (D_{50} of 0.1 and 0.15 mm, respectively, and an overall specific gravity of 4.8–5.2).

Uncompacted silty sand (d_{50} = 0.22 mm, 0.42% magnetite) was spread to a depth of about 0.04 m along the walls of valleys A and B and along the main channel. Thirty-four kilograms of the iron ore mixture were mixed with the silty sand and spread along the walls of valleys C and D to simulate a "disseminated deposit."

Precipitation was applied to the upper basin for durations ranging from 30 to 90 min at rates of 65 and 107 mm/hr, generating flows down the main channel of 1.26 and 2.8 L/sec, respectively. The high intensities were necessary to mobilize the sediment and heavy minerals. There were 26 runs, with a total duration of 1152 min (Table 3.1).

During each run, sediment and water samples were taken at the mouth of the main channel. The samples were sieved and the heavy-mineral content determined using a FRANZ magnetic separator. Following each run, seven cross sections (1–7) were surveyed along the main channel and seven (8–14) within the upper

TABLE 3.1 Summary of Experiments

Run Numbers	Elapsed Time (minutes)	Characteristic Events
0–11	0–505	Rainfall highest over valleys A and B. Baselevel constant.
12–13	506–591	Sprinkler systems adjusted. Rainfall highest over valleys C and D. Baselevel constant.
14–15	592–672	5-cm drop of baselevel at mouth of upper basin.
16–26	673–1152	13.3-cm drop of baselevel at mouth of main channel.

basin (Fig. 3.18). A BISON magnetic susceptibility meter was used to record magnetite concentrations at the same cross sections.

Spatial Distribution

Upper Valleys

The early runs were marked by aggradation and deposition in the upper valleys. After baselevel lowering at the beginning of run 16, stream incision, rejuvenation, and intermittent reworking of valley fills occurred, but by run 25 the drainage system stabilized.

In general, erosion and deposition events in the upper valleys strongly influenced sediment production and placer formation. Placer formation (heavy-mineral concentration) was minimal during aggradation and during the period of high sediment production and sediment discharge that followed basin rejuvenation. Then, as sediment discharge waned, placer formation increased.

Placer formation will be discussed for the upper reaches of valleys C and D. During the first two runs the very fine-grained iron ore was removed as wash load from the upper basin and, despite its high specific gravity, was flushed through the main channel. The remaining, coarser-grained heavy minerals formed placers.

Valley D. The longitudinal profile of valley D was initially steeply concave; it flattened during run 11 and then assumed a concave-convex shape by run 25 (Fig. 3.19). A preferred site for heavy-mineral enrichment developed at cross section 11 following deposition and repeated reworking of the sediment. This cross section was located at an inflection point in the profile.

A large amount of magnetite moved into this reach during run 20 as the channel incised and narrowed following baselevel lowering at the start of run 16. This magnetite augmented the existing deposit, increasing its concentration to 3.7% by weight by run 24. However, an episode of incision of 0.02 m during run 25 partially destroyed the placer, reducing the average concentration to 3.0%.

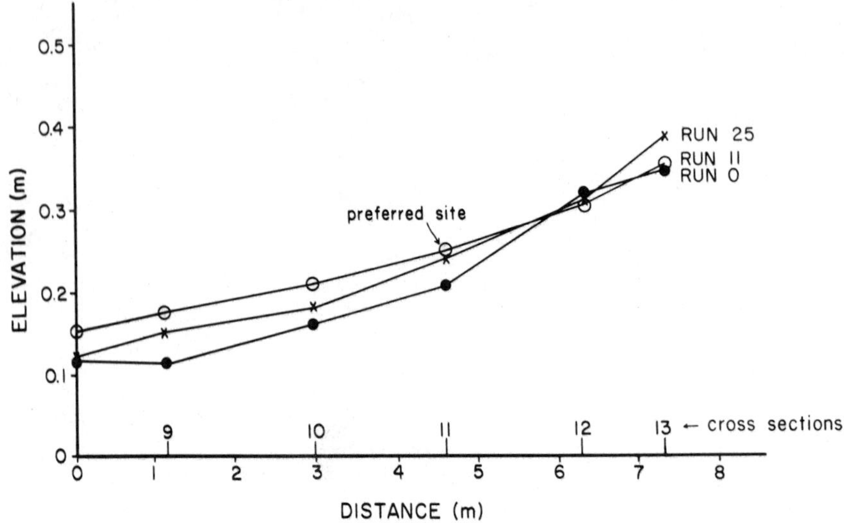

FIGURE 3.19. Longitudinal profiles of valley D (from Lidstone, 1981).

Valley C. The head of valley C was narrow and confined. Incision was nearly continuous from run 0 to run 25 and placers did not form. Cross section 11 was an enrichment zone, with concentrations of magnetite up to 3.8%. As in valley D, this site was an inflection point on the longitudinal profile, where the upstream slope of 4.7% decreased to 2.9% downstream (Fig. 3.20). The channel upstream generally incised from run 0 to 25. Downstream it initially aggraded, yet the sediments were regularly reworked and magnetite concentrations reached their peak by run 16. Channel incision had destroyed the placer by run 22.

Lower-Basin Channels

During the early runs the lower cross sections (8–10) were distinguished by channel shifting and development of an extensive valley fill deposit. A within-valley fan formed and the channel aggraded until run 7, with distributary channels continually shifting across the fan surface. Fanhead trenching started at cross section 10 after run 7, reworking dominated runs 10–16 (Fig. 3.21), and peak magnetite concentrations were recorded at the reworked fanhead (cross section 10) following run 15.

During the development and the early stages of reworking of the lower valley fill (cross section 8 and 9), no placer deposits were formed. Rapid aggradation prevented winnowing of the deposits. However, at cross sections 8 and 9, enrichment started after run 15 (Fig. 3.21). At the end of run 9 the average percentage of magnetite was 1.1%, but by the end of run 20 it was 4.0% at cross section 9. During runs 10–17, placers formed near the center of the channel and along the longitudinal bars and then, as the channel incised and terraces formed, the channel

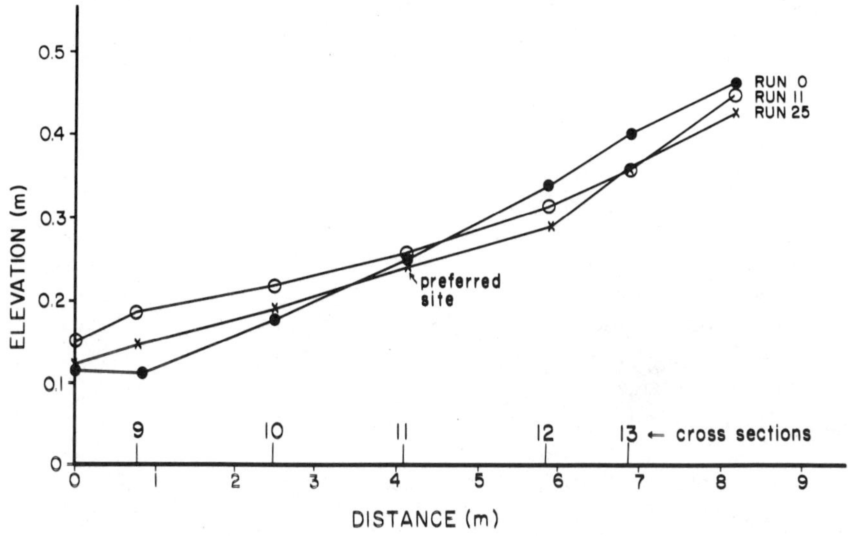

FIGURE 3.20. Longitudinal profiles of valley C (from Lidstone, 1981).

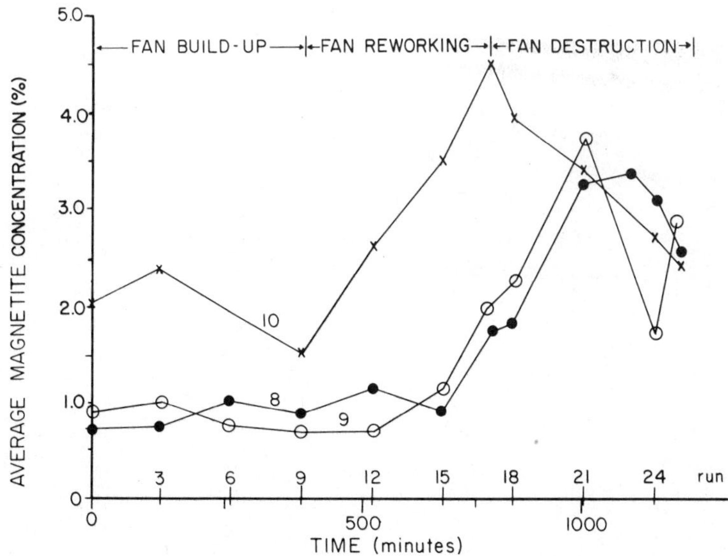

FIGURE 3.21. Magnetite concentration at cross sections 8, 9, and 10 in valley C (from Lidstone, 1981).

margins and the zones where terrace sediments were being reworked became the dominant sites of placer accumulation.

Temporal Distribution

Sediment discharge in Lidstone's experiment was initially about 50 g/sec, but by run 2 (100 min) it had stabilized at about 30 g/sec (Fig. 3.22). Baselevel lowering and basin rejuvenation after run 15 at 700 min increased sediment discharge, which reached 90 g/sec at 900 min. As the basin stabilized, sediment discharge once again decreased to about 25 g/sec. As with earlier studies in the REF, there was a high degree of variability about this general trend, with some particularly marked extreme values that are belived to be associated with sediment flushing.

Magnetite discharge varied in a broadly similar way to total sediment discharge (Fig. 3.23), again with considerable scatter about the general trend and the presence of many extreme values, which reflect episodic transport of magnetite from the watershed. However, there was only a weak relationship between magnetite discharge and sediment discharge (Fig. 3.24), and there were a number of significant differences in their trends with time (Figs. 3.22, 3.23). The increase of magnetite discharge after baselevel lowering (run 15) was much less than for sediment discharge, and the pattern of extreme values was different. There were more magnetite extreme values during runs 1–13, fewer during the immediate post-baselevel low-

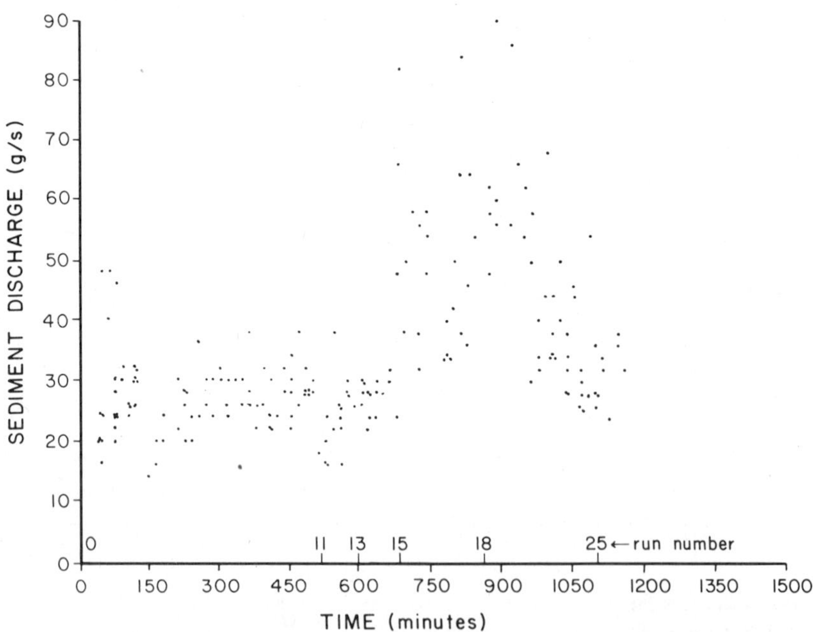

FIGURE 3.22. Sediment discharge through time (from Lidstone, 1981).

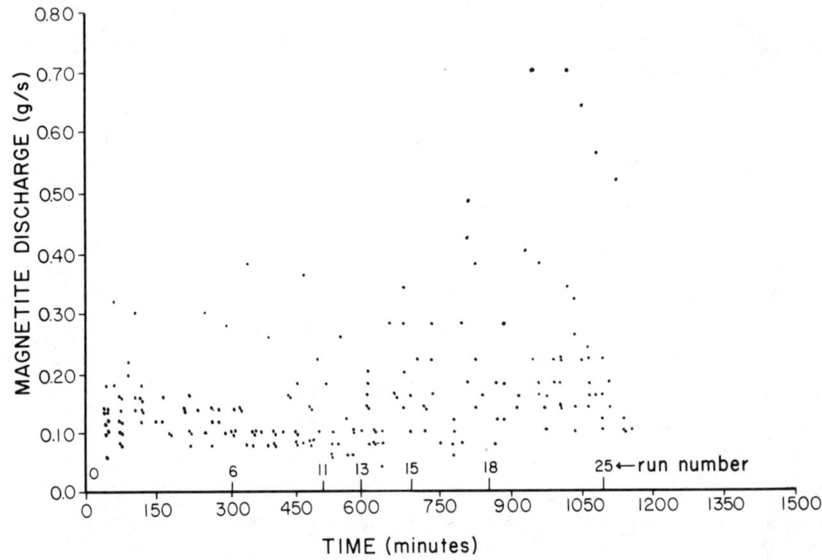

FIGURE 3.23. Magnetite discharge through time (from Lidstone, 1981).

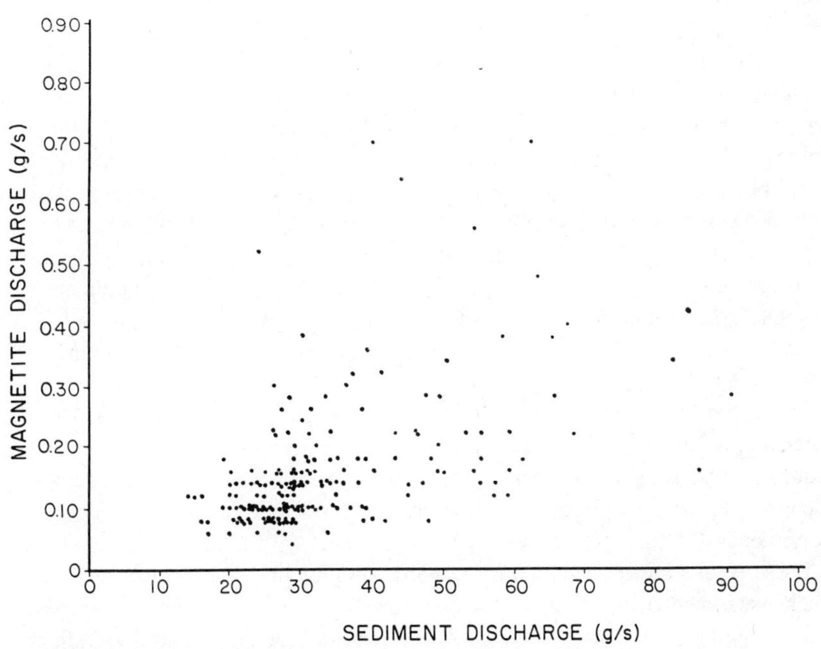

FIGURE 3.24. Plot of magnetite discharge against sediment discharge (from Lidstone, 1981). Correlation coefficient is 0.46.

ering period (runs 15 and 16, Table 3.1), and more during the period of stabilization following peak sediment discharge at 900 min. Rejuvenation of the basin induced bank slumping and incision of the upper-basin channels, which caused the increased total sediment discharges. In contrast, the highest values of magnetite discharge tended to occur during reworking of valley fills. This was particularly the case during runs 18–25 (850–1152 min), when channel incision triggered the release of magnetite from upstream placer sites.

In general, the period of steadily declining sediment discharge corresponded to a period of stability in the upper basin. Discrete events, such as bank failure or the removal of a valley fill or placer deposit, correspond to the extreme values on Fig. 3.22 and 3.23. In some cases, scour mobilized both barren sediment and heavy minerals, which were transported to the basin outlet. In these cases, peak sediment and magnetite discharges coincided. In other cases, the flow was unable to transport the heavy minerals and the magnetite was redeposited and stored. Later reworking triggered the release of pulses of magnetite-enriched sediment. The poor relationship between magnetite discharge and total sediment discharge (Fig. 3.24) indicates that they are not necessarily influenced by the same events or processes.

DISCUSSION

Although the difficulty of making accurate measurements of runoff characteristics posed serious problems for the hydrologic experiments, the results obtained were reasonable and have geomorphic significance. For example, the increased peak discharge for the same storm event when drainage density is high reveals that the flow is more efficient at transporting sediment and modifying the channel morphology (Fig. 3.1). With the higher peak discharge, previously stored sediment should be eroded and channel incision could convert a braided stream to a single-thalweg channel with a floodplain or convert a floodplain to a low terrace.

The effects of runoff source area pattern, storm intensity, and storm duration on the shape of the runoff hydrograph are more complex than was anticipated. It is clear that source area location is important during low-intensity storms, whereas at high intensity the effect is minimal under the conditions of this experiment (10% affected area).

The variations of source area location in a watershed as a result of land use (Hills, 1971; Lusby, 1970; Sharp et al., 1964) can significantly modify hydrograph characteristics. If outcrop areas of impermeable shale or other rock occur on divides, slopes, or floodplains, the effect should be similar to those produced by the experiments.

These results are consistent with research on the effects of landuse in the Caspar Creek watershed of north coastal California (Ziemer, 1979). Following selective tractor logging of second-growth redwood forests, Ziemer found a significant increase in the flood peaks of small storms but no significant change in moderate- or large-size storm flow peaks generated by higher-intensity rainfall.

The ability of the model drainage basins to evacuate sediment and precipitation appears to peak at a value of ruggedness number (relief × drainage density) of 1.0–1.5; increases beyond that had little effect on hydrologic response (Figs. 3.2, 3.12). Similarly, Zimpfer found that hydrograph characteristics changed as the experimental drainage basins evolved, until a point at which continued drainage pattern development had little further effect (Fig. 3.1). As Chorley and Kennedy (1971) explained, the additional channels at higher drainage densities are small, hydraulically rougher, and more intricately branched, so that at a certain point they cease to increase the speed at which runoff can be evacuated.

In spite of the constant rates of precipitation and runoff, sediment concentrations were highly variable as a result of valley-wall slumping, periods of lateral channel shifting, and episodic aggradation and degradation. Accordingly, it is difficult to relate sediment yields to geomorphic characteristics in the various studies. However, Mosley (1972a) achieved this by using sediment yields at maximum extension as the dependent variable in relationships with drainage density, surface shape, total channel length, and surface slope.

The studies consistently demonstrated a rapid exponential decline in sediment yields following initiation of erosion, commonly with a second peak following the initial rapid decline. This peak is believed to be the result of remobilization of sediments that were deposited as an alluvial fill in the lower main channel. Remobilization occurs because, as the drainage network reaches maximum extension, sediment concentrations decline and runoff increases its load by scouring the readily erodible alluvium. Subsequent, irregularly spaced peaks were frequently observed (Figs. 3.6–3.9) that were related to valley-wall slumping and to lateral attack of valley walls and of the toe of tributary alluvial fans, as well as to episodic aggradation and degradation.

The behavior of sediment yields from the experimental drainage basins, as a result of single initial impulse, such as initiation of precipitation onto a pristine surface or baselevel lowering, is highly relevant to natural watersheds. Whether caused by human interference or by some extreme climatic event, a newly initiated cycle of erosion may be expected to be associated with initially high sediment yield rates that rapidly decline, a high degree of variability of sediment yield, and periodic peaks that are inherent to the system and not caused by subsequent extreme events. Of course, the effects of floods will be superimposed on the natural variations observed in the experiments, so that the commonly observed variability of sediment yields in prototype drainage systems is hardly surprising (Rowntree, 1982). Obviously, predictions cannot be made without a full understanding of both basin morphology and basin dynamics, which include erosion, transport, storage and flushing of sediment, the effect of bank collapse, and the rate of nickpoint migration and network rejuvenation following baselevel lowering.

In the past, downstream dilution of sediment-bearing heavy minerals has been postulated to explain the distribution of fluvial placers. The experimental studies indicate that, rather than conforming to a pattern of simple downstream dilution, placer deposits are found at preferred sites such as at breaks in valley slope or below valley confluences.

In the experimental basin, the most important preferred site was the break in slope or inflection point on the longitudinal profile, which separated the steep-gradient upper reaches, where vertical incision predominated, and the lower-gradient lower reaches, where reworking of alluvial fills occurred. All such locations were enriched and placer values, while fluctuating, remained significantly above the background concentration.

Although the morphology of the experimental channels and valleys was continuously evolving, placers continually reappeared at the same locations, despite changes such as incision and aggradation due to variations in sediment production. Thus it appears that heavy-mineral concentrations may persist at a given location that is related to certain characteristics of valley or channel morphology and is, therefore, predictable.

Magnetite and sediment production from the REF were broadly related, but the general trends of both were interrupted by periods of exceptionally high production, caused by such processes as bank failure, episodes of scour or fill, or nickpoint retreat. Sediment and magnetite production commonly did not respond in the same way to such an event; in other words, the mechanisms that caused "pulses" of magnetite or sediment were not the same. Commonly, periods of high sediment production and infilling of the valley bottoms were periods of low magnetite production, because magnetite was being stored and concentrated in the valley fill. Periods of low sediment production, when the channels were incising, tended to coincide with periods of high magnetite production as placers were incised, reworked, and removed.

The formation of placers was also shown to result from a large-scale process of cyclic reworking, which is associated with nickpoint retreat or fanhead trenching downstream. In the absence of external influences such as uplift, baselevel lowering, or change in runoff, these processes were triggered by oversteepening of the valley floor or fanhead. This suggests that prototype placer formation may be a response to geomorphic processes intrinsic to the evolution of the drainage basin itself.

4 Basin Dynamics

The substantial variations of sediment yield from the experimental watersheds, under an essentially constant rate of water application, reflect a variety of geomorphic events, such as channel extension, nickpoint migration, and storage and remobilization of sediments in the valleys. This is obviously relevant to a major question of geomorphic research, which is how and at what rate a landform responds to changed controlling conditions. There is a vast literature dealing with this subject, but experimental studies that deal with these topics are few (see Chapter 1). A considerable amount of information was collected during the drainage basin experiments on basin response to rejuvenation by baselevel lowering, which can be used to explain the sediment yield variations that were described in Chapter 3.

REJUVENATION

Rejuvenation of a drainage basin by baselevel lowering produces a nickpoint that migrates up the main channel and its tributaries. Rejuvenation may be documented by monitoring nickpoint migration through the drainage network; and although nickpoints are discussed more fully in Chapter 6, they are considered here as an indicator of drainage basin response.

Parker (1977) measured nickpoint migration following baselevel lowering, but data were difficult to obtain because some nickpoints disappeared, others became multiple nickpoints, and multiple nickpoints at times coalesced to form a single nickpoint. Nevertheless, nickpoint migration rates measured in the main channel and in three tributaries (Fig. 4.1) decreased with time and with distance up the

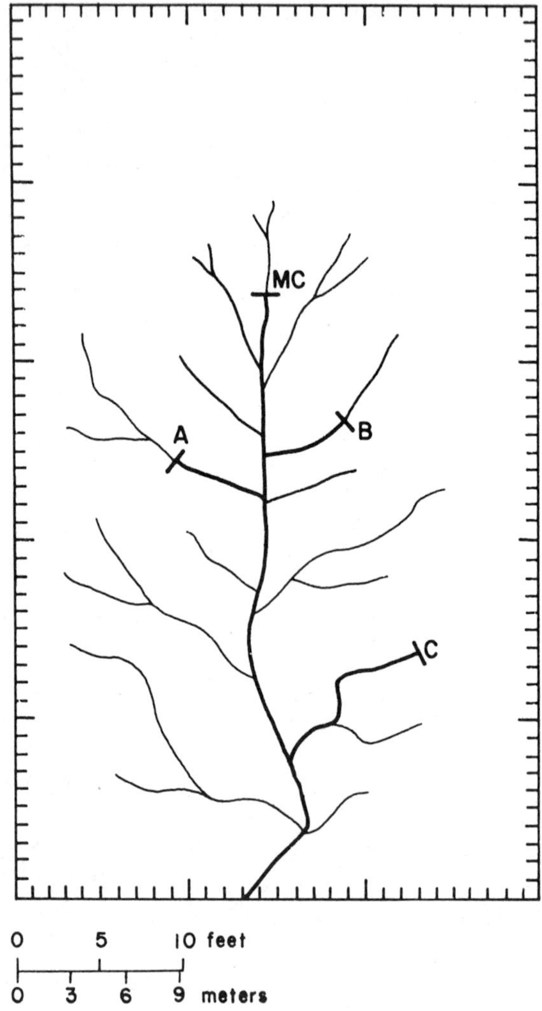

FIGURE 4.1. Position of nickpoints in Network 7 during Experiment 1 (from Parker, 1977).

main channel (Fig. 4.2) as the drainage area and discharge past each nickpoint declined.

Seginer (1966) found that drainage basin area (A), which is proportioned to discharge, could be used to predict gully advance (R) using an equation of the form

$$R = CA^b \quad (4.1)$$

He suggested that because headcuts in a gully network reach their positions by migration from a common origin, the constant C is the same for all gullies in the same network. Therefore, the relative rate of gully advance depends on the ex-

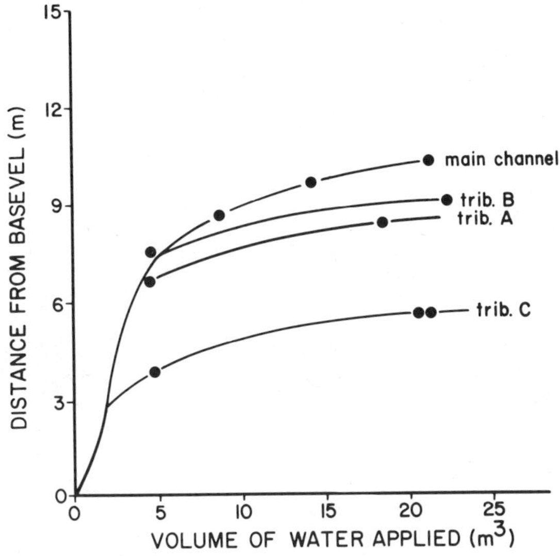

FIGURE 4.2. Nickpoint migration in Experiment 1 following baselevel change. Time is expressed as the volume of water applied to the REF (from Parker, 1977).

ponent *b*, which Seginer suggested is influenced by watershed size, and index of many primary hydrologic variables, and by soils, topography, land use, and land management.

Parker (1977) expressed this relation differently, with rate of nickpoint migration (K in ft/min) being inversely related to time, which is expressed as volume of water applied to the REF (V in ft^3).

$$K = 34.48\,V^{-0.985} \tag{4.2}$$

The parallelism of the lines representing the three tributaries (Fig. 4.2) indicates that there is a constant rate of nickpoint migration in each tributary, which implies a constant rate of nickpoint movement for a given Strahler order. Indeed, the rate of nickpoint migration increases as stream order increases (Fig. 4.3). As expected, the rates of nickpoint migration in Experiment 1 have an exponential relationship with watershed area above the nickpoint (Fig. 4.4). However, watershed area, distance from the basin outlet, and Strahler order are all highly correlated, and, of course, they are also related to discharge. Therefore, as a nickpoint moves farther into a watershed, its rate of migration is slowed because of the reduction of basin area and discharge.

If the nickpoints progress up the network in a regular fashion with respect to Strahler order, the number of streams along which nickpoints are migrating and the number of subbasins being eroded increase geometrically. However, the farther the nickpoints migrate up the network, the smaller is the discharge in each subbasin, the slower is nickpoint advance, and the lower and narrower is the nickpoint. These

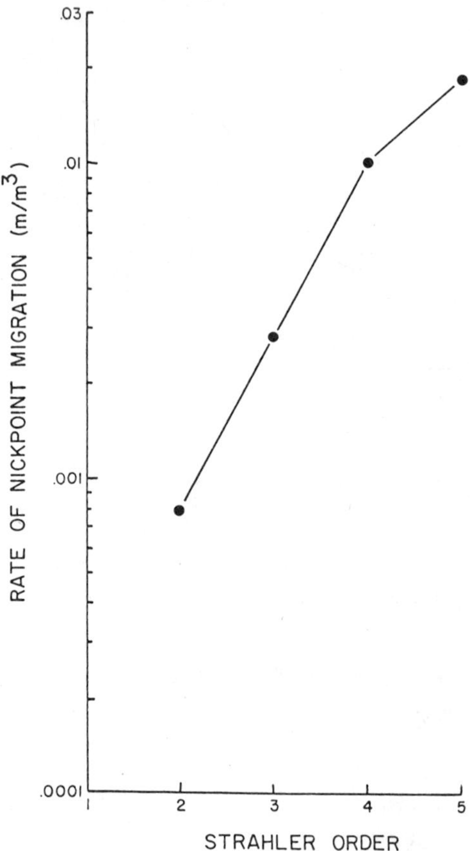

FIGURE 4.3. Rate of nickpoint migration (m/m^3 of water applied) and Strahler stream order (from Parker, 1977).

two tendencies interact to produce a sediment production curve that at some point during basin evolution reaches a maximum (Fig. 4.5). This sediment production curve is similar to the network growth curve that shows drainage density reaching a maximum and then declining (Fig. 2.22). After maximum extension, sediment production due to tributary extension essentially ceases and total sediment production declines.

However, a station that is recording sediment yield at some point along the main channel shows a rapid increase and then an exponential decline of sediment yield (Figs. 3.6, 3.7). Sediment yields are low before a nickpoint migrates past a station, but as the nickpoint passes, sediment yields increase sharply to a maximum (Fig. 4.5). As the nickpoint progresses upstream, there is an exponential decrease of sediment yield.

Although both the point sediment yield and basin sediment production curves are shown in Fig. 4.5, the relation in time between the two curves cannot be

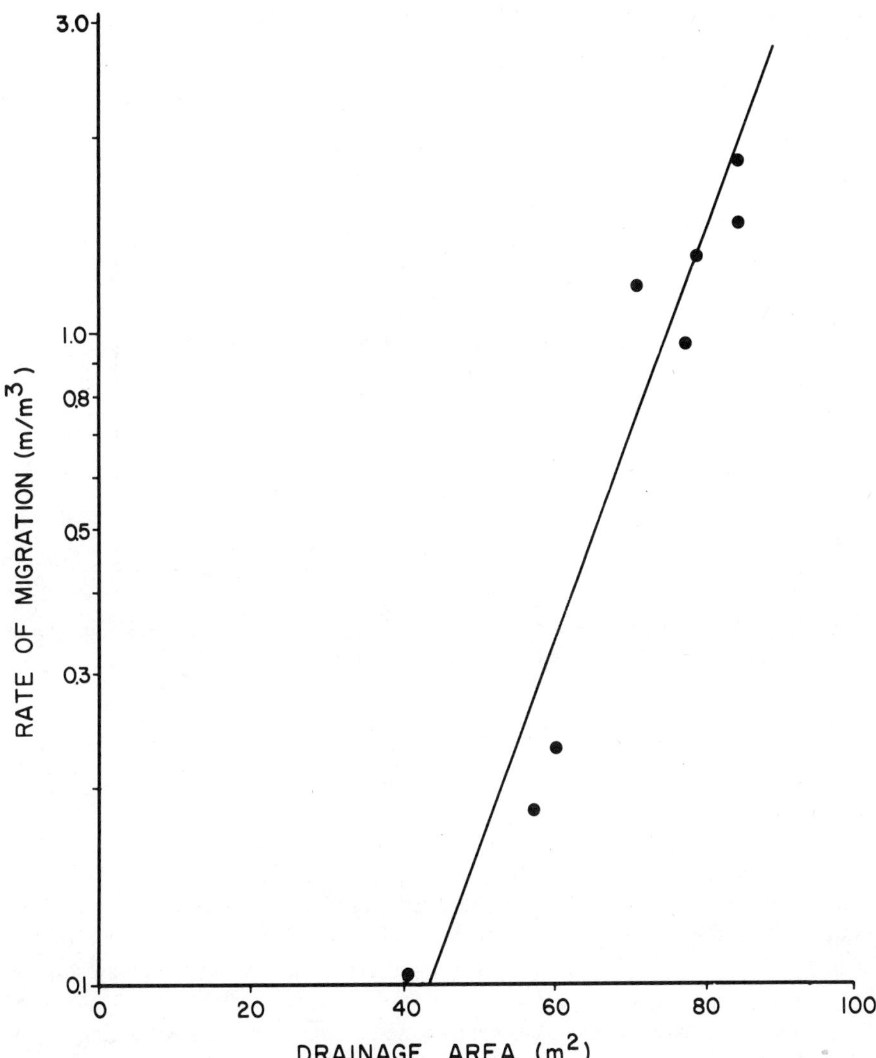

FIGURE 4.4. The rate of nickpoint migration (m/m^3 of water applied) as a function of the approximate drainage area above the nickpoint.

obtained from the experimental data. It is assumed, however, that at or near the peak of sediment production the variability noted in the sediment yield graphs (Figs. 3.7, 3.8) is produced. The difference in basin sediment production and point sediment yield reflects storage of material in the channel upstream from the measuring station. For a measuring station in the lower main channel, point sediment yield has already begun its exponential decay before basin wide sediment production has reached its maximum rate, because large amounts of sediment are being

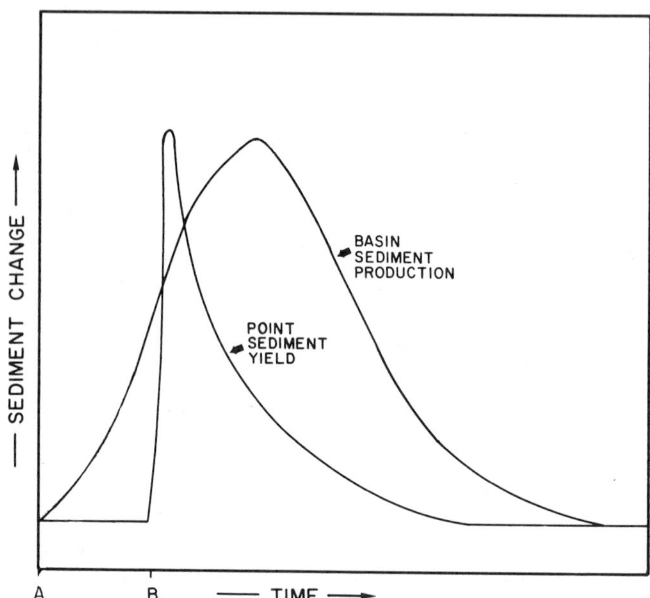

FIGURE 4.5. Hypothetical diagram of changes of sediment yield measured at a point at time B and sediment production after a change of baselevel (time A).

stored upstream. This alluvium may be flushed periodically past the measuring station, causing considerable sediment yield variability (Fig. 3.13).

Sediment Storage

During basin rejuvenation, the increasing sediment loads from the eroding channel system cause the main channel to widen and braid and to deposit some of this sediment. Thus initial degradation causes aggradation as erosion progresses up the network.

Although it was not possible to measure sediment production within the experimental basin, it was possible to document changes in the main stream profile that presumably reflect changes of sediment delivery to the main channel (Fig. 4.6). The first profile was surveyed 2 hr after baselevel lowering. The second profile, surveyed after 4 hr, shows continued degradation of the channel. The third profile, surveyed after 6 hr, shows continued degradation above 4 m but deposition of alluvium in the lower 4 m of channel, near the mouth of the basin. Deposition in the first 3–7 m of the profile was noted following baselevel lowering in both experiments.

Lateral migration of the channel caused extensive bank erosion, which also contributed to the alluvial fill. Both the main channel profile (Fig. 4.6) and cross sections (Fig. 4.7) provide evidence of natural aggradation in the main channel. The period of aggradation was sometimes followed by channel narrowing and

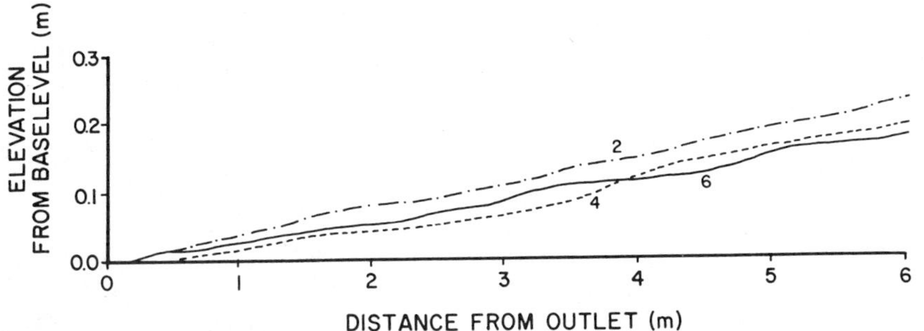

FIGURE 4.6. Changes in the profile of the main channel after baselevel lowering. Numbers refer to hours of run time (from Parker, 1977).

erosion into the alluvial fill as sediment production from upstream decreased. When this occurred (Fig. 4.7*C, D*), the channel became underfit as a result of decreased sediment load.

Observations of these changes in the long profile and cross sections of the main channel, coupled with the regularity of nickpoint migration and the variability of

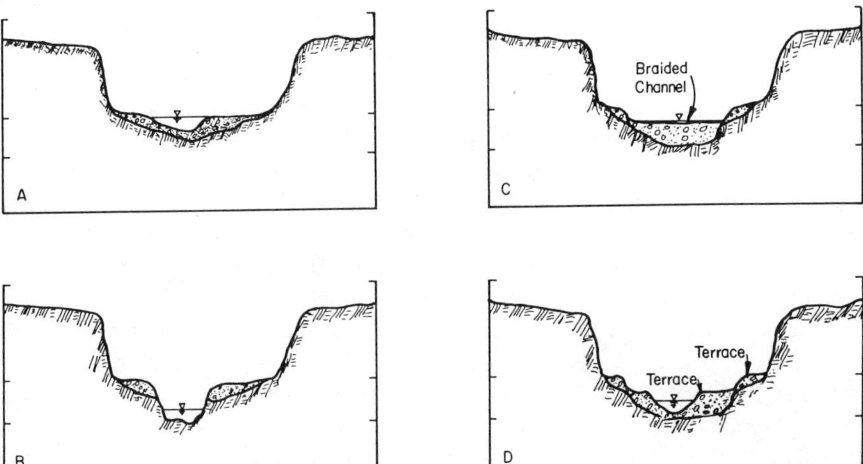

FIGURE 4.7. Diagrammatic cross sections of experimental channel 1.5 m from outlet of drainage system (baselevel) showing complex response of channel to a single lowering of baselevel (Schumm and Parker, 1973). (*A*) Valley and alluvium, which was deposited during previous run, before baselevel lowering. The channel flows on alluvium. (*B*) After baselevel lowering of 10 cm, the channel incises into alluvium and the bedrock floor of valley to form a terrace. Following incision, bank erosion widens the channel and partially destroys the terrace. (*C*) An inset alluvial fill is deposited as the sediment discharge from upstream increases. The channel with high width–depth ratio is braided and unstable. (*D*) A second terrace is formed as the channel incises slightly and assumes a low width–depth ratio in response to reduced sediment load. With time, channel migration will destroy part of the lower terrace and a floodplain will form at a lower level.

sediment yield with time, led to the formulation of a model of complex response of drainage basins to channel incision (Fig. 4.7; Schumm and Parker, 1973). In this model, initial degradation in response to baselevel change is followed by a period of deposition that results from increased sediment production upstream. As sediment production declines, the main channel narrows and incises into the previously deposited alluvial fill. This conclusion was foreshadowed in an experimental study carried out by Lewis (1944). Lewis observed the following sequence of events in a stream table 3.4 m long: (1) rapid supply of sediment to the streams in the early stages, when slopes were steep and new channels were being cut; (2) floodplain aggradation as main channel slope steepened sufficiently to transport this sediment; (3) a reduction in sediment supply as slopes upstream declined and stream courses stabilized; (4) incision and a corresponding reduction in gradient of the main channel in response to the reduced sediment load; and (5) creation of terraces in the upper reaches of the floodplain. Lewis regarded the fact that terraces were built with no change in baselevel, tilt, or discharge as the most significant outcome of his experiment.

The complex response as described by Schumm and Parker (1973), resulted from a baselevel change with channel incision and tributary rejuvenation (Fig. 4.7*B*), but the response was probably similar to that to be expected in any area of high sediment production. (Fig. 4.7*C, D*)

Channel Response to Watershed Erosion

A major difficulty with geomorphic research has been that, in practice, processes on hillslopes and in channels have generally been studied separately and the interactions have rarely been considered. This has been characteristic also of model studies, whether mathematical or physical, but the REF was used by Harvey (1980) to take a first step in remedying this omission.

Slope erosion by rapid mass movement and the consequent delivery of large volumes of heterogeneous debris to stream channels has been widely reported in mountainous regions of the world, where high relief, tectonic activity, and high-intensity rainstorms occur. In these regions, much of the landscape development is related to large-magnitude, low-frequency mass wasting. Apart from the obvious physical modification of the slopes, these events lead to changes in stream network morphology, and they can have serious effects on downstream transportation systems, soil, and land resources on the floodplains.

Although rapid mass movement is a principal mechanism for the introduction of large volumes of debris to the channel system, rejuvenation of tributaries also introduces large quantities of sediment. Regardless of the mode of sediment introduction to the drainage system, the net result is a series of aggradational and degradational events within the channel network. The nonuniformity within the system is reflected in the variability of sediment yields following the introduction of sediment. Aggradation and degradation at different sites within the channel network result in episodic sediment yields, which are manifested as pulses or waves of sediment moving downstream through the channels (Hayward, 1978), and they cause major changes in channel morphology with time.

In an attempt to document these changes in channel morphology following rapid mass movement and introduction of slope materials, Harvey (1980) designed an experiment in the REF to study sediment movement in a long channel (Figs. 4.8–4.10). Unlike the earlier experiments, he constructed the drainage network and divides. For purposes of discussion, the drainage basin is divided into three separate zones (Figs. 4.8, 4.9): Zone 1 (headwaters; channels G, H, D, C, B); Zone 2 (intermediate zone; channels F, A); Zone 3 (the main channel, M). The distinct zonation of channel and valley morphology into three zones has been reported by

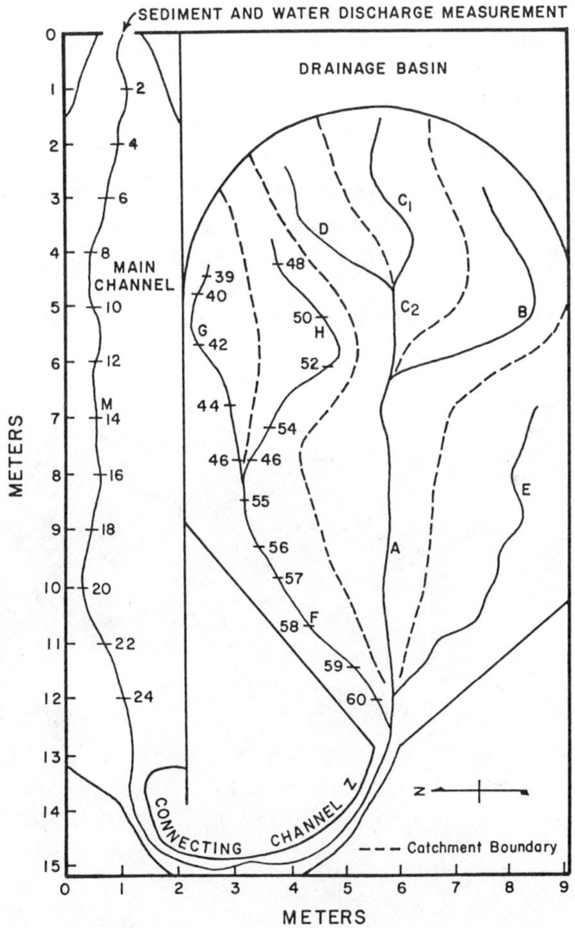

FIGURE 4.8. Plan view of REF showing the locations of cross sections during Harvey's experiment. Note the scale distortion between the horizontal and vertical axes. Numbers identify cross section locations (from Harvey, 1980).

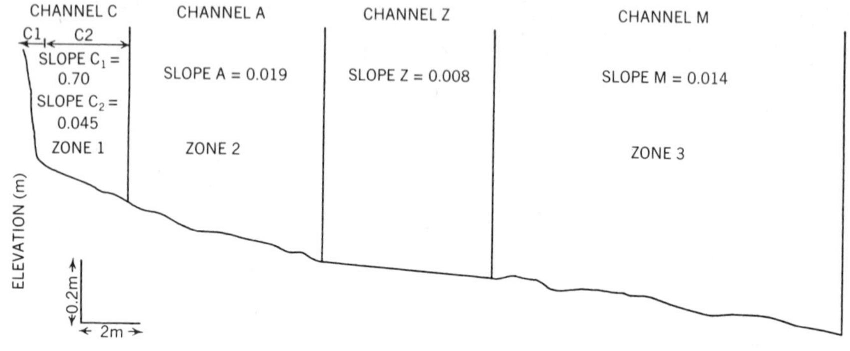

FIGURE 4.9. Longitudinal profile of main channel (Fig. 4.8).

O'Loughlin (1969), Kelsey (1977), Hayward (1978), and Bergstrom (1980). The zonation is based on: (1) the supply of sediment; (2) the availability of energy to transport the sediment, and (3) the presence or absence of confining valley walls. The bulk of the sediment is delivered from Zone 1, where the channels are cut into and flow on "bedrock." Energy is high and the channels are supply limited. The Zone 2 channels have wider valleys and flow on alluvium. Channel M is unconfined except by the wooden walls on the north and south sides of its "valley."

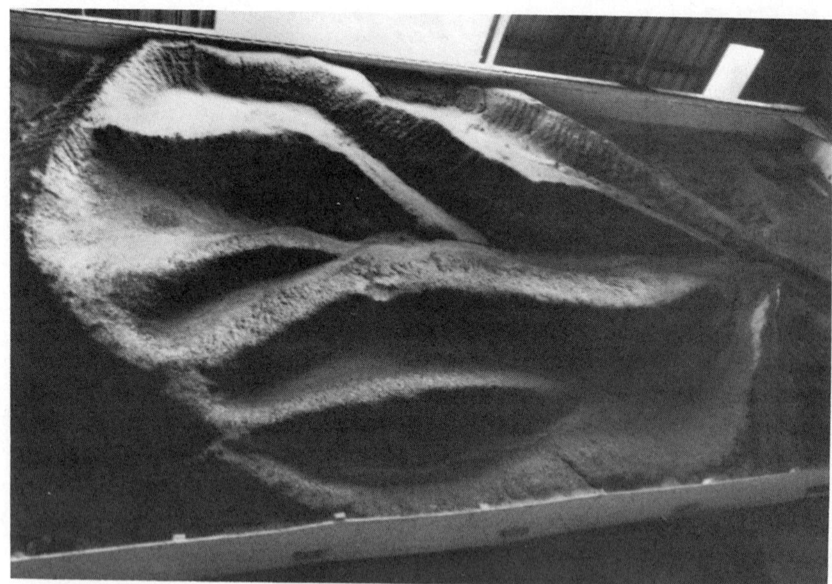

FIGURE 4.10. Initial topography of Zones 1 and 2 in the REF (from Harvey, 1980).

Twenty-one runs were conducted. During each run, precipitation was applied at 350 mm/min until outflow stabilized at about 1.2 L/sec. A 5-min burst of precipitation at an intensity of up to 600 mm/min followed, giving outflow rates of 2.3–4.9 L/sec; then precipitation was reduced until outflow again stabilized at 1.2 L/sec.

Studies prior to Harvey's tended to neglect channel changes because the emphasis was on pattern growth and hydrologic characteristics (Chapters 2 and 3). Harvey's observations are presented in some detail because they demonstrate complex-system behavior under conditions of high sediment load. For the sake of brevity, only the morphologic changes and development of subcatchment channels M, F, H, and G (Fig. 4.8), which are representative of the Zone 1 and Zone 2 channels and the main channel, are described.

Zone 1. Channels G and H represent the Zone 1 headwater channels (Fig. 4.8; Tables 4.1, 4.2). Relative to existing stream power, both channels were undersupplied with sediment during the first five runs. Loose material left in the steep channels as a result of construction of the drainage basin was rapidly transported from Zone 1. The experimental "bedrock" channel was eroded as headcuts moved upstream, and as they reached the upper cross sections, sufficient sediment was delivered to the confluence of channels G and H to form a fan during run 3 (Table 4.1). In the next two runs this fan was alternately cut and filled and a series of unpaired terraces was produced.

During run 6, slumps in incised sections of channels G and H caused a general aggradation of both channels. During the recessional flow, paired terraces were formed in both channels. Gullying of the slump sites resulted in further delivery of sediment to channel G, and the entire channel incision filled during run 7. During run 7, an additional slump occurred in channel H, and the upper reaches filled, but the lower reached scoured.

During runs 8–11, the channels alternately scoured and then filled progressively from the lower reaches to the upper reaches as sediment was moved out of storage and transported downstream. Scour of the stored sediment formed paired terraces in both channels. By the end of run 10, the fan at the confluence of G and H was deeply trenched, which indicated that the sediment supply from both channels had been reduced.

During runs 13–18 (Table 4.2), channel H progressively degraded. During this degradation, individual cross sections scoured and filled as the coarser fraction of the sediment was transported downstream and paired and unpaired terraces were formed.

Various reaches of both channels, G and H, were out of phase during most runs, but the two channels were dramatically out of phase during runs 12 and 13 (Tables 4.1, 4.2). During run 13, the slump sites gullied severely and produced a large quantity of sediment, which resulted in a general aggradation of the channel. During runs 14–19, the two channels were again in phase and generally degraded. One slump occurred in each of the drainages during run 20, which resulted in temporary reversal of the degrading trend.

TABLE 4.1 Morphologic Changes in Channel G, Runs 1–20[a]

Run Number	Aggrading Reaches (cross-section numbers)	Degrading Reaches (cross-section numbers)	Slope (m/m)	Comments
1	39–42	43–46	0.0808	Model adjusting to the effects of construction. Headcut moved up the channel to cs 43.
2	40–46	39	0.0815	Headcut moved up to cs 39. Channel filled below this. Sheet and rill erosion on the divides.
3	39; 42	40–41; 43–46	0.0867	Channel scoured in lower reaches and a fan formed at confluence with channel H.
4	39–43	44–46	0.0927	Headcut moved up to cs 44. Filled in upper reaches.
5	None	All cross sections scoured; 39–46	0.0912	Entire channel scoured. Unpaired terraces formed.
6	All cross sections filled; 39–46	None	0.0869	One slide into channel. Entire channel filled. Paired terraces formed at recessional flow.
7	All cross sections filled; 39–46	None	0.0839	Entire channel filled. Paired terraces formed in lower reaches at recessional flow.

8	41–46	39–40	0.0839	Upper reaches scoured. Lower reaches filled.
9	39–45	46	0.0924	Upper reaches filled. Fan at confluence with channel H forced flow to undercut slope in G—minor slide.
10	39–42	43–46	Unmeasured	Filled in upper reaches and scoured in lower reaches. Paired terraces in upper, unpaired in lower.
11	39–44	45–46	0.0856	Upper reaches filled. Scour into fan at lower reaches—unpaired terraces
12	39–40; 46	41–45	0.0870	Filled in upper reaches and at fan (confluence H). Scoured in lower reaches.
13	All cross sections filled; 39–46	None	0.0652	Entire channel filled. Paired terraces formed at recessional flow.
14	None	All cross sections scoured; 39–46	0.0950	Entire channel scoured. Unpaired terraces formed. Small headcut moved up to cs 42.
15	39–41	42–46	0.0960	Upper reaches filled. Lower reaches scoured. Headcut moved up to cs 42.
16	39–44	45–46	0.0965	Upper reaches filled. Headcut washed out.

TABLE 4.1 Morphologic Changes in Channel G, Runs 1–20[a] (*Continued*)

Run Number	Aggrading Reaches (cross-section numbers)	Degrading Reaches (cross-section numbers)	Slope (m/m)	Comments
17	39–45	46	0.0950	Entire channel filled except cs 46 on the fan that scoured.
18	39–40	41–46	0.0965	Upper reaches filled. Lower reaches scoured. Headcut formed at cs 45, moved up to cs 44.
19	39–42	43–46	0.0950	Upper reaches filled. Headcut moved up to cs 43.
20	39–45	46	0.0812	One slide in channel. Unpaired terraces in lower reaches.

[a] For location of cross sections (cs), refer to Fig. 4.8 (from Harvey, 1980).

TABLE 4.2 Morphologic Changes in Channel H, Runs 1–20[a]

Run Number	Aggrading Reaches (cross-section numbers)	Degrading Reaches (cross-section numbers)	Slope (m/m)	Comments
1	47–51	52–54	0.0886	Model adjusting to the affects of construction. Headcuts moved up channel to cs 52.
2	48–54	47	0.0896	Most of channel filled with debris from sheet and rill erosion on the divides.
3	47–48	49–54	0.0954	Channel scoured in middle and lower reaches. Headcut moved up to cs 51. Fan on confluence.
4	None	All cross sections scoured; 47–54	0.1012	Entire channel scoured. Headcut moved up to cs 48.
5	47–51	52–54	1.1001	Upper reaches filled. Lower reaches scoured. Headcut washed out. Unpaired terraces.
6	All cross sections filled; 47–54	None	0.0951	Two slides into channel. Paired terraces formed at recessional flow.
7	49–54	47–48	0.0920	One slide into channel. Fan at confluence with channel G enlarged.

TABLE 4.2 Morphologic Changes in Channel H, Runs 1–20[a] (*Continued*)

Run Number	Aggrading Reaches (cross-section numbers)	Degrading Reaches (cross-section numbers)	Slope (m/m)	Comments
8	47–48	49–54	0.0933	Lower and middle reaches scoured. Fan at confluence with channel G trenched. Unpaired terraces.
9	None	All cross sections scoured; 47–54	0.0998	Entire channel scoured. Headcut moved up to cs 52.
10	47–50	57–54	Unmeasured	Filled in upper reaches and scoured in lower reaches. Fan deeply trenched.
11	47–52	54	0.0938	Upper and middle reaches filled. Scour into fan at confluence with channel G.
12	All cross sections filled; 47–54	None	0.0854	Entire channel filled. Paired terraces formed at recessional flow. Fan enlarged.
13	None	All cross sections scoured; 47–54	0.0984	Entire channel scoured. Unpaired terraces formed in lower reaches. Fan trenched.

14	47–50	51–54	0.1000	Upper reaches filled. Lower reaches scoured. Small headcut formed at cs 52.
15	57–52	53–54	0.1005	Upper reaches filled. Lower reaches continued to scour.
16	48–53	47; 54	0.1000	Upper cross section scoured. Middle reaches filled. Fan at confluence with G deeply entrenched.
17	47–52	53–54	0.1025	Upper reaches filled. Lower reaches scoured. Headcut formed at cs 53.
18	None	All cross sections scoured; 47–54	0.1050	Entire channel scoured. Headcuts at cs 52, 49.
19	47–53	54	0.1040	Entire channel except cs 54 filled. Fan trenched. Unpaired terraces in lower reaches.
20	51–54	47–50	0.0771	One slide into channel. Unpaired terraces in lower reaches.

[a]For location of cross sections (cs), refer to Fig. 4.8 (from Harvey, 1980).

TABLE 4.3 Morphologic Changes in Channel F, Runs 1–20[a]

Run Number	Aggrading Reaches (cross-section numbers)	Degrading Reaches (cross-section numbers)	Slope (m/m)	Comments
1	55–59	60	0.0465	Model adjusting to the effects of construction. Headcuts moved up channel. Sheet and rill erosion of drainage divides.
2	58–50	55–57	0.0445	Headcuts reached cs 55 and stalled.
3	58–60	55–57	0.0406	Channel filled. Headcuts disappeared. Large fan formed at cs 60 confluence with channel A.
4	All cross sections filled; 55–60	None	0.0348	Channel filled. Fan enlarged at confluence with A. Multichanneled.
5	60	55–59	0.0345	Channel scoured. Fan at confluence with channel A forced flow in A to south.
6	55–59	60	0.0350	Two slides in channel H, one slide in channel G. Channel filled. Paired terraces along the length of channel.
7	None	All cross sections scoured; 55–60	0.0349	One slide in channel H. Headcut formed at cs 59 and moved upstream to cs 56 at recessional flow.

8	55–60	None	0.0337	Channel filled. Fan at confluence with A enlarged. Aggradation at high flow and scour at low flow.
9	55–60	None	0.0324	Channel filled. Fan truncated by flow down channel H. Terraces formed in channel.
10	None	All cross sections scoured 55–60	Unmeasured	Channel scoured. Fan enlarged at confluence with A. Unpaired terraces formed in channel.
11	57–60	55–56	0.0332	Channel filled in lower reaches, scoured in upper reaches. Unpaired terraces formed in lower reaches.
12	None	All cross sections scoured; 55–60	0.0362	Entire channel scoured. Fan prograded into channel A. Multichannel system on fan.
13	56–60	55	0.0205	Channel filled. Flow divided into two equal channels on fan. Paired terraces formed at recession of flow.
14	55–58	59–60	0.0194	Filling in upper reaches. Scour in lower reaches. Flow in fan all on west side. Single channel.

TABLE 4.3 Morphologic Changes in Channel F, Runs 1–20[a] (*Continued*)

Run Number	Aggrading Reaches (cross-section numbers)	Degrading Reaches (cross-section numbers)	Slope (m/m)	Comments
15	All cross sections filled; 55–60	None	0.0192	Channel filled. Paired terraces formed at recessional flow.
16	59–60	55–58	0.0200	Upper reaches scoured. Flow on fan in many small channels.
17	None	All cross sections scoured; 55–60	0.0225	Channel scoured. Channel floored with gravel along entire length. Terrace formation.
18	56–60	55	0.0195	Channel filled. Paired terrace formation. Channel floored with sands. Flow on fan in single channel.
19	59–60	55–58	0.0165	Large headcut moved up channel to cs 55. Fan extended out into channel A. Flow in one channel.
20	56–60	55	0.0118	One slide in channel H, one slide in channel G. Channel filled. Fan at A enlarged into Z.

[a] For location of cross sections (cs), refer to Fig. 4.8 (from Harvey, 1980).

Zone 2. Channel F is representative of Zone 2 channels (Table 4.3). During runs 1 and 2, nickpoints formed in the lower reaches of the channel and progressively moved upstream, reflecting a shortage of sediment from Zone 1. A large fan formed at the confluence of channels F and A (Fig. 4.8) during run 3. Degradation of channels G and H during runs 3 and 4 resulted in aggradation of channel F and elimination of the nickpoints. During run 5, all the cross sections except number 60 scoured, which produced a series of unpaired terraces along the length of the channel.

There was general aggradation of channel F during run 6, as fine sediment from slumps in basins G and H was rapidly transported downstream. During run 7, the entire channel degraded and a series of unpaired terraces were formed. These were destroyed during runs 8 and 9, when the entire channel again aggraded. Flow over the fan was confined to a single channel in the west side. The channel degraded and aggraded during runs 10 and 11, respectively. During run 12, the entire channel again degraded and a multichanneled pattern developed on the fan.

The overall trend of degradation in channels G and H caused net aggradation of channel F during runs 13–16 (Table 4.3), although individual reaches filled and scoured and flights of unpaired and paired terraces were successively formed and destroyed. This progressive aggradation oversteepened the channel, and as a result it scoured deeply during run 17. An armored layer of course sand and granules covered its bed and unpaired terraces were formed along its length. During runs 18–20, the channel filled again, the armor was buried, and the channel was covered with sand.

Figure 4.11 shows the cumulative changes of slope for runs 1–20 in channels F, H, and G. An increase in slope indicates channel filling, and a decrease indicates channel scouring. The cyclical nature of the fill and scour in channels G and H is evident, as is their degradation after runs 12 and 13, respectively. The corresponding aggradation of channel F in response to erosion in channels G and H is also apparent.

Zone 3. The main channel (M) is Zone 3. The floodplain was wide (approximately 2.0 m) and the flows were unconfined and therefore less competent than the Zone 1 and Zone 2 channels. There were generally three distinct reaches in the channel. These have been designated upper (cross sections 24–17), middle (cross sections 16–9) and lower (cross sections 8–1) (Fig. 4.8). After any run two of the reaches might have aggraded while the other degraded, or vice versa. Therefore, the behavior of a single reach could not be used to predict the overall behavior of the channel.

Three cross sections were selected to demonstrate the changing channel behavior with time as aggradation and degradation took place (Table 4.4). They are cross section 2 (lower reach, 1.5 m from outlet), cross section 14 (middle reach, 7.0 m from outlet), and cross section 20 (upper reach, 10.0 m from outlet). In general, any two of the cross sections were in phase but the third was out of phase, which reinforces the view (Andrews, 1979) that an entire channel cannot be regarded as a single entity when aggradation and degradation are being considered.

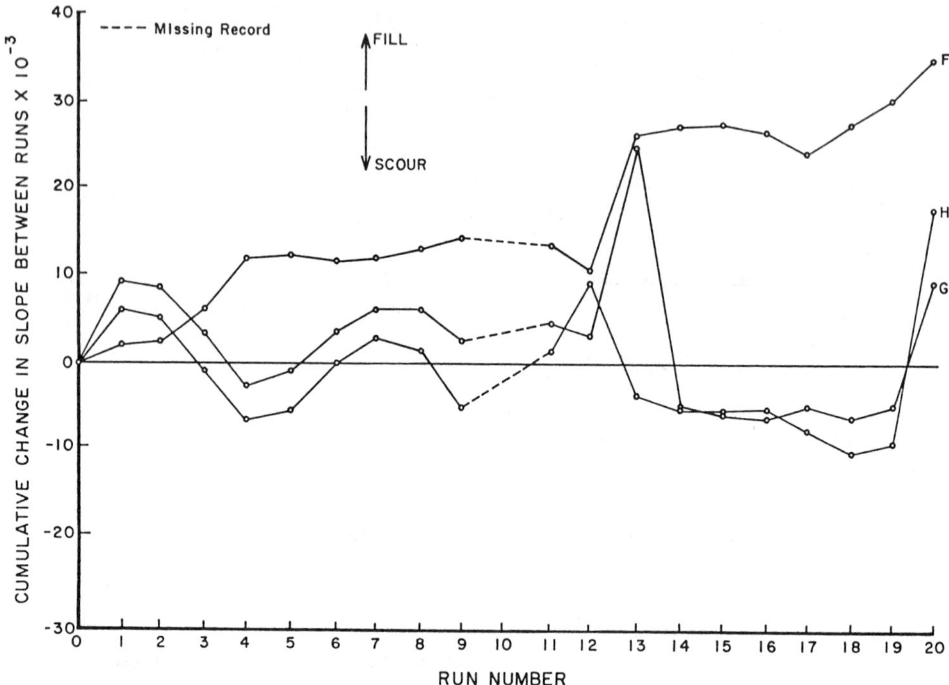

FIGURE 4.11. Cumulative changes in slope between runs for channels F, H, and G (from Harvey, 1980).

Sufficient coarse sediment had been transported from Zone 2 by the end of run 11 to form a large valley fan (i.e., a wedge-shaped body of sediment resembling an alluvial fan) in the upper reach. The fan lengthened 1 m downstream during run 12 and the channel braided. The beds of the channels on the fan were covered by coarse sand and granules. During run 13, the fan expanded 1.25 m farther downstream (cross sections 24–17), but the upper reach of the fan trenched (cross sections 24–21), thereby concentrating the flow. The fan continued to expand downstream (1 m) during run 14.

During run 17, the fan trench extended down to cross section 15, cutting off the side channels that fed the distributaries on the south side of the floodplain. During run 19, the toe of the fan reached cross section 12 and it trenched from its head to cross section 18 as a result of reduced sediment supply from Zone 2. During run 20, the upper fan trench filled (cross sections 24–18).

Wave or pulselike transport of temporarily stored bed materials has been reported by Kelsey (1977), Hayward (1978), Griffiths (1979), and Bergstrom (1980). However, the observations in the REF (Harvey, 1980) suggest that the wave or pulse like transport phenomenon is the result of erosion in locally aggraded channel reaches, which moves the material downstream to a new site of temporary storage. It is not, therefore, due to a progressive surge of sediment down the channel (Griffiths, 1979; Meade, 1985).

TABLE 4.4 Qualitative Summary of the Changes in Mean Elevation of Cross Sections (cs) 2, 14, and 20 during Runs 4–19[a]

	Cross section		
Run Number	2 Lower Reach	14 Middle Reach	20 Upper Reach
4	+	+	+
5	−	−	+
6	+	+	+
7	−	0	−
8	+	+	+
9	−	−	+
10	+	+	0
11	−	−	+
12	0	+	+
13	−	−	0
14	−	0	+
15	+	+	+
16	−	+	+
17	+	+	+
18	−	0	−
19	−	−	−

[a]From Harvey (1980).
+ = Increase in mean elevation (i.e., aggradation);
− = Decrease in mean elevation (i.e., degradation);
0 = No change.

In summary, Harvey's experiments indicated that the Zone 1 tributaries, Zone 2 channels, and Zone 3 channel behave distinctively because of differences in sediment supply, energy availability, and degree of confinement by valley sides. The headwater tributaries generally had an excess of energy over that required for sediment transport, but sediment supply was limited except when large volumes of sediment were periodically delivered by mass movement. At these times there was temporary storage of sediment in the channel bottoms because sediment availability exceeded transport capacity; this sediment was reworked in succeeding storm events and moved downstream. Aggradation tended to promote slumping, as it caused lateral stream shift and under-cutting of the valley walls. The clearest expression of sediment storage was the valley fans at tributary confluences; degradation and trenching of these fans indicated that the stored sediment was being exhausted and that the earlier balance between sediment supply and available energy was being restored.

The Zone 2 channels also responded to the balance between available energy and the quantity of sediment being delivered from the tributaries by storing sediment when incoming loads were high and degrading when loads were low. Repeated aggradation and degradation led to the formation and destruction of flights of both paired and unpaired terraces, and exhaustion of sediment was indicated by degradation, the formation of coarse lag deposits, and initiation and headward retreat of nickpoints.

Temporary storage of sediment in large valley fans at the confluences of Zone 2 channels was an important process. Channel patterns across the fan depended on sediment supply rates, with braiding when supply was high and incision into a single channel when the supply was low.

The Zone 3 channel generally had insufficient energy to transport the quantity of sediment supplied from the Zone 2 channel, but its response was also conditioned by sediment type. Following slumping in the headwaters, sediment was initially fine and a single well-defined channel could be maintained. However, with time the sediment became increasingly coarse and deposition formed a large valley fan. The fan was repeatedly trenched and filled; as this occurred, it advanced downstream as trenching provided sediment for progradation of the toe of the fan. Similar filling and trenching at other localized aggradation sites was responsible for temporal variations in sediment yield from the lower channel; variability in net sediment transport rate from the basin was therefore a function of the progressive reworking of these depositional features, rather than of the downstream movement of a sediment wave.

KRAFT BADLANDS

Hypotheses about drainage basin dynamics that were developed by Schumm and Parker (1973) in the REF have already been applied to management of drainage basins in semiarid areas (Begin and Schumm, 1979; Patton, 1973; Womack, 1975), and a field study (Bergstrom, 1980) of sediment movement in the Kraft Badlands, northeast of Lusk, Wyoming (lat. 42°58′ N, long. 104°0′ W), extends Harvey's observations in the REF to a small natural watershed with high sediment production. Although this is not an experimental study as defined in Chapter 1, it is an important study of small landforms because previous investigations in the area (Mosley, 1973) indicated marked seasonal changes of channel morphology and sediment movement. During one year, the area might experience changes that require decades or centuries in large watersheds. Hence the Kraft Badlands appeared to be an excellent place to study basin dynamics associated with sediment production, storage, erosion, and transport.

The Kraft Badlands (Figs. 4.12, 4.13, 4.14; Table 4.5) is a small area of high sediment production. The siltstone bedrock of the Brule Formation breaks down readily into coarse sand and granule-size particles (Fig. 4.15), which are transported into headwater drainage channels by winter winds, creep, and raindrop impact. The sediment is then transported to the main channel during summer thunderstorms, producing an annual cycle of sediment production, storage, and erosion. The wide, braided valley fills and steep, straight slopes resemble larger steepland drainage basins (Fig. 4.14).

The basin selected for study (Fig. 4.13) has an area of 24.82 ha and a drainage density of 0.21 m/m^2. It is composed of three main drainages; seven tributaries were studied in detail.

FIGURE 4.12. Aerial photograph of Kraft Badlands. See Fig. 4.13 for scale.

The climate is semiarid, with 70–80% of the 40-cm mean annual precipitation falling as spring or summer thunderstorms. During the June–July 1978 observation period, 119 mm of precipitation fell during a series of short-duration thunderstorms each of which delivered up to 20 mm of rainfall (Fig. 4.16).

Channel Dynamics

As in the REF, three channel zones were identified (Fig. 4.17*B*). The headwater Zone 1 channels are high sediment producers, and debris flows are common in the first- and second-order channels. The Zone 2 channels are marked by an abrupt increase in valley width and a decrease in slope and are characterized by both sediment storage and production. Sediment transported by debris flows down steep first- and second-order channels is deposited in lobes in lower Zone 1 and at the Zone 1/Zone 2 boundary, and at tributary junctions. Below second- and third-

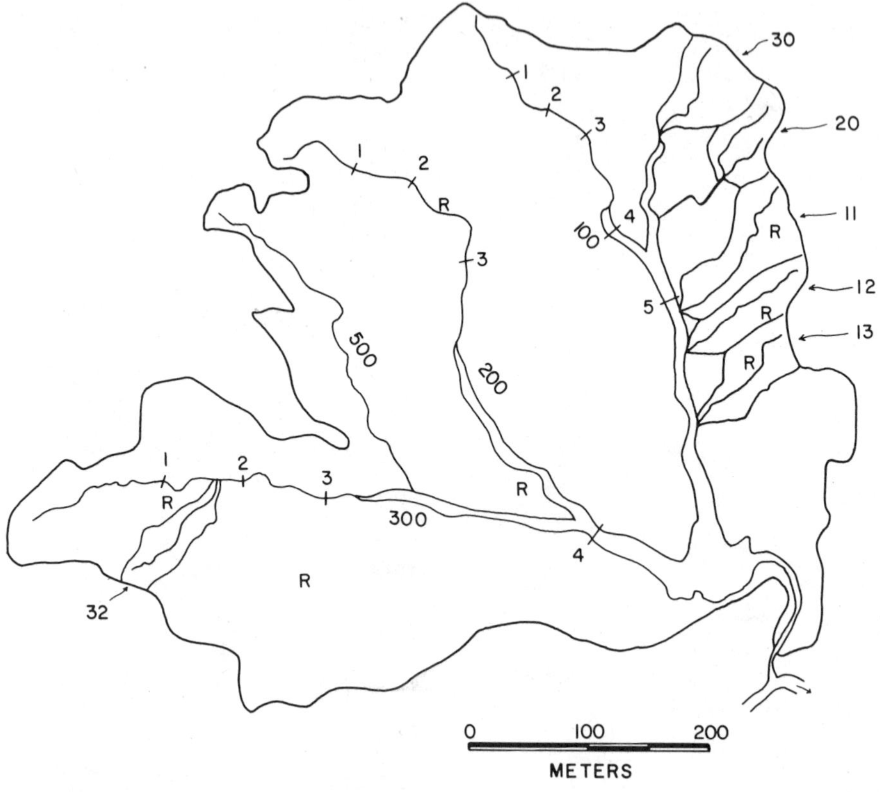

FIGURE 4.13. Kraft Badlands showing location of cross sections on main channels, rain gauges (R), and channel identification numbers (from Bergstrom, 1980).

TABLE 4.5 Geomorphic Characteristics of Four Tributary Basins in the Kraft Badlands[a]

Drainage Basin	Relief *H* (m)	Basin Length Parallel to Main Stream (m)	Length of Main Stream (m)	*A* Area (m^2)	Total stream Length *L* (m)	Drainage density *Dd* (m/m^2) = *L/A*	Relief Ratio (Rh) = *H/L*
11	8.7	112.4	143.0	4,574.9	867.0	0.19	0.08
12	7.3	100.7	107.4	2,662.1	492.9	0.18	0.07
13	9.2	99.9	102.1	2,661.3	658.2	0.25	0.08
32	9.4	90.6	99.5	2,031.4	444.7	0.22	0.10

[a] From Bergstrom (1980).

FIGURE 4.14. Characteristic landforms of Kraft Badlands. (*A*) Bedrock slopes and low-order channel in upper basin, Zone 1. (*B*) Braided main channel and tributary. (*C*) Multiple terraces in alluvium of Zone 2.

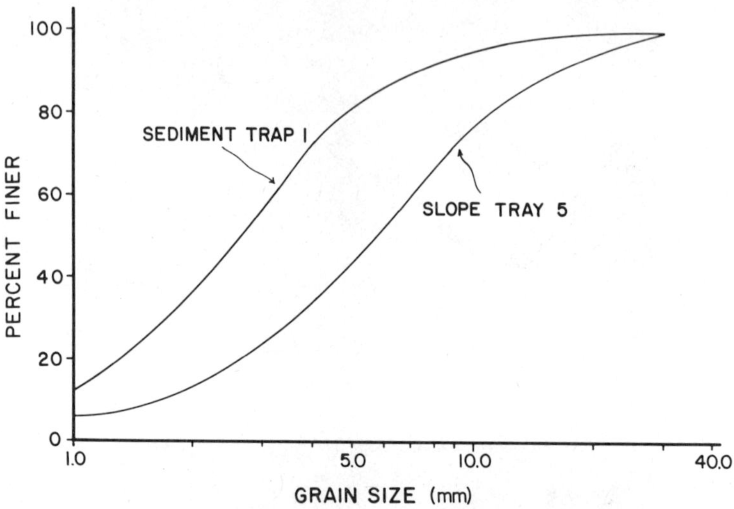

FIGURE 4.15. Grain-size analysis of material produced from slopes (slope tray 5) and in channel (sediment trap 1) (from Bergstrom, 1980).

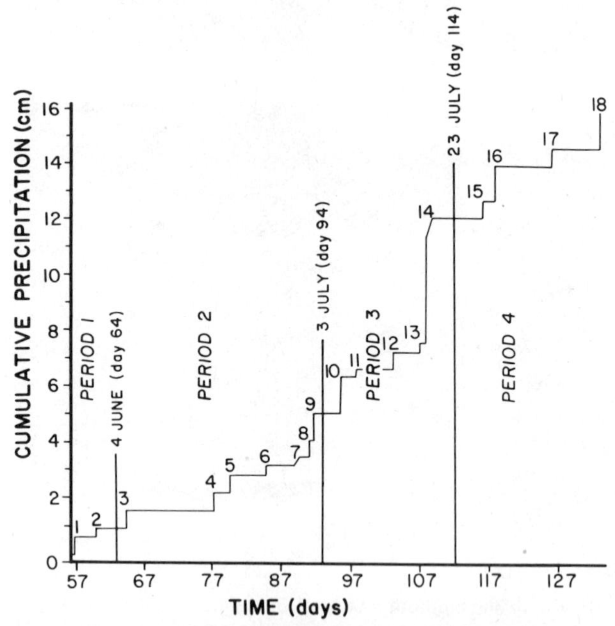

FIGURE 4.16. Cumulative precipitation during the study period (summer 1978). Individual storms are numbered and the four storm periods are indicated (from Bergstrom, 1980).

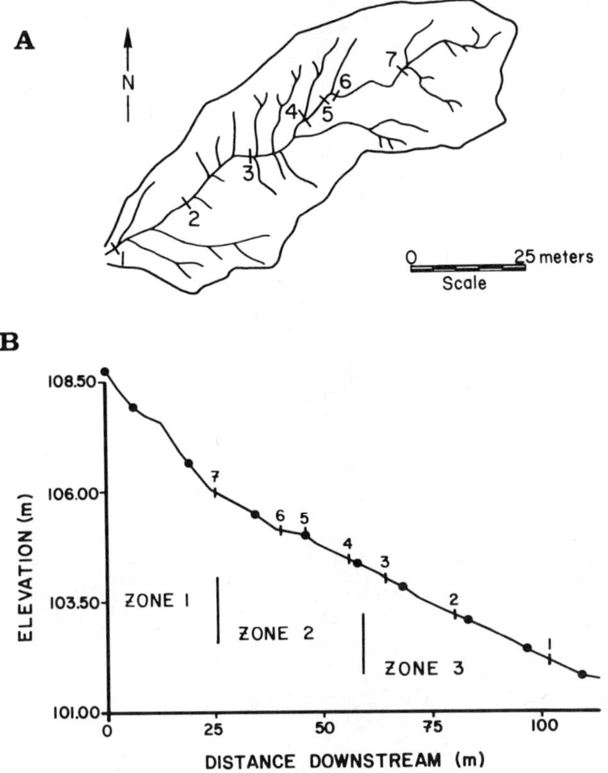

FIGURE 4.17. Map (*A*) and logitudinal profile (*B*) of basin 12. Numbers refer to cross section locations (from Bergstrom, 1980).

order junctions in Zone 2, sediment moves as bedload in a straight incised channel or braided channel that is confined by valley walls. Zone 3 has a wide, permanently braided channel.

Observations in basin 12 are typical of the seven subbasins (Fig. 4.17). During winter, the steep and narrow headwater channels filled with granule-size sediment, and snowmelt runoff caused incision in Zone 2; but this incised channel became shallow in Zone 3 and was not present in Zone 1. During the period April 2–June 3, precipitation produced no runoff, but during the period June 3–July 3, debris flows were initiated by high-intensity storms. The smaller headwater channels incised to bedrock and a large volume of sediment was delivered to the Zone 1/Zone 2 boundary, causing 9.7 cm of aggradation at cross section 7 (Fig. 4.18*A*) and somewhat less downstream at cross section 5 (Fig. 4.18*B*). During the period July 3–July 23, the channel incised to bedrock at cross section 7, while the channel downstream at cross section 5 at first filled and then incised in several episodes to create a series of unpaired terraces (Fig. 4.18*B*). Bedlevel changes along the profile during these periods were highly irregular due to reworking and storage of sediment in valley fans (Fig. 4.19).

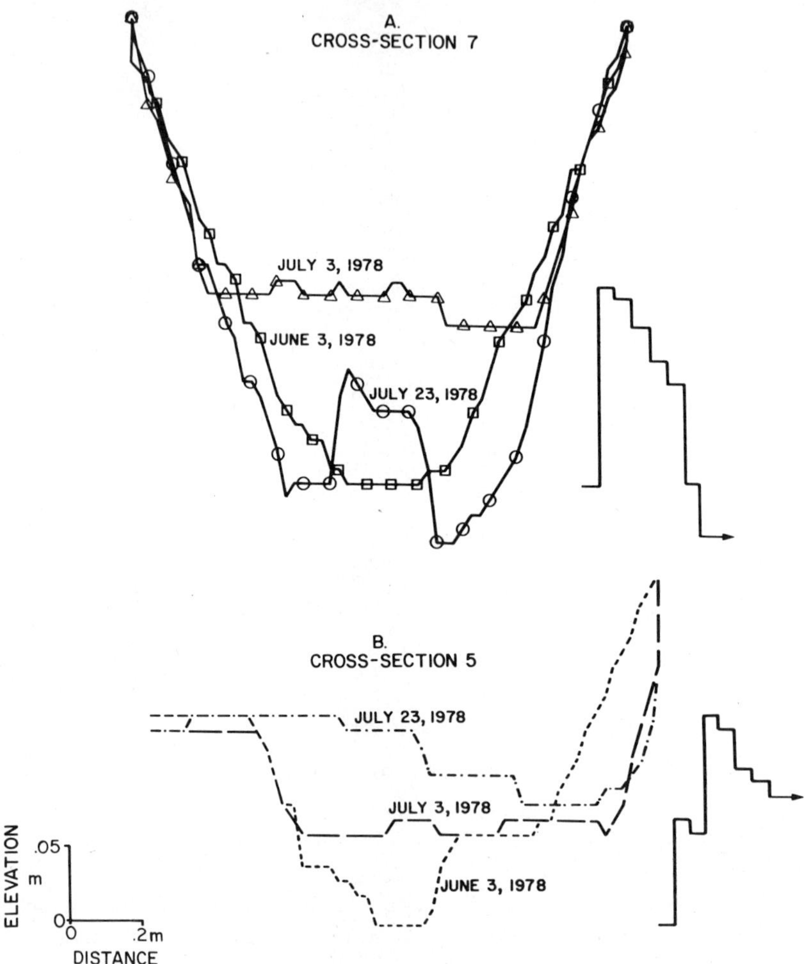

FIGURE 4.18. Cross sections 7 and 5 in basin 12. Schematic diagrams at the right illustrate channel bed behavior during the study period (from Bergstrom, 1980).

Only the heaviest storms produced runoff throughout the channel system. During small storms, Zones 1 and/or 2 were actively producing and transporting sediment, but there was no substantial flow in Zone 3, and sediment arriving from upstream was deposited. When runoff occurred in Zone 3, the storage sites were trenched and sediment was removed from the basin.

The channel bed elevations changed markedly during the study (Fig. 4.19). In upper Zone 1, the fill was scoured discontinuously between June 10 and July 4, and the sediment was deposited at the Zone 1–Zone 2 boundary. Cross section 7 illustrates this deposition, although the dates of survey differ slightly (Figs. 4.18*A*, 4.19). The initial U-shaped channel inherited from the winter period is shown by the June 3 survey. Between June 3 and July 3, sediment supplied by debris flows

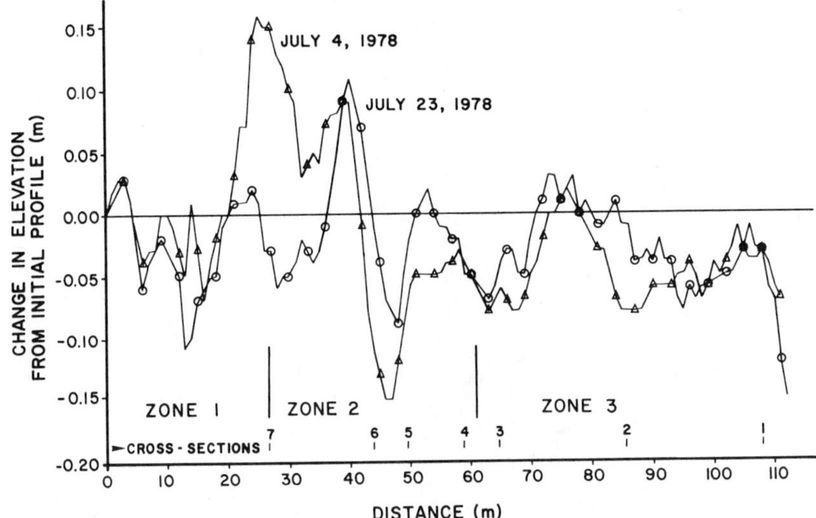

FIGURE 4.19. Change in thalweg bed elevation relative to the June 10, 1978, profile, which is the horizontal line at 0.0 m in basin 10. Numbers refer to cross section locations (from Bergstrom, 1980).

aggraded the bed 9.7 cm. The irregularity of the July 3 profile indicates reworking, minor deposition and erosion, and channel development on the right side of the profile (Fig. 4.18*A*).

Two broad peaks at 50–60 m and 70–85 m (Fig. 4.19) and a third, more subdued peak at 100–110 m in lower Zone 2 and Zone 3 are the result of sediment storage in valley fans. During period 3 (July 3–July 23), storms moved the stored sediment at cross section 7 downstream (Figs. 4.18, 4.19). This sediment filled and widened the channel, between 50 and 60 m at the point of valley widening and elsewhere downstream (Fig. 4.19).

Between July 3 and July 23, the channel at cross section 7 degraded, finally encountering bedrock at both valley sides (Fig. 4.18*A*). The overall changes during periods 2 and 3 were thus characterized by aggradation, followed by discountinuous incision (Fig. 4.18*A*).

In contrast, the Zone 2 valley, exemplified by cross section 5 (Fig. 4.18*B*), aggraded during the same period and subsequently incised episodically, forming unpaired terraces. However, there was an overall increase in sediment storage in Zone 2. Therefore, cross sections 5 and 7 were out of phase, with deposition at 5 while erosion was occurring at 7.

There were significant channel changes during the period of observation, and Bergstrom mapped channel conditions in basin 12 (Fig. 4.13) on July 2 and July 26 (Figs. 4.20, 4.21). These dates correspond to the end of storm periods 2 and 3 (Fig. 4.16).

Conditions of maximum sediment storage existed on June 4 and there was no change in the winter colluvial accumulations. However, by July 2, erosion extended

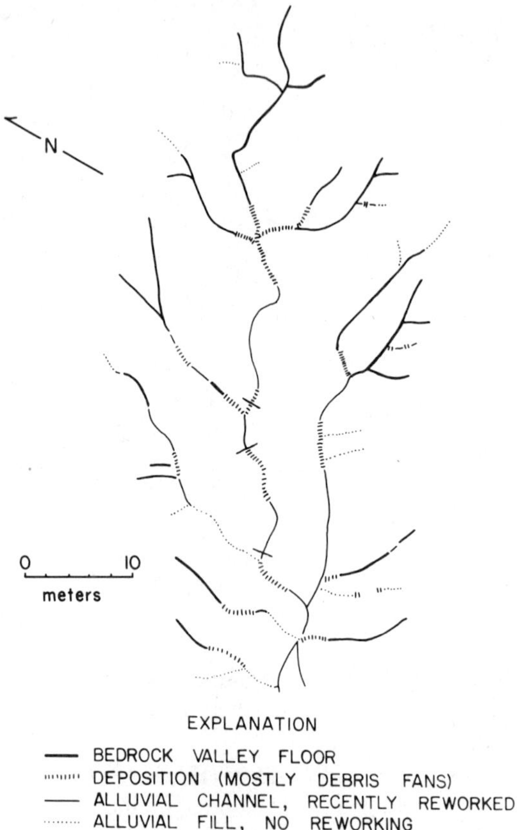

FIGURE 4.20. Channel morphology of Zones 1 and 2 of basin 12 at the end of storm period 2 (July 2, 1978) (from Bergstrom, 1980).

from the main divide downstream with abundant temporary sediment storage at fans in lower headwater channels (Fig. 4.20). Several channels contained a series of debris fans. By July 26 there was nearly complete flushing of the headwaters of the main channel and all tributaries (Fig. 4.21). In addition, many of the storage sites had been incised to form terraces.

Behavior of Steep Basins

In steep terrain, as exemplified by the Kraft Badlands and the REF, basin behavior and sediment production are not uniform in time or space. This episodic behavior (Bergstrom and Schumm, 1981) may be inherent in steepland areas of high sediment production.

The observations in basin 12 provide a basis for this model of episodic behavior of the channels in the three zones (Fig. 4.22). An idealized sequence of cross sections in lower Zone 1 is shown in Fig. 4.22*A*. The numbers 1–4 identify the altitude of the valley floor and channel at different times during the year. Winter

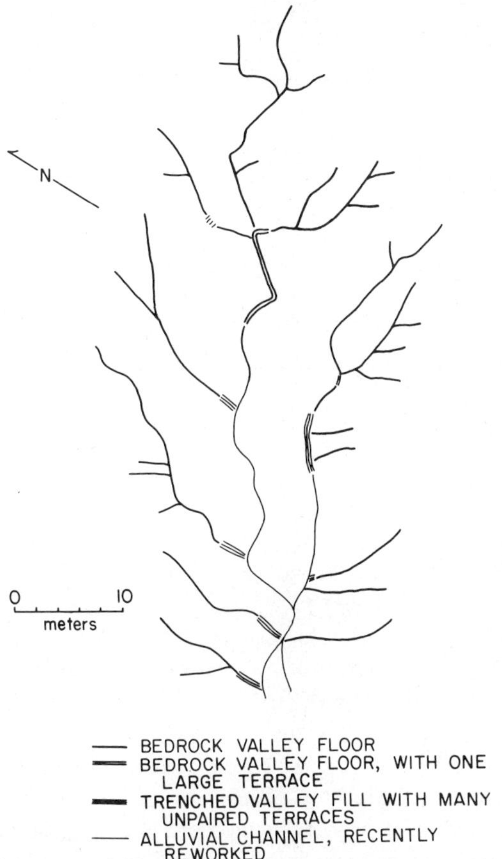

FIGURE 4.21. Channel morphology of Zones 1 and 2 of basin 12 at the end of storm period 3 (July 26, 1978) (from Bergstrom, 1980).

sediment production covers the bedrock valley floor (1) and partly fills the valley (2a). Spring rains cause further aggradation (2b); but eventually precipitation on bare slopes and flow in upstream bedrock channels yield flashy runoff with low sediment loads, and this causes incision and removal of much of the stored alluvium (3). Subsequent runoff events cause further incision down to the bedrock valley floor (4), leaving a valley-side terrace (3).

At stage 1 in Zone 2 (Fig. 4.22*B*) a valley fill is present (1). Early spring runoff incises the alluvium (2) and produces a terrace or multiple terraces. As sediment is introduced to Zone 2 as a result of major incision of the Zone 1 valley fill, aggradation raises the valley floor (3). When Zone 1 sediment production decreases as the channels reach bedrock, the Zone 2 alluvial fill is eroded episodically (4a, 4b, 4c). Zone 2 is clearly out of phase with Zone 1. Zone 1 aggradation coincides with Zone 2 degradation and vice versa.

Only during the largest hydrologic events did runoff occur in Zone 3. During

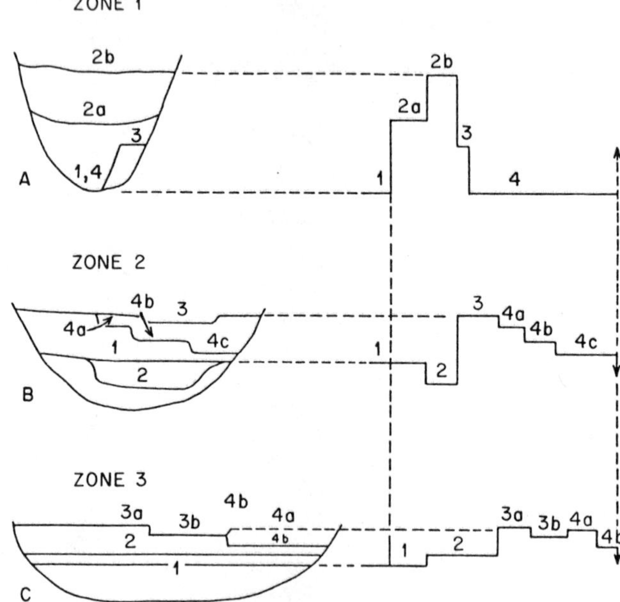

FIGURE 4.22. Idealized cross sections showing changes in Zones 1, 2, and 3 during one season in the Kraft Badlands. Diagram to right of each cross section illustrates channel behavior in each zone (from Bergstrom and Schumm, 1981).

small storms, when Zone 1 and/or Zone 2 were actively producing and transporting material, Zone 3 received little flow capable of modifying the valley fill (Fig. 4.22*C*). Therefore, sediment was stored in Zone 3 (1, 2, 3a), and valley fans, which are sites of maximum sediment storage, were formed. When runoff occurred in Zone 3, the fans were incised and sediment was moved out of the basin (3b). As sediment continued to be contributed from Zone 2, a particular cross section would alternately aggrade and degrade (4a, 4b), depending on whether a fan was being constructed or eroded. Therefore, reaches of channel in Zone 2 were also frequently out of phase at any given time.

In summary, in steep terrain sediment production is not uniform during time or in space. Widespread mass wasting of slopes and incision through channel-stored sediment in Zone 1 produces large quantities of sediment that is delivered to Zones 2 and 3. The main channel is overloaded and channel filling and storage are followed by erosion and sediment transport. Deposition and erosion alternate episodically through time within the three zones. The zones are usually out of phase with one another at a given time. While deposition is storing sediment in Zone 1, Zone 2 is a source of sediment and Zone 3 is a sink (Fig. 4.22*B*) time 2). As Zone 1 begins to produce sediment (time 3), Zones 2 and 3 both aggrade. Finally, Zone 2 acts primarily as a conveyor of sediment that was temporarily stored at the Zone 1–Zone 2 boundary, and Zone 3 acts as both a sink and source of sediment as valley fans form and are incised.

These conclusions are consistent with observations of steepland catchments in New Zealand and California (Hayward, 1978; Kelsey, 1977; Mosley, 1978; O'-Loughlin, 1969), where a longitudinal zonation of channel form and process has been observed.

DISCUSSION

The experimental studies by Parker and Harvey show how complex is the behavior of a small, steep drainage basin, and Bergstrom's observations in the Kraft Badlands permitted development of an episodic-behavior model of drainage basin behavior (Fig. 4.22), which stresses the out-of-phase condition of components of the drainage networks and their episodic response. This model of episodic erosion and deposition, which was developed from observations in the badlands, fits the observations by Harvey (1980) in the REF and has close similarities to the simpler complex-response model proposed by Schumm and Parker (1973) on the basis of the first experiments in the REF and field observations elsewhere (Araya and Higashi, 1983; Bettis and Thompson, 1982; Schumm and Hadley, 1957; Schumm et al., 1984).

In the Kraft Badlands, the model of episodic behavior describes sediment supply and transport on an annual basis. Other examples indicate that episodic erosion occurs widely over longer periods of time, in fact, whenever extreme events supply large quantities of sediment to and overload the drainage system (Grant, 1977; Janda et al., 1975; Pearce and Watson, 1983). Womack and Schumm (1977) and Wildman (1981) found clear evidence for episodic erosion on a time scale of decades as sediment was removed from the Douglas Creek channel (Fig. 4.23) and the Bear and Yuba valleys of California (Fig. 4.24). Douglas Creek in northwest Colorado displays a complex series of up to seven discontinuous, unpaired terraces, which in some places may be dated by tree ring counts, aerial photographs, and historical records. In 1882, the first settler described the valley as "the best cattle country you ever seen . . . no brush and deep gullies like today, but lush grass up to the stirrups of a horse. The creek was right on top of the ground. You could dip out water with a bucket." The homestead site is now cut by a vertical-walled channel 10 m deep and the stream is intermittent.

Channel erosion began after, and it is probably related to, the introduction of large numbers of cattle in 1882, and the terrace sequence demonstrates that incision has been episodic, with periods of incision separated by periods of infilling at different locations (Fig. 4.23). On the basis of the complex-response and episodic-erosion models developed from work in the REF, it is concluded that when erosion began in the valley, the headward erosion of the main and side valleys in response to major storms produced sediment that temporarily overwhelmed the channel's ability to transport sediment, leading to aggradation and oversteepening of the valley floor. Repeated trenching of such deposits produced flights of unpaired terraces that persist for only short distances. Much of the evidence may have been destroyed by more recent erosion, but the differences between terrace flights at

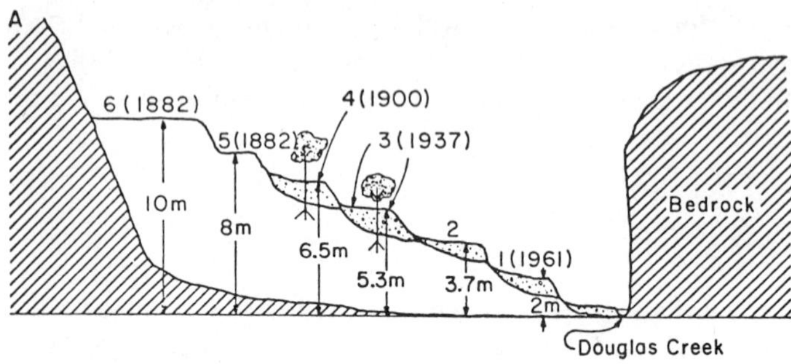

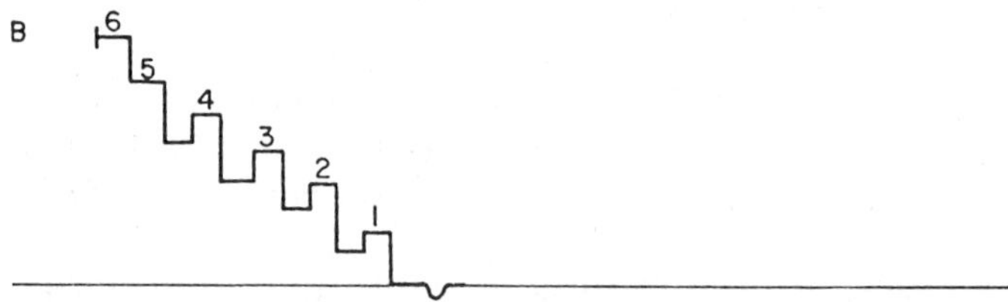

FIGURE 4.23. (*A*) Idealized cross section of Douglas Creek valley. Terraces 6 and 5 represent pre-1882 valley floor and probable channel of Douglas Creek, respectively. Terraces 1–4 were formed after 1900 (from Womack and Schumm, 1977). (*B*) Diagrammatic representation of channel behavior at this cross section.

different locations suggest a different erosional history at each location and that erosion and deposition along Douglas Creek was out of phase. It is not possible to correlate the terraces purely on the basis of elevation, nor to relate the terrace remnants throughout the valley to specific stimuli such as climatic variations or baselevel changes. Moreover, there is a lack of similar terraces in nearby valleys. Thus it appears that the controls are inherent to the geomorphic system, even though the initial impulse was the external one of overgrazing.

Other examples of terraces that were formed due to system behavior rather than in response to several external stimuli (tectonics, climate change) have been provided by Born and Ritter (1970), Gage (1970), and Hoyer, (1980a, b). Born and Ritter mapped six discontinuous and unpaired terraces at the mouth of the Truckee River, where it enters Pyramid Lake, Nevada. A single lake-level lowering induced discontinuous downcuttig by the river, which created the terraces. The Waiho River, New Zeland, aggraded many meters in response to a major storm, and then incised through the deposits over a period of several weeks (Gage, 1970). This erosion produced a flight of 3-m-high terraces, which, in the absence of observation of their formation, might have been attributed to several periods of erosion and deposition over a long period. Hoyer (1980a, b) reported that the formation of numerous terraces in Little Sioux River Valley, Iowa, are the result of a drainage change during Wisconsin time that caused episodic incision of pre-Wisconsin gla-

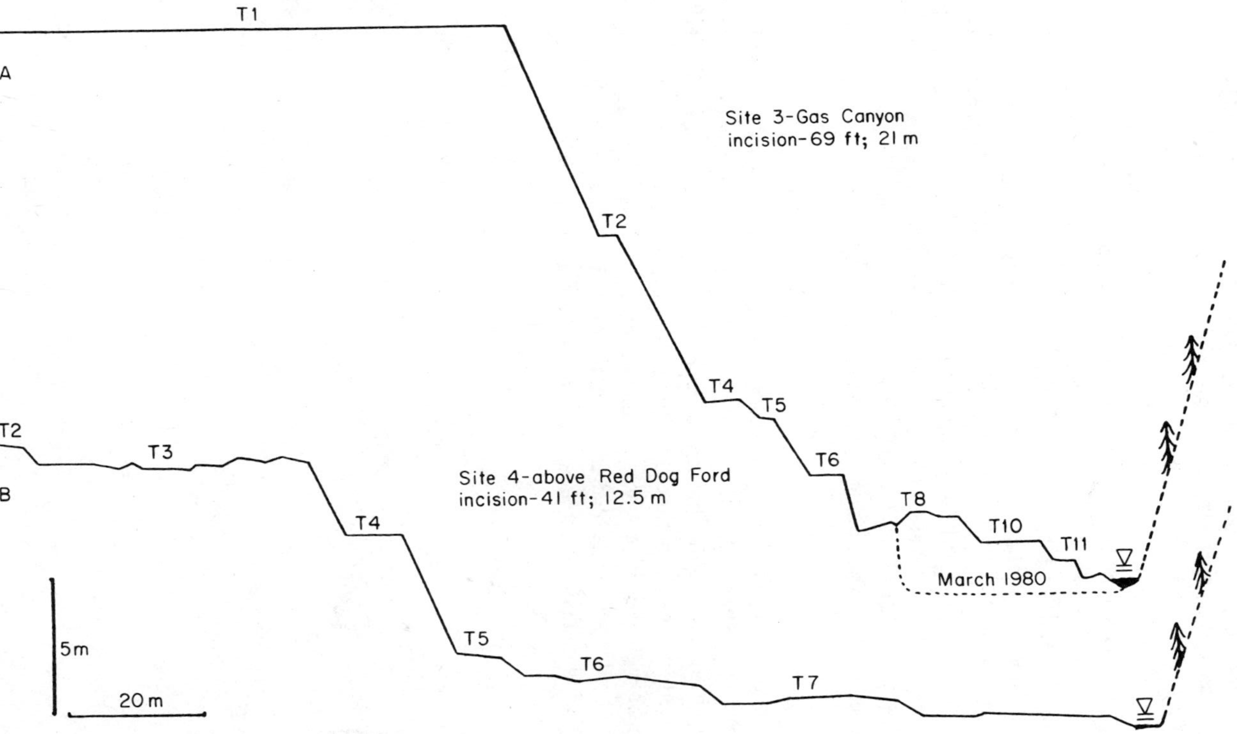

FIGURE 4.24. Terraces formed during the past 100 years as sediments produced by hydraulic mining were removed from the Bear River valley, Sierra Nevada, California (from Wildman, 1981).

cial deposits. Finally, the incision of upper Toutle River, near Mount St. Helens, Washington, as a result of pumping from Spirit Lake, led to similar episodic behavior. At a constant discharge, multiple terraces formed as a result of sediment storage and removal that reflected bank collapse, tributary contribution, and general variability in sediment load (Paine, 1985). Clearly, sediment storage is an important component of any fluvial system (Ashida et al., 1976; Graf, 1983; Kelsey, 1980; Meade, 1982; Trimble, 1975), as is its subsequent remobilization.

Perhaps the most important result of the experimental studies is the recognition of complex response (Fig. 4.7), episodic erosion (Figs. 4.23, 4.24), and episodic behavior (Fig. 4.22). The resulting complexity of alluvial stratigraphy and the out-of-phase characteristics between valleys will lead to problems of alluvial and terrace correlation in many areas (McDowell, 1983). Obviously, this also greatly complicates attempts to develop sediment yield predictions based upon short hydrologic records.

Also of great significance are the relatively large magnitude of changes involved and their nonuniformity in time. For example, engineering designs that are based on short-term sediment yield records and existing channel morphology can prove to be wholly inadequate if the drainage system is responding complexly. It should be remembered, however, that the dynamic drainage basin behavior that has been described in this chapter is probably a function of high energy and high sediment production. Gentle, well-vegetated drainage basins will not behave in this dynamic fashion unless denuded or subjected to major climatic, tectonic, or baselevel changes.

II RIVERS

River channels comprise only a few percent of the total surface area of the continents, but they have received a high proportion of the attention of geomorphologists and members of the other disciplines involved in the study and management of the earth's surface. This may partially be a result of man's eternal fascination with water, but it is also because rivers are of great practical importance, and an understanding of their behavior is needed for management and utilization of this resource. For example, one class of rivers that has received particularly extensive study is the large, low-gradient rivers such as the Mississippi and the Rhine, which are used for commercial navigation. A prime need is to establish a stable channel along which navigable depths can be maintained without incessant dredging.

In addition, irrigation canals such as those in the Indian subcontinent have been the subject of extensive study in order to facilitate the design of stable channels that will neither lose capacity by being choked with sediment, nor scour their beds and thus undermine their banks and structures such as bridges, nor migrate laterally and destroy valuable riparian land. Because many rivers flow through settled areas or are bordered by highly productive floodplain soils, the need for river control has prompted much research into river behavior, with the goal of developing design methods to estimate such channel characteristics as width, slope, and meander wavelength.

The purpose of this section is to review experimental studies of river morphology (Chapter 5), incised channels (Chapter 6), valleys and bedrock channels (Chapter 7), and the effect of tectonics on alluvial channels (Chapter 8). In all cases the walls of the channels are not the rigid walls of the flume (Soni, 1981); rather, they are erodible boundaries that permit changes of channel morphology and position.

The many flume studies of sediment transport (Kennedy, 1983) will not be examined, because in general they relate only obliquely to channel morphology as a whole. The reader is referred to texts such as those by Graf (1971) and Raudkivi (1967) for a thorough treatment of such work. Much of the relevant work on natural rivers has been comprehensively reviewed by Richards (1982), Knighton (1984), and Morisawa (1985).

5 Alluvial River Channels

As recently as 1983 Kennedy, in his Rouse Memorial Lecture to the American Society of Civil Engineers, remarked on the slow progress that has been made in river hydraulics and sediment transport engineering, and there is a continuing discussion as to whether the shape of alluvial channels is indeterminate (Kennedy and Brooks, 1965; Maddock, 1970), determinate (Hey, 1978), or semideterminate (Mosley, 1981). It seems that, in principle, the form and behavior of alluvial channels should be explicable in terms of the laws of physics, although in practice there are limitations to data gathering that may prevent this (Watson, 1969; Schumm, 1984), and the fact that there are many different types of channels (Schumm, 1977; Brice, 1984) makes generalization difficult. There have been many formulations of the relationship between river channel form and the variables which control it. These range from Richards's (1982) ''speculative representation of the alluvial channel system'' (Fig. 5.1), through Hey's (1979) qualitative ''process-response model of river channel development,'' to the type of dimensional analysis used by hydraulicians such as Ackers and Charlton (1970a). However, much research into river morphology has been empirical and descriptive (Henderson, 1966, p. 455).

There may be some question whether the list of dependent and independent variables included in Fig. 5.1 is exhaustive or whether all the variables are of equal importance, but the variables that should be included in any given situation are, in principle, specifiable. Another problem that is associated with the question of indeterminacy is the direction of causality—what are causes and what are effects? Kennedy and Brooks (1965) and Schumm and Lichty (1965) have demonstrated that different variables may be dependent or independent, depending on the time scale at which a river is being considered (Table 5.1). Streams are seldom in a static state, and transitory adjustments are accomplished by storage of sediment and water. In the short term, water storage occurs by a rise of river stage or

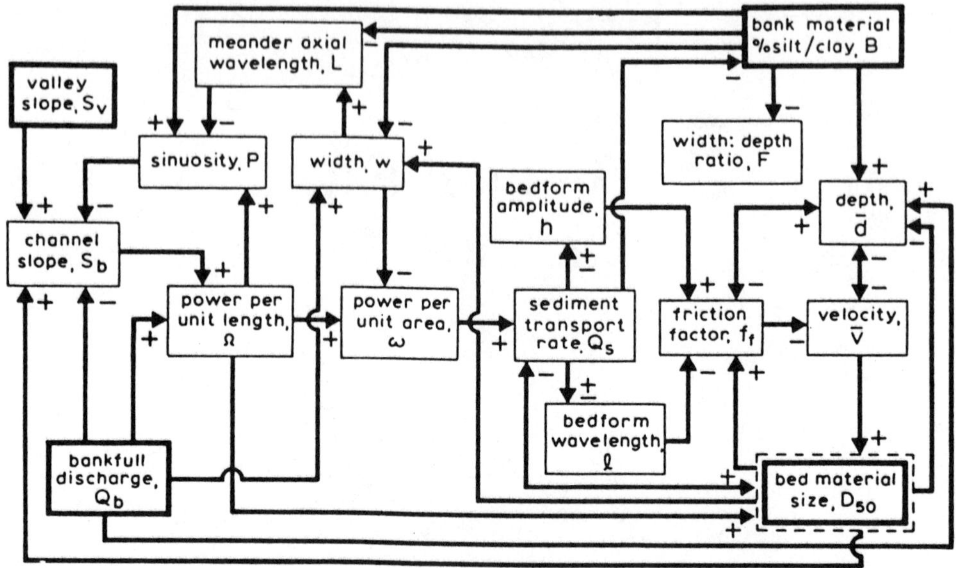

FIGURE 5.1. Richards' conception of the alluvial channel system. Independent variables have heavy outlines; bed material size, though ultimately controlled by lithology, is semi-independent as it is affected by sediment transport. Direct relationships are shown by +, inverse by −; arrows show direction of influence. Some links are reversible; for example, as friction increases, so does depth of flow, but increasing flow depth drowns resistance and reduces friction (from Richards, 1982).

overbank flooding. Departures of the sediment inflow from the equilibrium transport rate are absorbed by scour or fill. However, in the longer term, the river will assume a profile and other characteristics for which the outflows of sediment and water equal the inflows—that is, they are both independent. Over the very long term, a river responds to the geology and climate of the area, so that every characteristic of the river, including water and sediment discharge, depends on something else (Table 5.1). The form of a river channel is, on the very long ("cyclic") or long ("graded") time scales, dependent on a complex set of variables. Isolation of the effect of any one of these variables may prove exceedingly difficult because they cannot be manipulated to facilitate the observation of the response of a natural river to changing environmental conditions.

Schumm (1977) noted that these changes in the status of variables may cause confusion in the interpretation of geomorphic processes because a temporary reversal of cause and effect may occur at the "steady," or short, time scale at which rivers are commonly studied. An example is the relationship between secondary circulation in a river bend and meandering; observations during steady time may indicate that secondary circulation is caused by the bend (Bathurst et al., 1979), while some theories of meander causation conclude that the bend is caused by secondary circulation (Einstein and Shen, 1964). Furthermore, Rubey (1952) pointed out that certain variables are mutually adjusted: "if discharge, load, grain

TABLE 5.1 The Status of River Variables during Time Spans of Decreasing Duration[a]

Variables	Status of Variables during Designated Time Spans		
	Cyclic	Graded	Steady
1. Time (stage)	Independent	Not relevant	Not relevant
2. Initial relief	Independent	Not relevant	Not relevant
3. Geology (lithology, structure)	Independent	Independent	Independent
4. Paleoclimate	Independent	Independent	Independent
5. Paleohydrology	Independent	Independent	Independent
6. Relief or volume of system above baselevel	Dependent	Independent	Independent
7. Valley dimensions (width, depth, slope)	Dependent	Independent	Independent
8. Climate (precipitation, temp., seasonality)	—	Independent	Independent
9. Vegetation (type and density)	—	Independent	Independent
10. Hydrology (mean discharge of water and sediment)	—	Independent	Independent
11. Channel morphology	—	Dependent	Independent
12. Momentary water and sediment discharge (at-a-section)	—	—	Dependent
13. Hydraulics of flow (at-a-section)	—	—	Dependent

[a] After Schumm and Lichty (1965).

size, and sorting are considered the controlling factors, then velocity, slope, width and depth of channel are dependent variables that are affected not only by the independent variables but also by one another." This view led Maddock (1970) to suggest that it is to some degree impossible to achieve unique or determinate relationships among velocity, depth, slope, and the rate of sediment movement, because the relationships are influenced by other constraints.

A diversity of approaches to explanation of channel morphology has been used. Conceptual models of the type proposed in Fig. 5.1 are only a starting point, even though they are a necessary stage in development of a physically based, deterministic theory of river behavior. To progress beyond the conceptual models, studies of rivers have followed three main directions that have involved theoretical and semitheoretical treatments, empirical investigations of natural channels, and physical models of actual or idealized rivers.

Empirical investigations of natural rivers have taken two major routes. Commonly, data from a number of rivers having different environments have been collected and simple relationships developed between channel morphology and indices of environment, notably sediment grain-size characteristics, mean annual flood, and bankfull discharge (Brush, 1961; Carlston, 1965; Leopold and Maddock,

1953; Leopold and Wolman, 1957; Mosley, 1981; Schumm, 1960; Wolman, 1955). This "hydraulic geometry" approach is similar to the "regime theory" developed by river engineers such as Blench (1969), Inglis (1949), and Lacey (1930), as based on data from artificial canal systems. Both approaches emphasize the influence of water and sediment discharge on the river channel and make little reference to the other variables listed in Table 5.1.

More recently, there has been increasing opportunity to study, in a moderately controlled fashion, the response of river systems to changes in some variables such as discharge, flood frequency, sediment yields, vegetation cover, and climatic fluctuations (Grant, 1977; Gregory, 1976; Lisle, 1982; Mosley, 1975a, 1978; Petts, 1979; Williams and Wolman, 1984). Most of these opportunities have arisen because of human modification of the land surface and dam construction with consequent impacts on flow regime and sediment yields.

On the short ("steady") time scale, channel morphology may be regarded as controlled primarily by the physical characteristics and quantities of water and sediment supplied to the river (Table 5.1). At this level, formulation of physical theories of river morphology begins to be possible, and there have been many efforts to explain channel form in a theoretical or semitheoretical manner (e.g., Lane, 1957; Parker, 1976). Nevertheless, such theories must be tested against the physical world, and extensive use has been made of laboratory flumes to provide the necessary quantitative information (e.g., Gilbert, 1917; Simons and Richardson, 1966).

In some cases, rivers with controlled discharges have been utilized to study river morphology under quasi-experimental conditions (Meade et al., 1981; Mosley, 1982a, b). Field studies of fluvial processes are, however, increasingly expensive because of large manpower requirements, the need for robust but sophisticated instrumentation, and, often, remoteness of field sites. Furthermore, during the conditions of steady flow under which these studies are generally carried out, channel morphology is itself an independent variable that controls the phenomena that the scientist is able to measure, particularly the values of water depth, velocity, and sediment concentration. During floods, when channel morphology is a dependent variable and is changing in response to variation in the controlling variables, water and sediment discharge measurements are most hazardous, difficult, and slow, processes are most rapid, and the least time is available to make measurements.

Studies of small-scale laboratory rivers under controlled conditions have been widely used to elucidate the relationship between channel form, flow processes, and variations in certain controlling variables—primarily valley slope, water discharge, sediment characteristics, and sediment discharge (variables 7 and 10, Table 5.1). These variables are the immediate controls upon river form, and flume studies have been little used to consider the effect of the other independent variables listed in Table 5.1, although Chapter 2 describes physical model studies of the influence of variable 2 on the character of a drainage network.

A fact that is frequently ignored by experimenters, model builders, and field investigators is that there exists a considerable variety of alluvial river types (Brice,

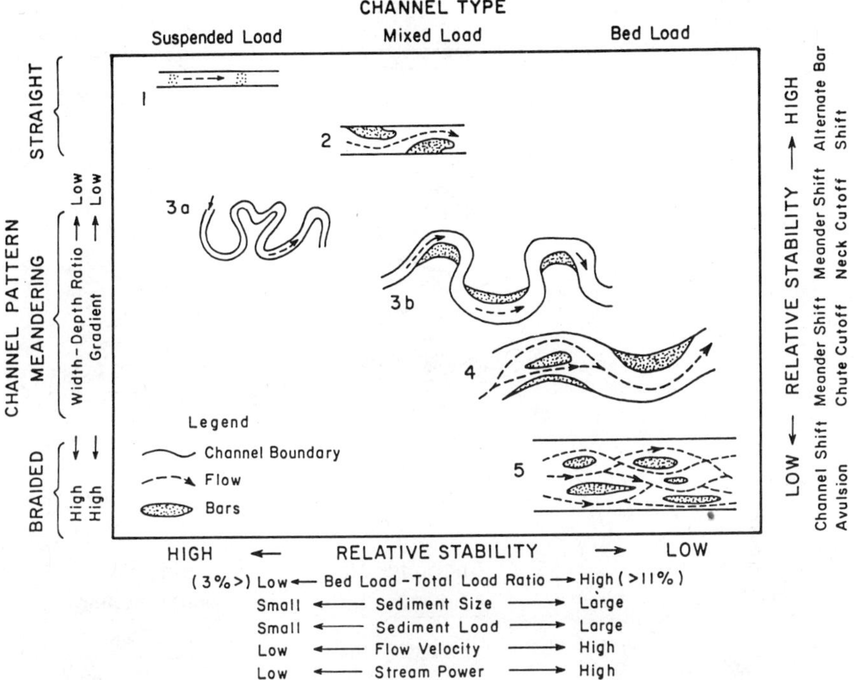

FIGURE 5.2. Classification of alluvial channels based upon type of sediment load (from Schumm, 1981).

1984; Mollard, 1972; Schumm, 1981). There appear to be about 5 basic river types (Fig. 5.2), although Schumm (1981) has suggested that there are 14, and Brice shows that within one type there can be a considerable range of morphologic characteristics (Fig. 5.3). Herein lies the greatest problem: any single field or experimental study will provide information on only one type of river and the results cannot be extrapolated with impunity to other channels (Schumm, 1984). In spite of this complication, rivers have a regularity of form that both is intriguing and holds out the promise that explanation is feasible.

Recent efforts to explain river behavior have concentrated on understanding the relationship between river morphology, the materials of which the channel is composed, and the distribution and expenditure of energy or power to do work. Stream power can be expressed as the product of discharge and slope, the other two independent variables that immediately control channel morphology. Various attempts have been made to use concepts such as the minimization of rate of work or energy expenditure (Yang et al., 1981), maximization of friction along the channel perimeter (Davies and Sutherland, 1980), and minimization of the variance in rate of energy expenditure (Leopold and Langbein, 1962). However, Richards (1982, p. 203) remarks that "teleological and anthropomorphic arguments that rivers tend to minimize power loss, or its variance, cannot replace an explanation

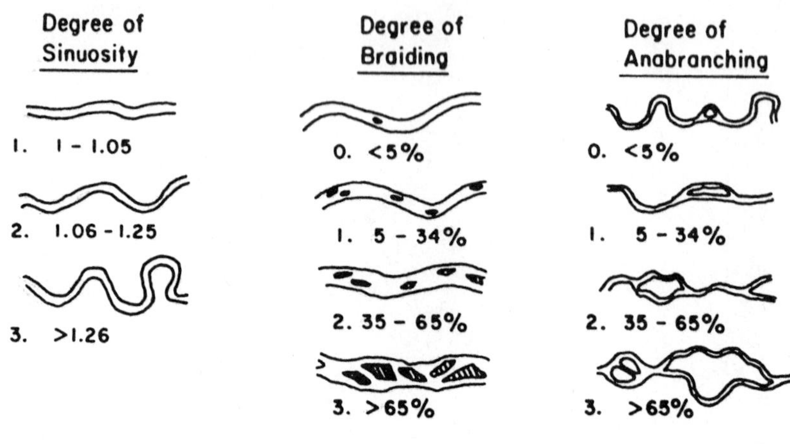

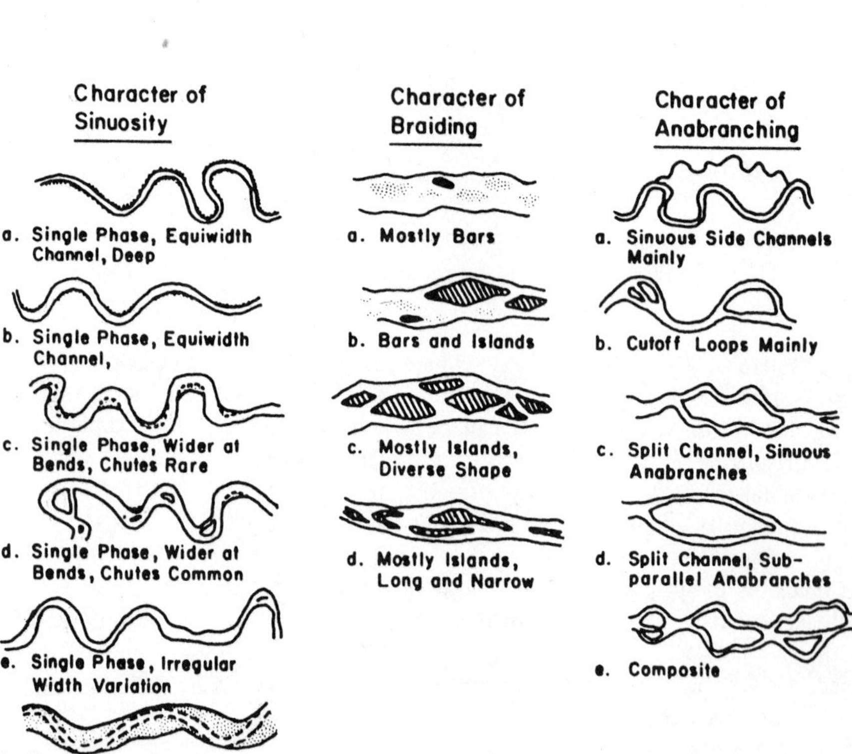

FIGURE 5.3. Range of channel types (from Brice, 1984).

based on the physical mechanisms of meander development'' or any other aspect of river morphology.

RIVER MORPHOLOGY

A number of books provide information on natural channel morphology and channel change as a result of altered conditions (Gregory, 1977; Leliavsky, 1955; Leopold et al., 1964; Schumm, 1977; Richards, 1982). The aspects of river morphology that received most attention are channel width [W, closely equivalent to wetted perimeter (R) in wide channels]; mean and maximum depth (D and D_{max}); water surface slope (S), which fixes channel sinuosity (P) if valley slope (S_v) is given (S/S_v); meander wavelength (L) and amplitude (H); and some index of channel roughness (f or n), which fixes mean velocity (V) when water discharge (Q_w), slope, and cross-section shape (W/D) are given.

The relationships can often be described by simple power functions, the constants and exponents of which vary between areas, in response to variations of discharge, sediment load, and bed and bank sediments, which in turn are dependent upon lithology, climate, vegetation cover, and the intensity of denudation. To take account of sediment characteristics, Blench (1969) and Lacey (1930) both incorporated ''silt factors'' (d_{50}) into their regime equation sets, but their equations were based largely on data from Indian irrigation canals with a rather restricted range of sediment types.

It can be demonstrated that the size of a river channel (W, D) and the size of its pattern (meander wavelength, amplitude) are largely the result of water discharge through the channel (Dury, 1964a,b; Leopold and Maddock, 1953). These relationships, as well as channel shape (W/D) and sinuosity (P), are affected by the type of sediment load moved through the channels (Schumm, 1960, 1963a, 1977). This is expressed as the percentage of silt and clay in a channel perimeter (M). It was demonstrated by Schumm (1968) for five locations where total sediment load was sampled, that M is related to the ratio of wash load to bed-material load or to the percentage of total load (Q_s) that is bed-material load (Q_{sb}) as follows:

$$M = \frac{55}{Q_{sb}} \tag{5.1}$$

Simons and Albertson (1960) assembled data from a range of canals and identified the effect of sediments on canal morphology and hydraulics. For example, wetted perimeter is less and hydraulic radius is greater for a given discharge in a channel with cohesive bed and bank sediments than in a channel with sand bed and banks.

Water surface slope has been a major concern, particularly for engineers with the task of designing stable channels. Lane (1955) suggested that slope (S) is related to the dominant controlling variables of sediment load (Q_s), grain size (d_{50}) and discharge (Q_w) as follows:

$$S = \frac{Q_s d_{50}}{Q_w} \tag{5.2}$$

Channel slope appears to be closely related to channel pattern, although Carson (1984) attributes this to the relation between channel slope and bed-material load. Leopold and Wolman (1957) assembled data from a number of rivers in the United States and India and found that the relationship between channel slope (S) and bankfull discharge (Q_w in m^3/sec),

$$S = 0.288 Q_w^{0.44} \tag{5.3}$$

could be used to discriminate between meandering and braided rivers. Henderson (1961) refined their analysis by including bed-material size (d_{50} in mm) to obtain the discriminant function

$$S = 3.07 d_{50}^{1.14} Q_w^{0.44} \tag{5.4}$$

which is remarkably similar to an equation (Henderson, 1966, p. 454) for the limiting value of slope at which a movable-bed channel is just stable:

$$S = 2.27 d_{75}^{1.15} Q_w^{0.46} \tag{5.5}$$

Parker (1976) extended this work still further, developing a theory to predict the number of braids as a function of the variables S/Fr and D/W, where Fr is Froude number.

The characteristics of river planform, particularly of meandering rivers, have also received much attention from both geomorphologists and engineers, the latter because much river training work has for many years been based on the belief that a river can be guided into a stable, smoothly sinuous course if the variables channel width, meander wavelength, and water surface slope (which constrains sinuosity when valley slope is fixed) are chosen correctly.

Data for a vast number of waterways, ranging from 3-mm-wide streams of water flowing down a glass sheet (Davies and Tinker, 1984; Gorycki, 1973), through natural rivers flowing in both alluvium and bedrock, to the Gulf Stream (Leopold et al., 1964), have been assembled by different authors and indicate a strong relationship between meander dimensions and some index of discharge. For meander wavelength (L), numerous equations of the form

$$L = f(Q_w^x) \tag{5.6}$$

where the exponent x is close to 0.5, have been developed for different data sets (Shahjahan, 1970). Leopold and Wolman (1957) suggested that wavelength is more closely related to channel width (W) than to discharge, and again many equations of the form

$$L = f(W^y) \tag{5.7}$$

in which the exponent y is close to unity, have been fitted to different data sets.

Since channel width at a given discharge is related to the character of the bank sediments, it might be expected that meander wavelength also should be controlled by the character of the channel perimeter, and Schumm (1968) found that multivariate relationships with discharge and the channel silt–clay percentage (M) accounted for almost 90% of the variation in meander wavelength. These relationships indicate that the meander wavelength of a channel carrying fine sediment will be shorter than that of a channel carrying coarser material, discharge being the same. Similarly, sinuosity (P) was found to be closely related ($r^2 = 0.83$) to channel silt–clay percent (Schumm, 1963a) for stable alluvial rivers of the Great Plains, U.S.

$$P = 0.94m^{0.25} \tag{5.8}$$

This relationship expresses the observed tendency for channel slope to decline as the sediment load becomes finer.

The preceding brief summary indicates that the factors that seem to be most important for channel morphology are discharge, sediment load and character, and valley-floor slope. All of these can be altered during experimental studies of channel morphology.

EXPERIMENTAL STUDIES

Physical models of rivers have been used for at least a century, since Thomson (1879) examined the path of sediment moving through a bend. The first permanent hydraulics laboratory in Dresden commenced operations in 1898, and a strong tradition of experimentation grew up in several European countries. Leliavsky (1955) mentions, for example, experimental studies undertaken in engineering laboratories in Germany, Hungary, Russia, and France. Unfortunately, English-speaking scientists and engineers have frequently been unaware of such work. This has led to duplication, which perhaps is not altogether undesirable. Thus, for example, Leliavsky (1955, p. 147) remarked, with reference to Friedkin's (1945) studies of meandering at the U.S. Waterways Experiment Station, that "the only objection one may possibly raise against these experiments resides in the lack of originality in the choice of the problems, for with a very few exceptions all solutions could have been obtained at much lesser cost from a careful study of the literature."

Nevertheless, Friedkin's work is now widely regarded as a classic, and it has strongly influenced more recent studies. Using flumes varying in length from 50 to 150 ft and a wide variety of materials (loess, silt, fine and coarse sand, granulated coal, and haydite, which is an industrial material with a specific gravity of 1.85), Friedkin examined the processes of meander formation and the relationship of meander form to water discharge, sediment load, the composition of the banks, and slope.

Studies by Edgar (1973, 1984), Khan (1971), Schumm and Khan (1972), and Zimpfer (1975) at Colorado State University were directed toward the influence of valley slope, sediment type, and discharge on channel morphology. During these experiments other observations of interest were made, and Mosley (1975c) concentrated on the morphology and dynamics of tributary junctions.

Equipment

The experimental facilities used by different scientists have varied widely, and to some extent they have influenced the outcome of the investigations. Commonly, large beds of sand (often called stream tables) have been used, for example, by Ackers (1964), Ackers and Charlton (1970a), Hong and Davies (1979), and Tiffany and Nelson (1939). The size of these facilities has varied from 5 m × 2 m (Hong and Davies) to 91 m × 10 m (Ackers and Charlton, 1970b). Changing the valley slope in such a facility may be a substantial operation, so that other workers have used tilting flumes (Quraishy, 1973; Shahjahan, 1970; Stebbings, 1963). Several such studies have experienced difficulties because the flume was too narrow and the developing channels contacted the flume walls before equilibrium had been attained (Hickin, 1972; Wolman and Brush, 1961). The CSU studies were conducted in two nontilting flumes, the first 31 m × 7.5 m, the second 2.8 m × 1.3 m.

Flume 1 is constructed of concrete blocks, with double walls 1 m high (Figs. 5.4, 5.8). The inner wall is permeable to permit control of the groundwater level in the material filling the flume. The flume was filled with either a coarse, non-

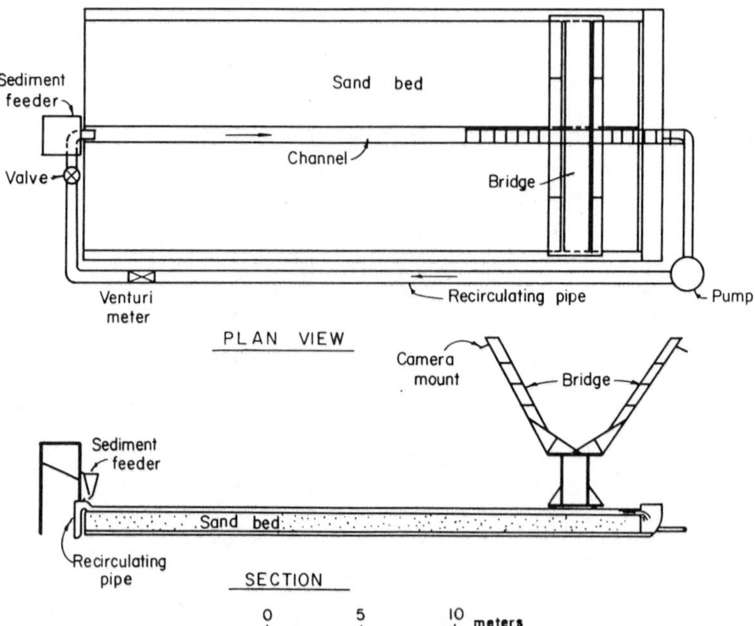

FIGURE 5.4. Flume used for alluvial channel experiments. Working area in flume is about 7 × 30 m.

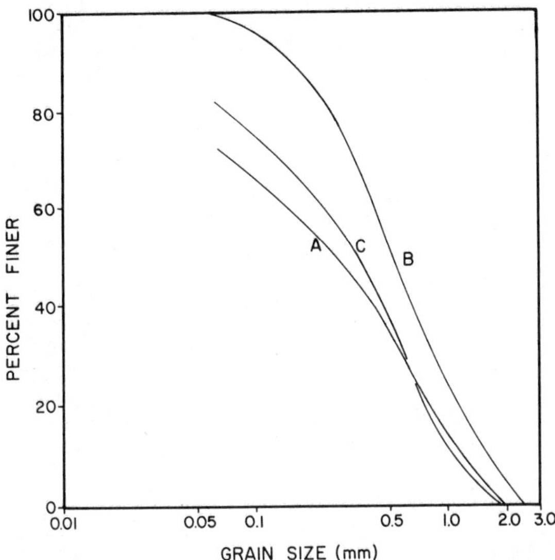

FIGURE 5.5. Grain-size distributions of materials used in channel studies. A = Sediment used in flume 1 (Fig. 5.4) by Mosley (1975c). B = Sediment used in flume 1 by Schumm and Khan (1972), Edgar (1973), and Zimpfer (1982). C = Sediment used in flume 2.

cohesive, poorly sorted sand (builder's plaster sand) with a median diameter (D_{50}) of 0.56 mm (sediment B), or an equal mixture of the first material and natural soil, which was passed through a 12-mm screen (sediment A) (Fig. 5.5). Sediment A had a D_{50} of 0.25 mm and, because of the presence of silt and clay, was cohesive.

The sediments in flume 1 were reworked after every run to ensure homogeneity; where sorting occurred (particularly in channels forming in sediment A), the sediment in the affected areas was removed and fresh sediment added from a stockpile. After compaction, the desired surface slope was established by placing stakes every 2–3 m at appropriate elevations along both sides of the flume. Two 8-m girders were buried in the sediment adjacent to the surveyed stakes, one on each side of the flume, and the surface was prepared by dragging a 6-m beam along the buried girders. Finally, straight channels 0.3 m wide and 0.08 m deep were cut down the centerline using a rectangular template.

Water was introduced into the channel through a headbox equipped with baffles to dampen turbulence and to ensure that flow was distributed evenly across the initial channel. A recirculating water supply powered by a 1 cm impeller-type electric pump that was capable of delivering up to 9 L/sec was used. Water was supplied to the flume from an underfloor sump maintained at room temperature, and water temperature throughout the experiments was 18–19°C. One series of runs used a second supply line to study flow processes at channel confluences; the second line could deliver up to 10 L/sec.

Sediment B was introduced at the head of the flume by vibrating sediment feeders, which were regularly calibrated and adjusted; but the feeders could not feed the cohesive sediment A, the fine fraction of which was in any case washed through the flume and deposited in the tailbox. Sediment was collected in the tailbox and was not recirculated with the water.

Measurements of channel characteristics were made with a point gauge mounted on the movable bridge (Fig. 5.4). Because the carriage rails were not perfectly level, a correction matrix was obtained by filling the flume with water and making point gauge readings to the water surface. All measurements were subsequently corrected.

Many experiments were carried out in flume 1 (Fig. 5.4), with various initial conditions, water discharges, and sediment loads. The operating procedure was generally the same in all runs:

1. The initial surface was prepared and the channel cut.
2. The water table in the flume was raised by adding water to the tailbox and side channels.
3. Discharge in the initial channel was slowly increased to the desired rate and the sand feeder was started; discharge was measured with a venturi meter and manometer board.
4. Water was allowed to flow through the initial channel until it had developed a pattern having uniform dimensions through the length of the flume.
5. Water surface and bed elevations were measured along the channel, vertical and oblique photographs were taken, and planimetric and cross-sectional data were collected with which to construct maps and cross sections of the channel. In some runs, this information was also collected at intermediate stages during the development of the channel.

Flume 2 consisted of a metal tray 2.8 m long, 1.3 m wide, and 0.12 m deep, within which a timber frame 2.45 m long by 1 m wide outlined the working surface. The flume was used by Mosley (1975c) for his tributary junction experiments. The working surface, which had a longitudinal slope of 0.016, was composed of the silt–sand–clay mixture A, which was additionally passed through a 2-mm mesh screen to produce sediment C (Fig. 5.5). Water was supplied to flume 2 from a constant head tank with a maximum discharge of 0.3 L/sec. The constant head tank was supplied from the public water supply. Two independently controlled supply lines fed water to small, movable headboxes; sediment was fed by hand in measured amounts, the sediment being the same as used for the working surface. Operational procedures were otherwise as in flume 1.

Initial Conditions

Initial conditions in flume studies have been of three main types:

1. An accurate representation of a prototype river, with the banks commonly molded in concrete and with an alluvial bed.
2. A generalized meander pattern molded in concrete or constructed of some

other nonerodible material, with an alluvial bed; the meander geometry has commonly been selected to conform to relationships observed in natural rivers, consistent with the area available (e.g., Hooke, 1975);

3. A smooth initial surface of erodible material into which is cut a straight channel of arbitrarily chosen shape and size. However, some investigators have also commenced work with a meandering channel cut into erodible material, which is free to erode its banks.

Stebbings (1963) carried out experiments in a 8.2 × 0.9-m flume in which water was applied to a planar, sloping sand surface with no initial channel. He found that a channel with defined banks would form at the upstream end of the flume if flow was introduced for long enough. The experiments in the REF (Chapter 2) are similar, except that creation of an initial surface composed of intersecting planes, plus baselevel lowering, encouraged development of a well-defined channel.

With the first two types of initial conditions, only the thalweg depth and disposition of bed forms are free to adjust and planimetric variables are fixed. This type of approach is most appropriate to studies related to maintenance of navigation channels, for example, and will not be considered further.

A major point of difference of studies with the third initial condition is whether flow should be introduced into the channel directly in a downflume direction or whether an initial bend is necessary or desirable to induce meandering. If the processes of meandering and braiding are instability phenomena (Engelund and Skovgaard, 1973; Parker, 1976), an initially straight flume channel should presumably develop a meandering form appropriate to the prevailing conditions of flow and sediment load. In practice, experience varies widely. Ackers and Charlton (1970a) make no mention of an initial bend, but several of their initially straight experimental channels developed meanders, meandering thalwegs, and braiding patterns (their Figs. 2, 11, 12, 13). Quraishy (1973) and Hickin (1969) similarly documented the growth of "skew shoals" and meander patterns from initial straight-entrance channels, while Wolman and Brush (1961) found that some channels developed "diagonal shoals" (equivalent to Quraishy's skew shoals) and pseudo-meanders. On the other hand, Tiffany and Nelson (1939) stated that "a preliminary test established the fact that an initially straight channel, without any disturbing factors to cause a deflection of the directive force of the water, would remain straight." Ackers (1964) concluded that there was a tendency for streams on steeper slopes to meander, which was influenced by gradient and sediment load; this suggests that Tiffany and Nelson's (1939) test may have been at a slope inappropriate for meandering. Certainly, Friedkin (1945), who continued Tiffany and Nelson's work at the U.S. Waterways Experiment Station, obtained a winding channel from an initially straight, straight-entrance channel.

The experiments by Khan (1971), discussed below, confirm that under the appropriate hydraulic conditions, an initial straight-entrance channel will tend to meander. On the other hand, if hydraulic conditions are not appropriate for stable meandering to occur, then bends that are induced by introducing the flow at an angle rapidly die out in the downstream direction. Therefore, the practice during the experiments at Colorado State University was to mold a bend at the head of the initial straight channel and to introduce flow at an angle of 40° to the flume.

Final Equilibrium Conditions

There has been a remarkable variation in the length of time different investigators have considered it necessary to continue their experiments, ranging from only a few minutes (Hickin, 1972) to 35 days (Ackers and Charlton, 1970a). This is not due solely to the effects of scale, because Friedkin (1945) reports run times of from 3 to 160 hr that were required for the development of channel stability. Similarly, there has been a wide range in operational definitions of *equilibrium*. Wolman and Brush (1961) defined *equilibrium* as the point when

1. the rate of bed-load movement became constant,
2. changes in channel shape approached zero, and
3. the longitudinal profile was regular and nearly constant.

Ackers and Charlton (1970a) ran their experiments until they judged that a stable channel pattern had formed; Hickin (1972), however, terminated his runs when the channel bends impinged on the walls of the relatively narrow flume. In the CSU studies, runs were terminated upon the development of a channel pattern "uniform in size and shape throughout the length of the flume" (Edgar, 1973), or when a stable (relatively unchanging) channel pattern had formed (Khan, 1971). In the latter case, it will be noted that stability and equilibrium are not necessarily identical. A laboratory stream flowing at a low slope and discharge may be stable but unable to rework the initial channel to an equilibrium form appropriate to the hydraulic conditions. On the other hand, a braided river may have a form that is in equilibrium with hydraulic conditions, but it is hardly stable.

Reproducibility

A prime reason for resorting to experimental work is that, under closely controlled constraints, there is a definite, unique relationship between the dependent (form) and independent (controlling) variables that is not influenced by other factors over which there is no control. In other words, under the same constraints and with the same values of the independent variables, channel form should be reproducible.

That this is so has been very satisfactorily demonstrated by Tiffany and Nelson (1939) and Zimpfer (1975) (Fig. 5.6; Table 5.2), for sets of runs with the same initial conditions, discharges, and sediment loads. Tiffany and Nelson's duplicate runs were with a constant (but unspecified) discharge; Zimpfer's runs were made with a varying hydrograph which averaged 5.66 L/sec.

CHANNEL DEVELOPMENT

There are three basic channel types, straight, meandering, and braided, but a river may exhibit combinations of these patterns along its length. These same basic

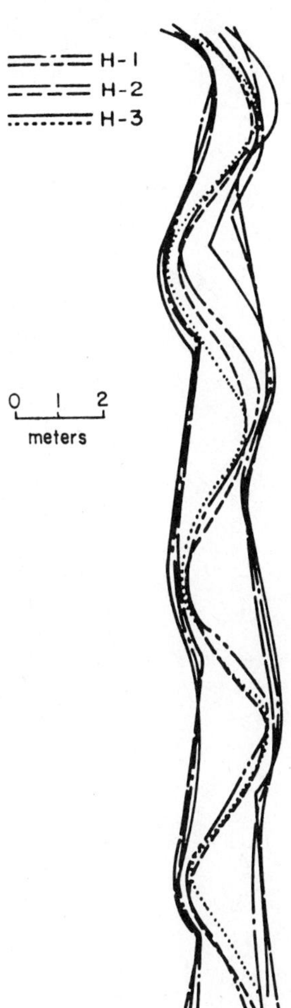

FIGURE 5.6. Repeat runs under similar initial and operating conditions, showing high degree of reproducibility of three experimental channels (from Zimpfer, 1975).

channel patterns were developed in the laboratory, but during most experiments a truly meandering channel was never obtained, only a pattern with a meandering thalweg. The thalweg is the deepest part of the active channel, where the majority of the discharge is conveyed. However, it may or may not correspond to the lowest point of a given cross section, as some backwater areas were actually deeper.

Straight Channels

Three channels with straight patterns were produced in flume 1 by Edgar (1973). All had a discharge of 2.83 L/sec and very flat slopes. The maximum valley slope was 0.003. Below this slope it was impossible to produce any pattern other than straight, although the flow passed through an initial bend upon entering the flume. The straight channels were the result of a low-energy condition. The rate of sediment transport was low, and the flow was unable to scour its bed.

TABLE 5.2 Reproducibility of Channel Form[a]

Run	Mean Discharge (L/sec)	Flume Slope	Sediment Feed Rate (gm/min)	Meander Wavelength (m)	Meander Amplitude (m)	Sinuosity	Cross-section Area (m^2)	Mean Depth (m)	Width–Depth Ratio
H-1	5.66	0.008	150	7.63	1.93	1.17	0.014	0.015	66.0
H-2	5.66	0.008	150	7.53	1.70	1.13	0.017	0.015	71.2
H-3	5.66	0.008	150	7.47	1.95	1.20	0.018	0.015	75.9

[a] Data from Zimpfer (1975, Appendix 1).

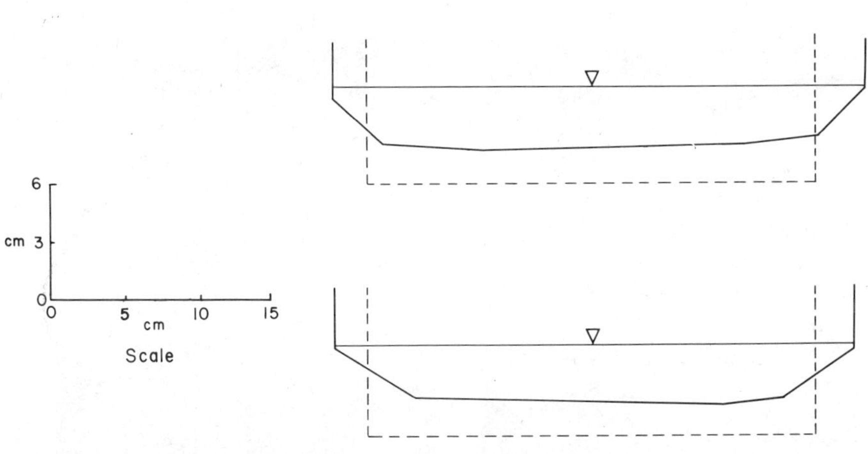

FIGURE 5.7. Cross sections of straight channel. Dashed lines indicate shape of initial channel (from Edgar, 1973).

The formation of straight channels was relatively uncomplicated, as only a slight widening and notable shallowing occurred. The net result was a change from an initially rectangular to a trapezoidal cross section (Fig. 5.7). Although the flow was incapable of exerting a large force upon the channel periphery, bank collapse nevertheless introduced small amounts of sediment into the channel.

The zone of sediment transport was confined to a narrow band along the center of the channel, where sediment moved very slowly as very low dunes or linguoidal sand waves. Tractive force or boundary shear stress (τ) is the primary force moving sediment, and Ghosh and Roy (1970) found that the maximum shear stress on the bed was along the center. Shear decreased away from the center and approached zero at the corners of a rectangular channel; therefore, the area of active sediment movement was in the center of the straight channel and secondary flow was absent.

Meandering-Thalweg Channels

As the slope of the sediment surface was increased, meandering-thalweg channels formed (Fig. 5.8). The minimum valley slopes for which this pattern was obtained by Edgar (1973) for discharges of 2.83, 5.66, and 8.49 L/sec were 0.006, 0.0044, and 0.0045, respectively.

Longitudinal profiles (Fig. 5.9) and cross sections (Fig. 5.10) indicated a series of alternating pools and crossings. The pools were found at and slightly downstream of the apex of each bend, whereas the crossings or shallow areas were located in the tangents connecting these bends. These correspond to pools in the bendways and crossings in natural meandering rivers.

There are many differences in sinuous-channel evolution described by different investigators, depending upon initial experimental conditions. Nevertheless, the development of alternate bars and meanders has been repeatedly described, often in very similar terms, and it is clear that these features form under a wide range

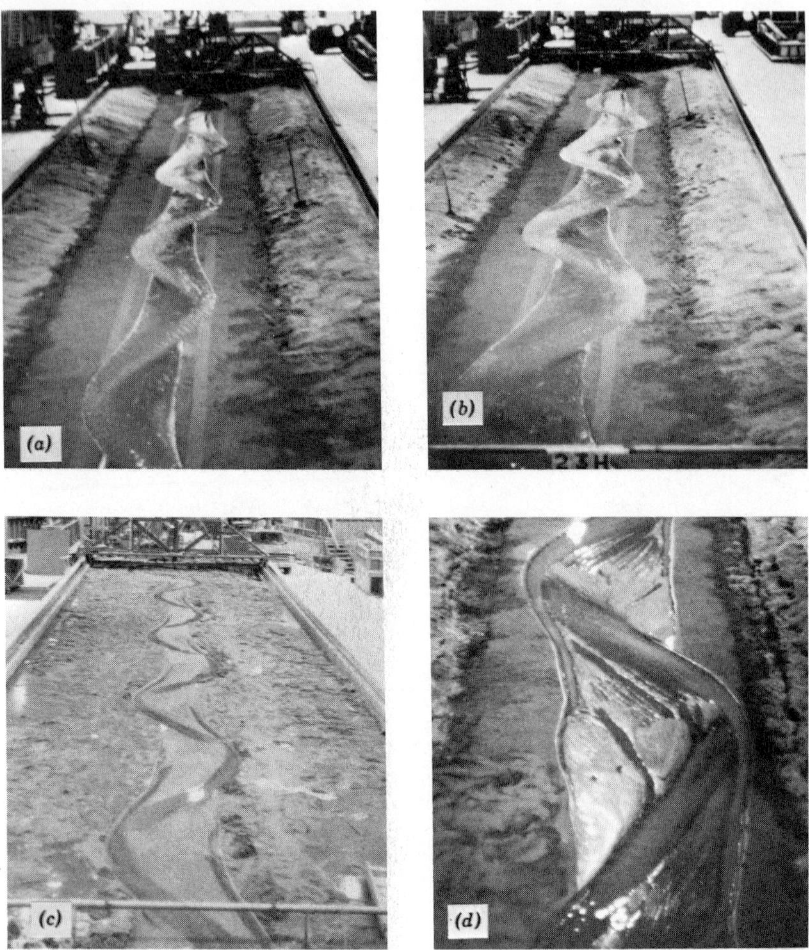

FIGURE 5.8. Sequence of photographs looking upstream and showing the development of a meandering-thalweg channel. Flow is towards the bottom of each photograph. (*A*) After 3 hr; (*B*) after 4 hr; (*C*) thalweg and backwater channels at low water as the flume was being drained; (*D*) dry channel at the conclusion of the test. All views are looking upstream except *D*.

of conditions. For example, Ackers and Charlton (1970b, p.16) described development of meanders in the following way:

> Immediately after the introduction of water and sediment at the upstream end of the prepared trapezoidal channel, the bed developed ripples which were symmetrical about a longitudinal line. After a period of about 6 hours for the higher discharges and about 24 hours for the lower values, the symmetry deteriorated and shoals began to form near the sides of the channel along its length. These were at fairly regular intervals, the shoals forming on alternate sides of the channel with deeps near the opposite bank. The shoals

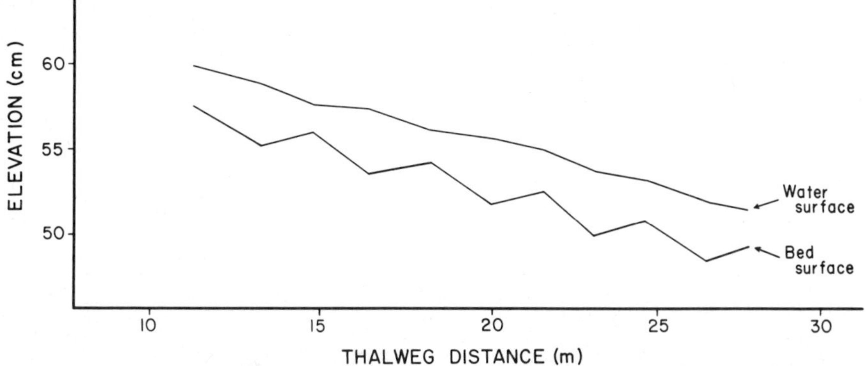

FIGURE 5.9. Longitudinal profile of meandering-thalweg channel (from Edgar, 1973).

> continued to increase in length and height while they migrated fairly rapidly downstream, increasing their pitch as the channel became wider through general bank erosion even though it remained straight. Then, almost simultaneously along the channel, embayments were eroded in the banks opposite each shoal. These expanded, producing a sinuous channel pattern in plan, and also migrated downstream, although more slowly than the shoals had done. The pitch of these features then remained constant, or tended to decrease slightly.

(Pitch is meander amplitude.) A similar sequence of events that is described by Hickin (1969, Fig. 1), Quraishy (1973, Fig. 6), Stebbings (1963, Fig. 6), and Wolman and Brush (1961, Fig. 124) bears a remarkable resemblance to the mode of channel pattern development by the River Ystwyth (Fig. 5.11) after artificial channel straightening (Lewin, 1976).

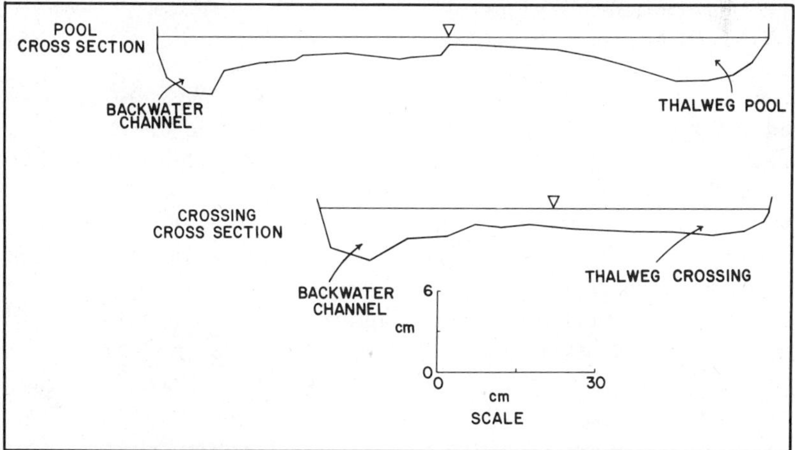

FIGURE 5.10. Cross sections of meandering-thalweg channel showing pool and crossing sequence (from Edgar, 1973).

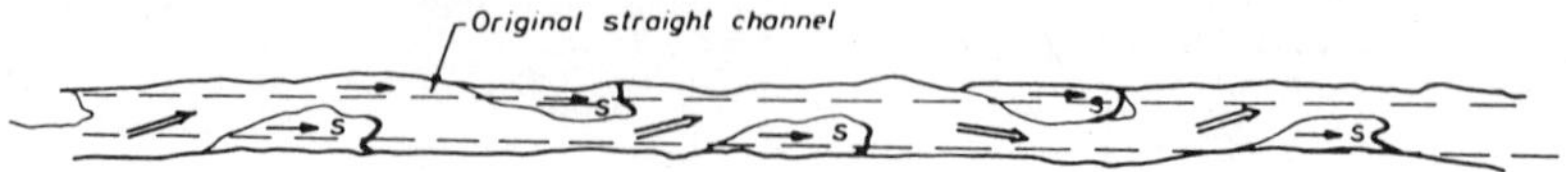

(a)- ENLARGING AND SHOAL FORMATION

(b)- BANK EROSION

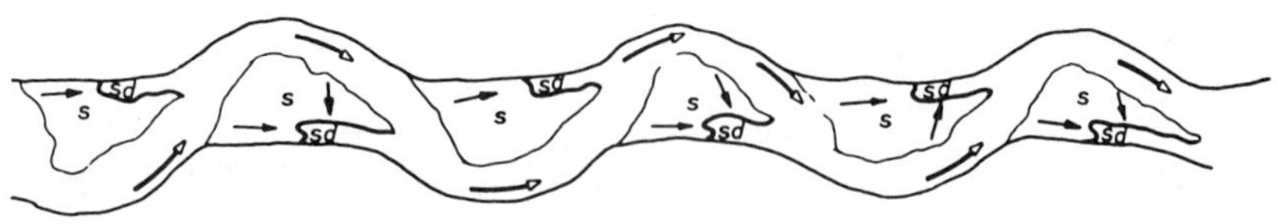

(c)- MEANDERING

SYMBOLS

S - Shoal

Sd - Stagnant water, deep

⟹ *Main water flow*

→ *Subsidiary flow due to difference in head between main flow and the stagnant deeps.*

FIGURE 5.11. Development of alternate bars in flume channel of Ackers and Charlton (1970b).

In tests with a bend at the top of the initial channel, the sequence of events tended to be different, although the end result was the same if hydraulic conditions were appropriate (Edgar, 1973; Zimpfer, 1975). With the introduction of flow into the initial channel, the banks began to cave and the channel widened. The initial bend directed the flow against the bank on the opposite side, the bank was eroded, and sediment was added to the channel. The upstream part of the thalweg began to develop a sinuous path, which extended progressively downstream and directed the flow against the outer banks, causing further erosion on alternate sides of the channel. The eroded sediment was deposited as alternate bars; sediment eroded from the outer (concave) bank at a developing bend was deposited on the inner (convex) bar next downstream (Fig. 5.12). The bars divide the actively outward-migrating thalweg from an area of still water, which was the dead slough of Wolman and Brush (1961) or the backwater channel of Edgar (1973). The sinuous thalweg grew downstream, followed by bank erosion, bar formation, and devel-

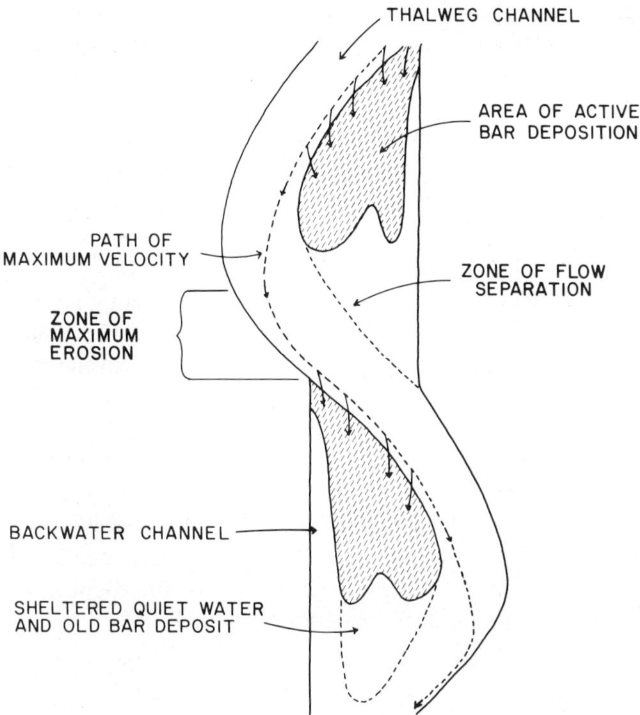

FIGURE 5.12. Flow conditions and bar locations in meandering-thalweg channel (from Edgar, 1973).

opment of a meandering thalweg (Fig. 5.8). Because lateral erosion began at the head of the flume, the upstream bends began to form earlier than the downstream bends. Bends eroded laterally at a uniform rate, so the amplitude of the bends initially was less downstream. Eventually, however, the upstream bends ceased to increase in amplitude, allowing the downstream bends, which continued to develop, to equal and in some cases surpass the amplitude of the upper bends.

The rate of bend development was not uniform through time. As sediment was transported into the area from upstream, it was concentrated in a low, elongated alternate bar on one side of the channel. As the bar increased in size, a very rapid increase in lateral growth, channel widening, and bend development followed. After a certain width and bend size was attained, the rate of growth decreased markedly. This was probably associated with the flow hydraulics, which was dictated primarily by the initial slope and discharge. The bend, however, continued to grow laterally at a reduced rate due to the ease with which the banks could be eroded.

Commonly, at this point investigators have terminated the run and collected data. Zimpfer (1975), however, pointed out that meandering-thalweg channels continued to develop past the stage at which a sequence of uniform bends formed.

Chute cutoffs began to form and progressively diverted more and more flow from the main thalweg, so that flow was eventually carried by a group of minor channels. In other words, the channel braided.

Zimpfer (1975) carried out eight runs with an initial straight channel and various initial valley slopes and discharge hydrographs, and in every case the channel pattern changed as follows: straight–meandering thalweg–meandering thalweg with cutoffs–braided (Table 5.3A). These experiments are discussed more fully below. However, it will be noted here that the rate of development increased with initial channel slope and discharge (Table 5.3B). Furthermore, Tiffany (1935) found that rate of development was reduced as sediment load increased, discharge and valley slope being held constant (Table 5.3C). Under constant flows, Tiffany and Nelson (1939, p. 644) and Friedkin (1945) noted a sequence of development similar to that observed by Zimpfer.

A meandering pattern is not stationary; it migrates laterally and down valley with time (Davis, 1902; Konditerova and Ivanov, 1969; Kondrat'yev, 1968). The dynamic nature of meanders is visualized as consisting of three basic types of motion (Daniel, 1971): translation (downvalley sweep), expansion (lateral growth), and rotation (rotation of the bend axis). These three basic motions can occur in any combination at a particular bend, depending upon the conditions of discharge and grain size of the bed and bank material (Daniel, 1971).

TABLE 5-3 Rate of Development of Channel Patterns[a]

				Elapsed Run Time (hr) to	
Run	Initial Slope	Mean Discharge (L/sec)	Sand Feed Rate (gm/min)	Meandering Thalweg	Braided
		A. Zimpfer's (1975) Runs			
1	0.006	5.1	150	2.0	6.0
2	0.006	5.4	150	3.0	6.5
3	0.006	5.1	150	2.5	8.5
4	0.012	5.4	150	0.5	3.5
5	0.018	5.1	150	< 0.5	1.5
6	0.008	5.7	150	1.0	—
		B. Edgar's (1973) Runs			
23A	0.006	2.8	60	14.75	runs
23C	0.006	5.6	180	4.0	terminated
25A	0.006	8.5	260	2.5	before
26E	0.012	8.5	210	1.75	braiding
27B	0.018	8.5	560	1.0	
		C. Tiffany's (1935) Runs			
1	0.005	5.1	0	15.0	runs
2	0.005	5.1	70	13.5	terminated
3	0.005	5.1	140	12.25	before
4	0.005	5.1	270	12.25	braiding

[a] From Zimpfer (1975, Tables 1, 2).

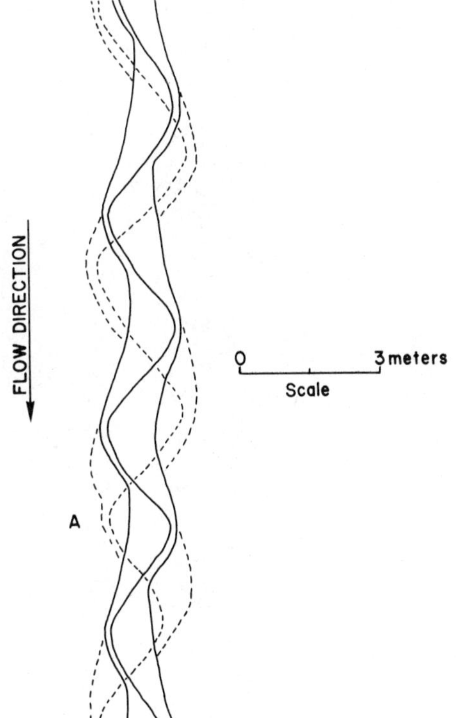

FIGURE 5.13. Meandering-thalweg channel migration with time. Solid lines represent channel after 3.5 hr of run time, dashed lines represent channel after 5.5 hr (from Edgar, 1973). The letter *A* indicates location of meander shift.

In the laboratory, bend migration is simplified because grain size is unchanged throughout the length of the channel and discharge is constant. Figure 5.13 demonstrates how the meandering-thalweg pattern changes with time. Two types of meander change can be detected, namely, a downvalley sweep or movement parallel to the valley or flume axis and an increase in amplitude or meander belt width.

The migration of meanders is generally quite uniform and meanders are not deformed, but in a natural situation sediments of differing erodibility would be encountered and pattern distortion would result. Furthermore, in the field, any portion of the bars or bed exposed for any significant length of time would be stabilized somewhat by vegetation.

Meander Shift

Normally there are two types of movement of a meander pattern on a floodplain. The first is meander sweep, which is a progressive downstream shift of the pattern; the second is meander swing, which is the lateral movement of the channel back and forth across the valley. Both of these types of river behavior are well known and have been described by Davis (1902) and others.

In Edgar's (1973) and Zimpfer's (1975) studies, the meandering-thalweg pattern

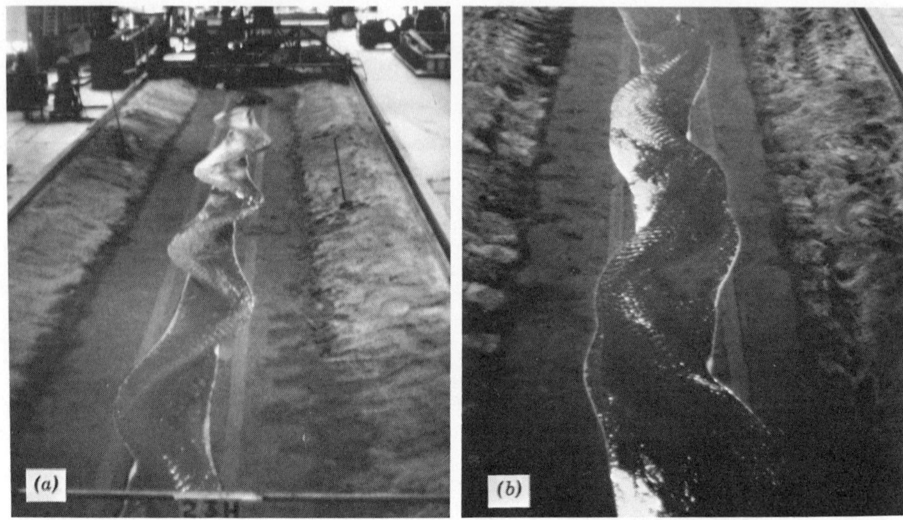

FIGURE 5.14. Photographs showing channel pattern (*A*) before meander shift and (*B*) after meander shift.

tended to develop and migrate in a uniform manner. However, on occasion, certain bends deviated from this uniform pattern (Fig. 5.13, point A). The scalloped form of the concave bank in plan (point A) was caused by an abrupt downstream shift in the thalweg through the bend (Fig. 5.14). This in turn caused a downstream shift of the thalweg bend apex in the next bend downstream and formed bars and cusps at the edge of the floodplain (Fig. 5.15). Usually meander shift occurred when the experiment was continued after the meander pattern had fully developed. During meander sweep in the flume, the downstream limb of the bend eroded into sediment that had not been eroded previously, and this sediment was slightly more resistant to erosion than that forming the upstream part of the point bar. Hence sweep of the downstream limb of the bend was slower than that of the upstream limb, which began to overtake it. This resulted in the development of an asymmetrical bend (Fig. 5.15), which was unstable.

When meanders are incised into bedrock or when the downstream limb of a bend is held by resistant materials or bedrock, the upstream limb, which is in alluvium, will sweep down on this fixed part of the bend, compressing the meander. This frequently is the explanation for deformed incised meanders (see Chapter 7).

In most cases, meander shift takes place in one bend. But after it has occurred there, the effect of this change in thalweg position is experienced both upstream and downstream, thereby inducing the meander shift at these locations also (Fig. 5.14). Similar results were obtained during Friedkin's (1945) experiments. Friedkin did not go into detail on this process, but it is obvious from his Plate 22 and the discussion on page 15 of his report that his results were identical, although he apparently considered the changes to be normal chute formation.

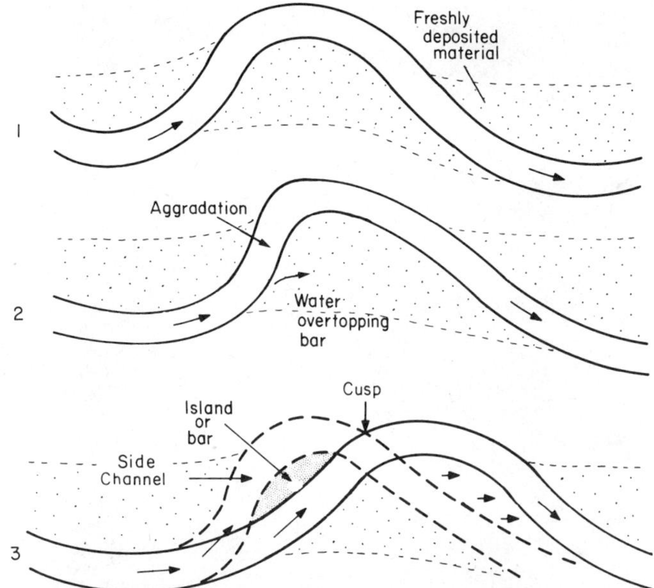

FIGURE 5.15. Diagram showing process of meander shift.

Transitional Meandering–Braided Channels

As the range of slopes was increased over that which produced meandering-thalweg channels, a pattern that was both meandering thalweg and braided in nature was produced (Fig. 5.16). The initial valley slopes required to produce this pattern for discharges of 2.83, 5.66, and 8.49 L/sec were 0.021, 0.0164, and 0.0180, respectively. This represents a further increase in flow energy over the conditions that produced the meandering-thalweg and straight patterns. The channel was much wider and had a larger meander wavelength and amplitude than typical meandering-thalweg patterns, which reflects the larger values of sediment transport, discharge, and slope.

The same initial bend and channel dimensions were used as those described previously, but due to the steeper slope, when the flow impinged on the bank of the straight initial channel, a bend formed quickly due to the rapid bank erosion and accumulation of the corresponding alternate bar. This effect progressed rapidly downstream, and only slightly more than 1 hr was required to produce the channel shown in Fig. 5.16.

Due to the momentum of the high-velocity flows, there was more flow over the bars than in the meandering-thalweg pattern. Before a uniform meandering pattern could be produced, the alternate bars were dissected. Numerous small channels were observed to flow in parallel and essentially straight courses from one crossing to the next. However, the primary area of water conveyance was still the meandering-thalweg portion of the channel.

FIGURE 5.16. Photograph of transitional meandering–braided channel. Flow direction is towards observer. Note meandering thalweg.

The longitudinal profile of a channel of this type exhibited alternating high and low points corresponding to the pool and crossing sequence of the meandering-thalweg pattern (Fig. 5.9). Flow depths were smaller and the pools and crossings of the combination channel were farther apart, due to the larger channel dimensions, and more subdued. This was due in part to the fact that more of the flow was channeled over the bar areas, whereas in the meandering channel practically all of the flow was confined to the thalweg channel. The combination pattern represented the intermediate phase of the transition from a meandering-thalweg channel into a braided pattern. The pattern was very unstable, and a change in the regimen of this channel will change its form to either meandering or braided, depending upon the direction of change of the controlling variables (Knighton, 1972). Hickin (1969) pointed out that a pattern of this type is the final stage in the development of what he terms a ''pseudeomeander pattern'' (see also Wolman and Brush, 1961) and immediately precedes a braided condition.

Braided Channels

The last channel pattern to be considered is the braided channel (Fig. 5.17). It is the product of the highest energy and steepest valley slope. Stream power and rate of sediment feed were almost five times greater than the values observed at the initiation of meandering.

Due to the presence of the initial bend, a series of bends formed at the start of this experiment. However, the bends were not uniform in size, and they were quickly destroyed. The most notable characteristic of the braided-pattern development was the rapid increase of channel width. The high-velocity flow produced a high shear, which quickly eroded the low-cohesion channel banks. While the channel was developing, it had much the same appearance as the combination pattern at various places along its length. However, as widening continued, the flow subdivided into individual thalwegs. The pattern was never stable; the positions of the individual thalwegs were constantly changing as the flow divided and rejoined and reworked the channel bed. However, these were minor changes, and the channel remained braided.

The primary characteristics of the laboratory braided channel are: (1) relatively straight channel banks, (2) division of the flow into numerous thalwegs separated by bars or islands, (3) a very wide, flat-bottomed, shallow cross section (Fig. 5.18), (4) a steep longitudinal profile, (5) a high concentration of bed load, and (6) a continuous shifting of the positions of the thalwegs. These same characteristics are found in natural braided streams (Fahnestock, 1963; Lane, 1957; Leopold et al., 1964).

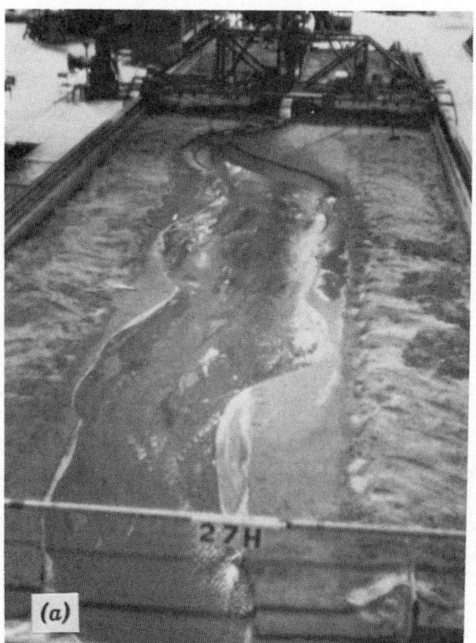

FIGURE 5.17. Photographs of a braided channel. (*A*) Channel with water flowing towards the observer and (*B*) same channel at end of test.

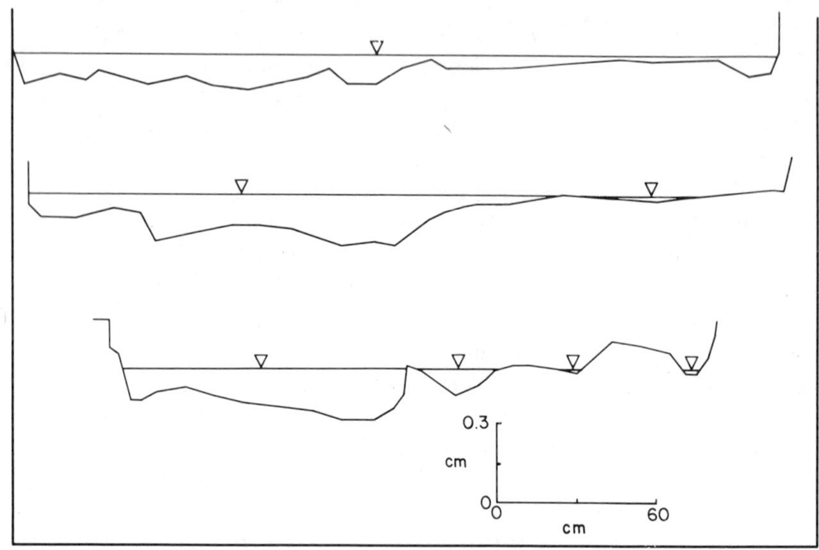

FIGURE 5.18. Cross sections of braided channel (from Edgar, 1973).

Working in a 5 × 2-m stream table, Hong and Davies (1979) studied braided channels formed by flows of 0.2–1.2 L/min and carrying sediment at rates of 1.5–5.0 g/min. Channels were 0.12 to 0.8 m wide, containing 4 to 20 channels with depths of about 1.5 mm. Despite the small size of the model channels, Hong and Davies found that they were geometrically similar to natural braided rivers hundreds of meters wide, although dynamic similarity, which would exist if Froude, Reynolds, Weber, and Euler numbers were the same in the model as in the prototype, was not attained. Froude number equality was observed, but Reynolds number was less than 200, that is, flow was laminar. Nevertheless, the gross discrepancies in channel form and behavior that might have been expected in the experimental channel did not occur.

The channels braided in precisely the same way as observed by Edgar (1973) and Zimpfer (1975), and also by Ashmore (1982) (Fig. 5.19). Alternate bars initially developed, but as amplitude of the thalweg meandering increased, chute cutoffs developed and the flow separated into a number of small channels. Only immediately downstream of the flume entrance did a central bar affect the widening process, which suggests a functional relationship with the entrance conditions. For example, at the upstream end of their channel, Ackers and Charlton (1970b, p. 16) documented the following mode of channel development:

> A shoal formed in the center of the channel near the inlet, and scour occurred on both sides. The shoal increased in length, width and height as the channel widened, until the banks on both sides of the channel became markedly concave. The resulting bulb-like shape at the entry to the channel was a familiar feature in almost every test of the series. It expanded fairly evenly, but flow was disposed to concentrate more to one side or the

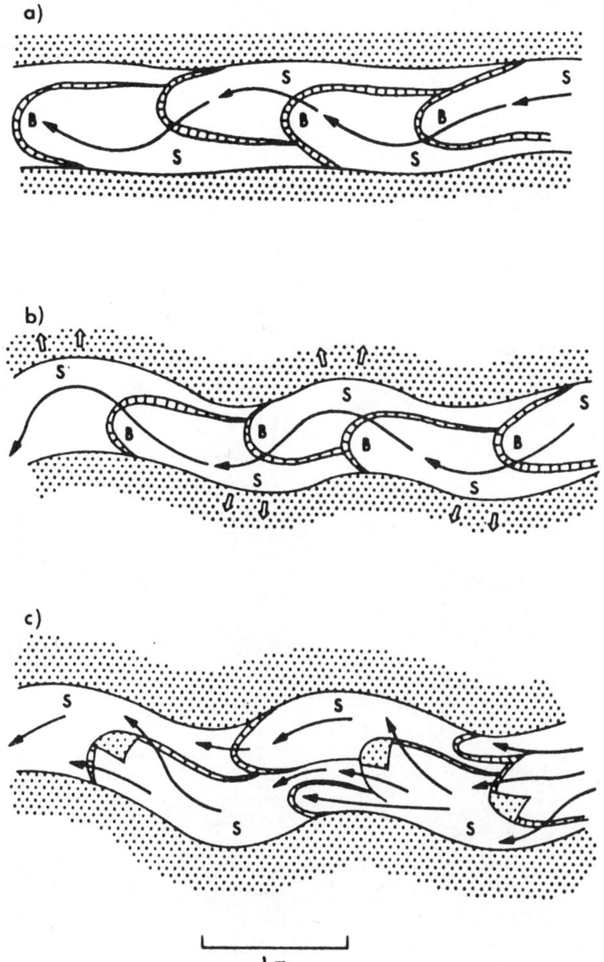

FIGURE 5.19. Development of braiding in Ashmore's (1982) laboratory channel (*a*) after 1 hr; (*b*) after 2 hr; (*c*) after 4 hr, showing bar (B) and scour (S) patterns.

> other where more marked bank erosion would occur. The deposition of sediment to form the central shoal at the inlet and the flanking shoals along the length of the channel usually began almost simultaneously.

This description is remarkably similar to that of Leopold and Wolman (1957), which has commonly been taken as a type example for the development of braiding (Fig. 5.20). That the central bar may be an artifact of the entrance conditions rather than an expected feature of all braided channels is indicated by the occurrence of similar features on the Rakaia and Waimakariri rivers, New Zealand, where they issue from bedrock gorges (Fig. 5.21).

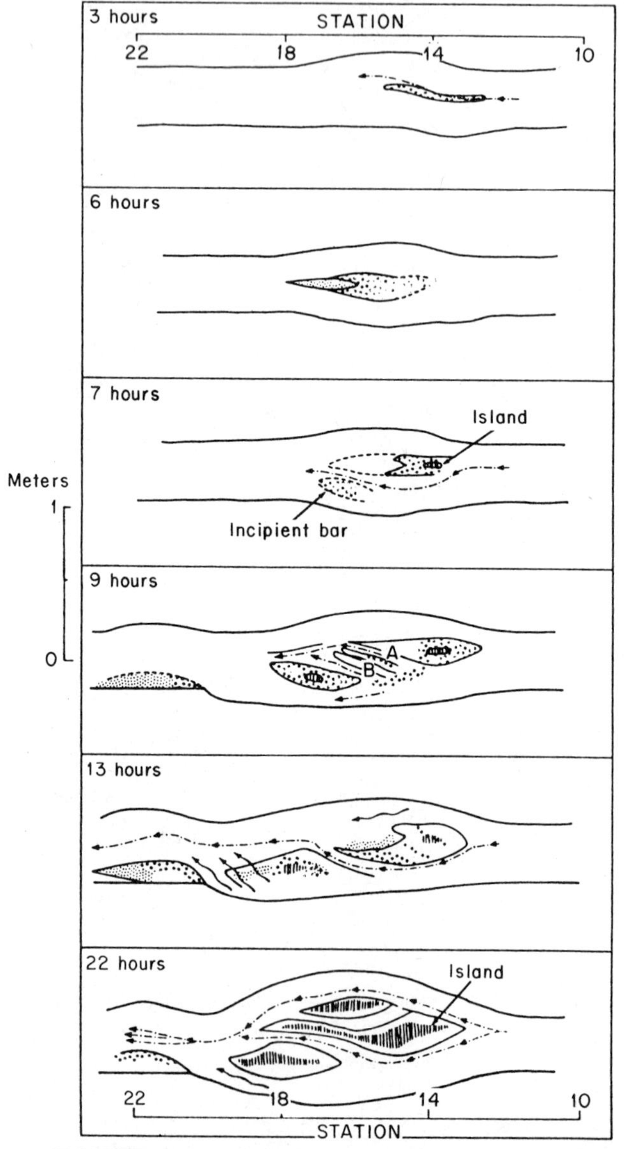

FIGURE 5.20. Development of central bar near the inlet of a laboratory flume (Leopold and Wolman, 1957).

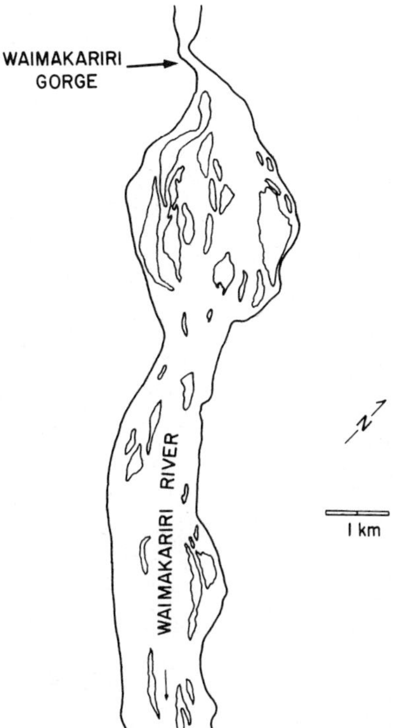

FIGURE 5.21. Bulb-shaped widening of Waimakariri River below a rock gorge (South Island, New Zealand).

Ashmore (1982) observed two further processes of channel division (Fig. 5.22). In the first of these, a bar develops in the center of the channel, the flow splits into two main currents on either side (Fig. 5.22*A*), and the center of the bar face emerges. Scour and outward channel migration occur downstream from the bar, the center of the channel downstream becomes inactive, and two effectively independent channels come into being. The two segments of the original central bar grow and deform, and pronounced flow convergence and divergence in the new channels cause thalweg scour and aggradation, respectively, in the center of the old channel. Each channel could then, in principle, evolve in the same fashion and split again, while the central bar may be further dissected as shown in Fig. 5.20.

Another process of branching occurs (Fig. 5.22*B*) when flow from one channel (a) spills as a diffuse sheet across a diagonal bar on a gravel flat, the water draining into a lower channel (b), creating a well-defined channel (c) that eventually meets the higher, original channel. Spillage is concentrated into channel c, water levels in channel a are lowered, and the bar becomes wholly emergent.

Ashmore's detailed observations allowed him to identify the major characteristics of constructional and erosional features in braided channels (Fig. 5.23); despite the fact that they were formed by a constant discharge, they bear a remarkable resemblance to features in many natural braided rivers (e.g., Bluck, 1976; Church and Jones, 1982).

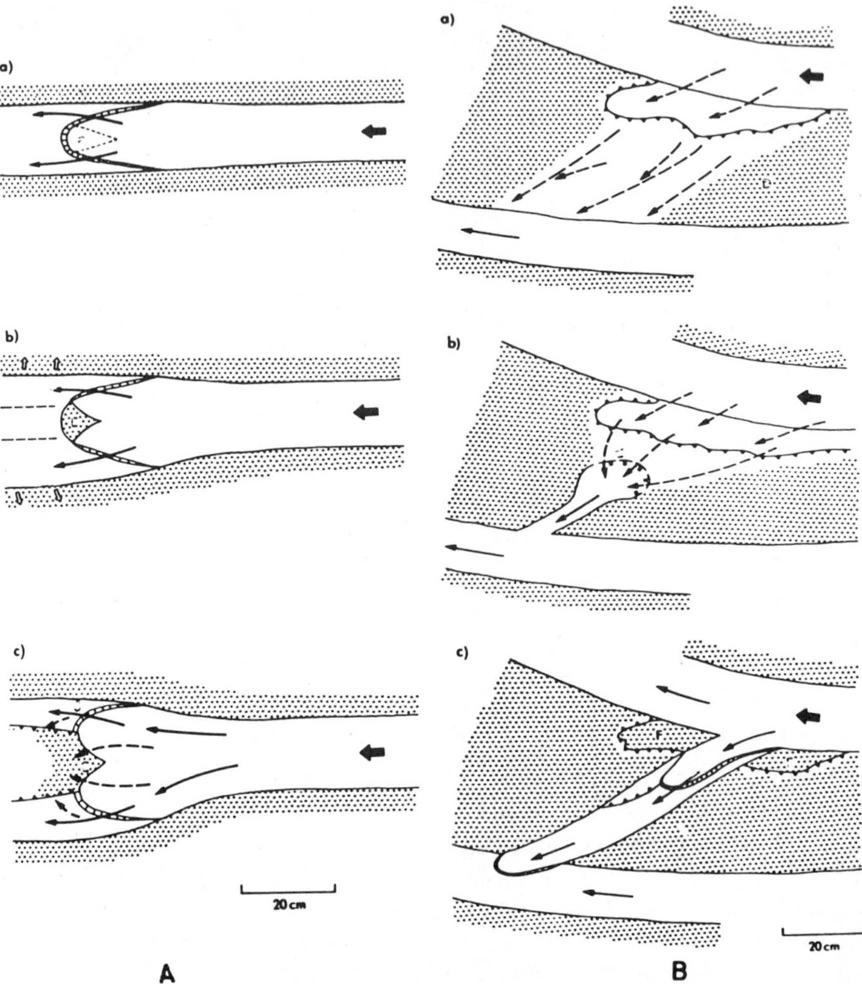

FIGURE 5.22. Processes of channel division in model channels (*A*) by separation around middle bar; (*B*) by incision of a new channel across a diagonal bar (from Ashmore, 1982).

CONTROLS ON CHANNEL MORPHOLOGY

The various experimental studies have, in aggregate, provided a great deal of information on the response of channel morphology to controlling variables. Commonly, the independent variables such as discharge, slope, and sediment load have been manipulated one at a time to develop bivariate relationships between the independent variable and some dependent variable such as meander wavelength. It should be noted that dimensional analysis, in which variables are combined into scale-free compound variables, is a powerful tool that has been used by Ackers and Charlton (1970b) and Shahjahan (1970) to analyze data in a more rigorous way.

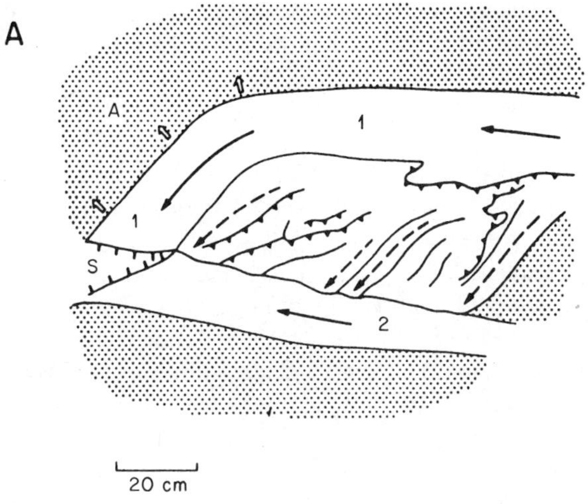

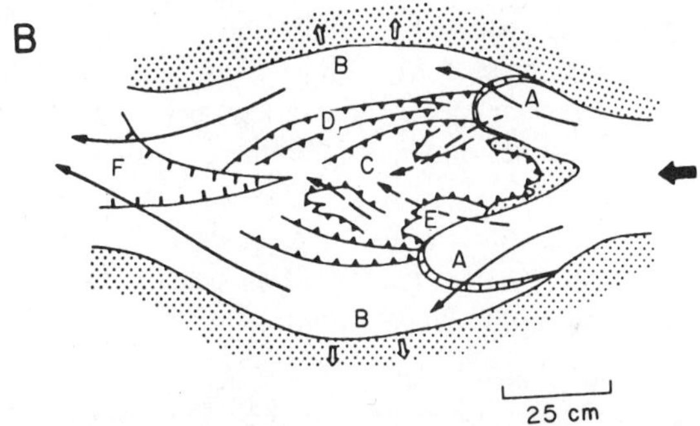

FIGURE 5.23. Typical constructional features in model braided channels (Ashmore, 1982). (*A*). A complex flat produced by erosion at A and resultant construction and abandonment of bars and channels. (*B*) A medial complex flat consisting of active channel bars (A), actively migrating channels (B), abandoned channels (C), overlapping gravel sheets (D), and abandoned bar lobes (E). A large scour hole is found where the two channels rejoin (F).

The effects of both hydrologic and sedimentologic variables on channel morphology are indeed significant, but, in addition, morphologic controls are important. They are related to experimental design and therefore can be considered as the influence of initial conditions on the experimental channels.

Morphologic Controls

Initial Width

Initial channel width in some studies has proved to be a significant control on final channel form because bank erosion has commonly been a major source of sediment load in the model streams, as indeed it is in natural rivers. Most investigators have chosen an initial channel "somewhat smaller than the expected size of the equilibrium channel" (Wolman and Brush, 1961, p. 185); this has commonly led to bank erosion, channel widening, aggradation, and a final channel form markedly different from the initial shape (Fig. 5.24).

Friedkin (1945) commenced runs with different initial channel cross sections having bottom widths of 0.1 and 0.23 m, but he found that the stable meander patterns were identical, with final top widths at a meander apex of 0.49 and 0.52 m. Mosley (1975c) conducted three runs at discharges of 2.8, 5.7, and 8.5 L/sec in flume 1. He used a straight entrance with channels that were 0.3 and 0.6 m wide and had an initial slope of 0.005. Channel evolution differed between runs, and he found that bank erosion, shoaling, and a tendency for the thalweg to wind (rather than to meander) were characteristic of runs with an initially wider channel in which unit stream power was lower (in the range of 1.9–6.6 $N/m^2 \cdot sec$, in comparison with 5.1–8.7 $N/m^2 \cdot sec$ for the runs with the narrower, 0.3-m-wide initial channel). This contrasts with observations of Ackers (1964), who found that model streams with steeper gradients (that is, higher unit stream power) tended to wind, as the formation of minor shoals on alternate sides in the upper channel led to bank erosion on the opposite banks, more prominent shoaling, and additional bank erosion. Channels that were initially too wide in Ackers's series of runs remained straight, with rippled beds. It is likely that the different experimental materials (grain size, percentage of silt and clay) behaved differently, and they had a greater effect than channel dimensions on these channels.

Initial Entrance Bend

Friedkin (1945, p. 10) remarks that "in each meandering river it is the direction of flow into each bend which largely determines its size and shape." A series of

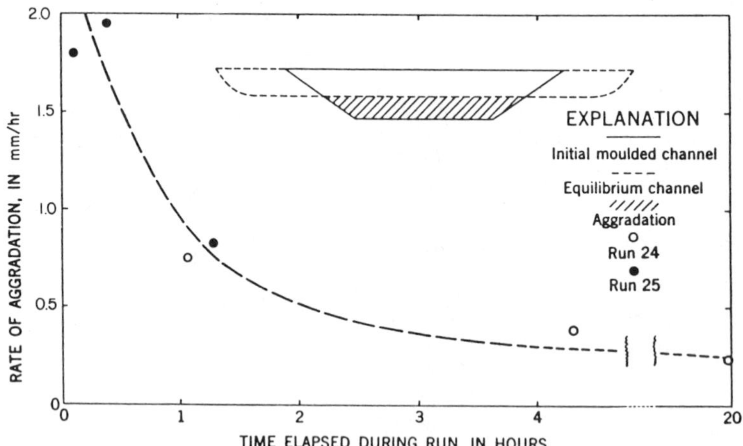

FIGURE 5.24. Widening and aggradation of an initially too narrow flume channel (from Wolman and Brush, 1961).

tests, all with an initial slope of 0.0075 and a discharge of 2.83 L/sec but with entrance bends of 30°, 45°, and 60°, showed that meander wavelength decreased and amplitude increased with angle of attack (Fig. 5.25). The bends were uniform down the full length of the flume, although Friedkin cautions that this is not the case for natural rivers because local changes in bank material change the angle of attack from one bend to another. Friedkin found that when the angle of attack approached 90°, the "lateral oscillatory motion" (meandering tendency) appeared to be destroyed by excessive turbulence, so that the flow did not deflect from the

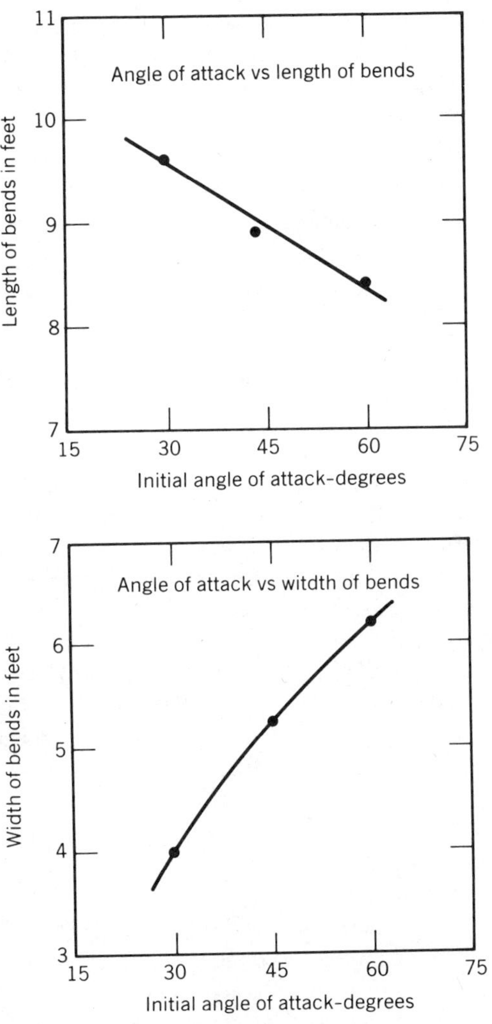

FIGURE 5.25. Effect of different initial angles of attack on flume channel morphology (from Friedkin, 1945). Slope was 0.0075, bed material was Mississippi River sand, and duration of run was 6 hr.

bank and develop a bend farther downstream on the opposite bank. On the other hand, with an entrance angle of 0°, an irregular, winding, and constantly shifting channel developed, with amplitude increasing downstream.

A series of runs by Mosley (1975) in flume 1 with discharges of 2.83, 5.66, and 8.5 L/sec and with an initial entrance angle of 40° may be compared with those described above. Initial slope was 0.005 and initial channel width was 0.3 m; the floodplain material was cohesive. With a discharge of 2.83 L/sec (Fig. 5.26), each bend developed singly, in response to the effect of bend development and changing flow exit angle from the bend upstream. Initiation of each bend occurred only after the bend upstream had reached a small radius of curvature of about 0.5 m and was directing flow at a high angle against the opposite bank. By plotting bend amplitude as a function of time for each bend, it is apparent (Fig. 5.27) that

1. the time intervals between initiation of successive bends were neither constant nor constantly increasing or decreasing,
2. the maximum amplitude generally decreased with each successive bend, and
3. the amplitude of each bend tended to decline after its maximum value was attained. Amplitude reduction tended to be achieved by chute cutoffs across the point bar.

At higher discharges of 5.66 L/sec, there was rapid evolution of the channel to an effectively straight final form. With a discharge of 8.5 L/sec, intense bank caving along the whole channel caused very rapid channel evolution (Fig. 5.28). Again, bends developed successively downstream, but chute cutoffs and downstream shift of bends caused the number and amplitude of bends progressively to diminish. The end result was a channel composed of a series of straight segments separated by bends at points where the flow contacted the undisturbed, more resistant bank material.

The effect of the entrance bend was in most cases to induce a meandering channel. However, the bends were not always stable. At low-energy conditions (low discharge, low valley slope) the bends quickly died out downstream. On the other hand, under high-energy conditions bend amplitude increased rapidly in the downstream direction, the bends were destroyed by lateral erosion, and the channel braided.

Effects of Tributary Entrance Angle

Mosley (1975, 1976) studied the effect of the entrance angle of a tributary or anabranch channel on the channel below the junction (Fig. 5.29). During 95 experiments, he not only varied tributary confluence angle (θ) but the ratio of tributary discharges (Q_1/Q_2), tributary width (w_1, w_2), and tributary sediment loads (Q_{s1}, Q_{s2}). Four runs in straight, unbranched channels with discharges of 0.25 L/sec were also made to obtain a value of mean channel depth (0.78 cm) to which the depth of scour (D_s) at tributary junctions was compared.

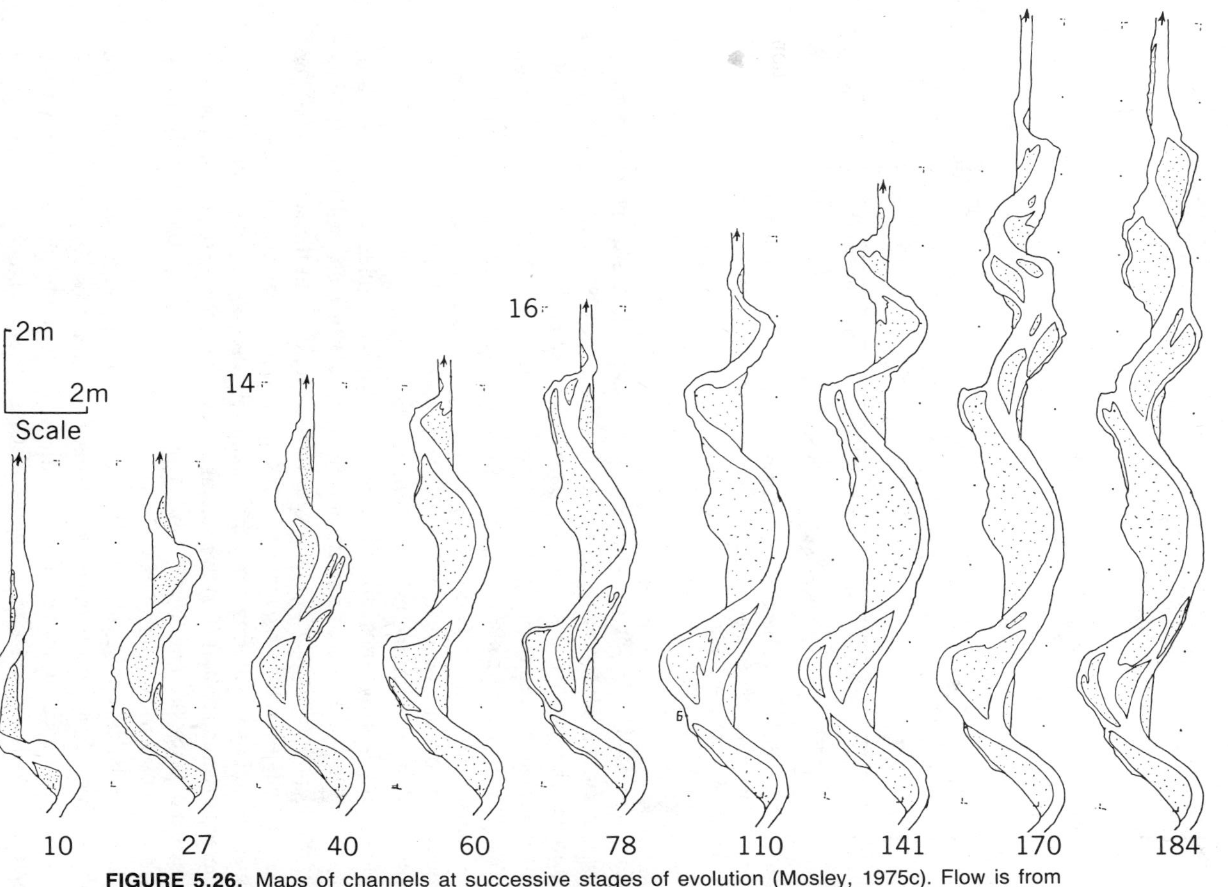

FIGURE 5.26. Maps of channels at successive stages of evolution (Mosley, 1975c). Flow is from bottom to top; number below each map is time in hours from start of run. Stippled areas represent point bar deposits.

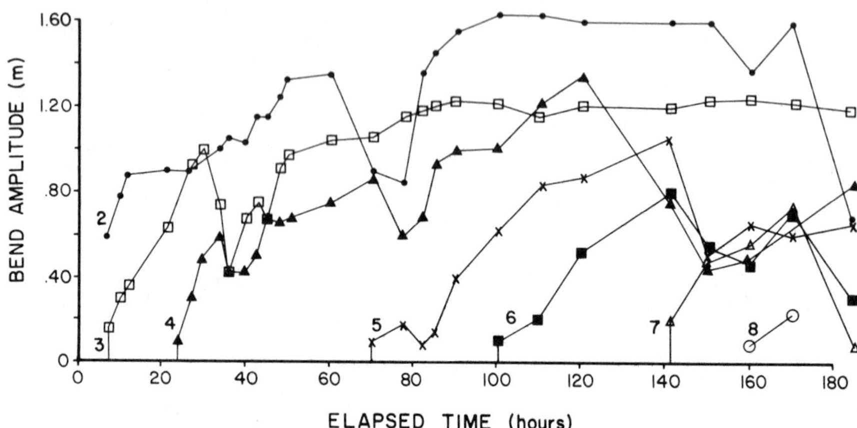

FIGURE 5.27. Ampltiude of bends as a function of elapsed run time (from Mosley, 1975c). Different symbols identify different bends; bends are numbered at their time of initiation (e.g., bend 5 was initiated after 70 hr).

As sediment was fed at the head of the tributary channels, the sand formed dunes that moved downstream. When the dunes reached the confluence, they rotated to parallel the flow in the lower channel. The scour hole that formed below the junction was therefore partly depositional in nature (Fig. 5.29). Dye injection demonstrated that flow patterns in the confluence were very complex. Shear along the boundary between the two converging flows at the confluence set up vertical vortices, which were swept downstream. They were revealed by a distinct, scalloped boundary between the two emerging tributary flows.

Flow from the tributaries crossed to the center of the confluence at the surface, plunged to the bottom, then returned to the surface along the walls of the scour hole. Two opposed helicoidal cells were thus established in the scour hole. The helicoidal cells prevented movement of sediment into the center of the scour hole, so that the sediment entering from the tributaries was transported along the base of the scour walls and left it in two lateral zones. These converged some distance downstream from the confluence, and a band of magnetite was usually found between the two zones (Fig. 5.29).

The scour holes were evidently a result of the shear and turbulence generated as the confluent flows were turned into the main channel. Description of the precise nature of turbulence in such a complex situation is virtually impossible, but the degree of turbulence and rate of energy dissipation are related to the change in flow momentum in the confluence, that is, to the forces exerted on the flows as they pass through the confluence (Chow, 1959). The momentum equation shows that these are in turn related to confluence angle, water discharge, water density (constant in the experiment), and flow velocity (approximately 0.3 m/sec in all runs). Figure 5.30 shows relative scour depth D_s/D as a function of tributary confluence angle. Runs with straight, unbranched channels and discharges of 0.25

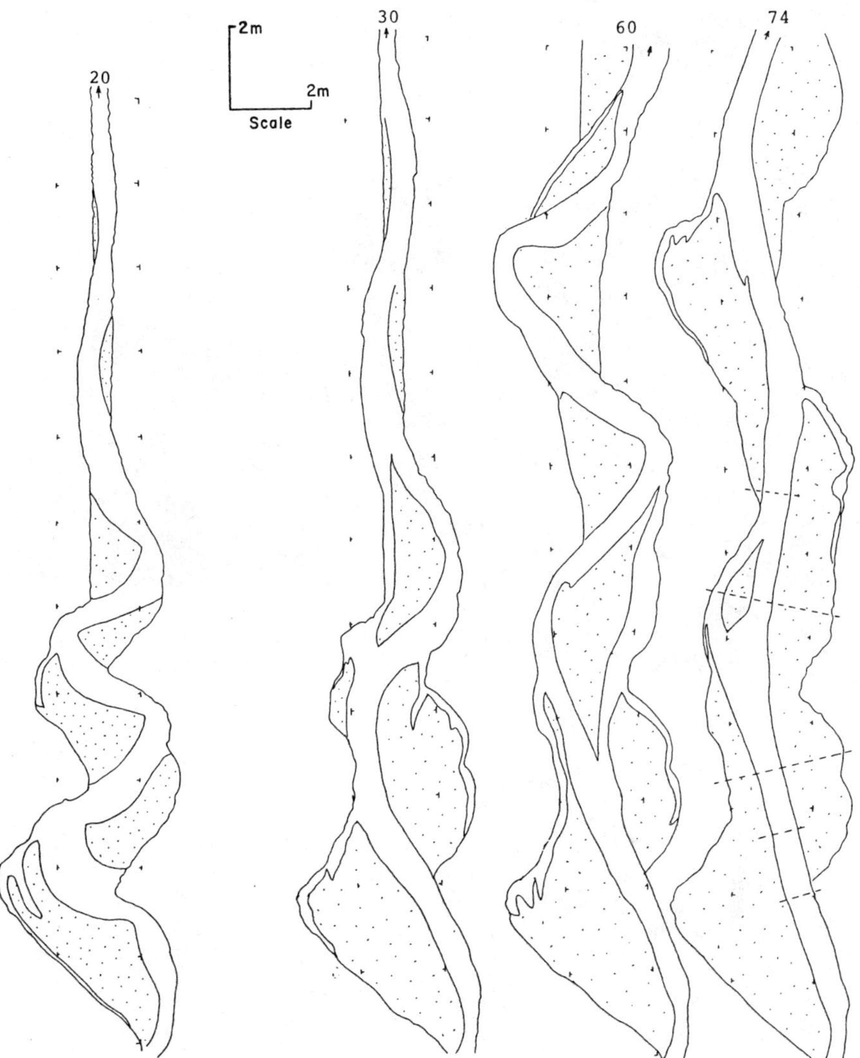

FIGURE 5.28. Maps of channels at successive stages of evolution (Mosley, 1975c). Numbers indicate hours of run time.

L/sec are also included. At confluence angles of 15°, there was little shear along the boundary of the converging flows and hence little tendency for a scour hole to form. This corresponds to engineering experience with generation of turbulence in rigid boundary confluences (U.S. Army Corps of Engineers, 1970). Qualitative observation indicated that shear and turbulence increase rapidly above angles of 15° to 90°, and more slowly thereafter; scour hole depths show a similar variation with confluence angle.

FIGURE 5.29. Scour hole in a typical channel confluence, looking upstream. Note the concentration of magnetite (dark band in center of channel) downstream from the confluence.

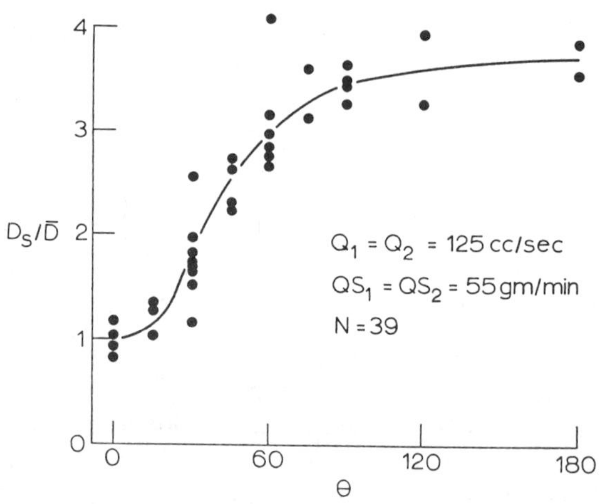

FIGURE 5.30. Relation between relative scour hole depth $D_s/\bar{D}$ in the confluence and confluence angle θ (from Mosley, 1975c). Mean depths for four straight channels ($\theta = 0°$) are included.

Valley Slope

Valley slope is, on the steady or graded time scales, an independent variable (Table 5.1) that partially controls the erosive capacity of a watercourse through its inclusion in the variables shear stress ($\tau = \gamma Rs$) and stream power ($\omega = \gamma Qs$). Friedkin (1945) demonstrated the effect of varying flume slope, other factors being held constant, on meander pattern (Fig. 5.31). He explained the tendency for wavelength and amplitude to increase with increasing slope as a function of the inability of the channel to make sharp turns at steeper slopes. For a given bank material, the river cuts its banks to a radius of curvature and a wavelength at which the erosive ability of the flow is balanced by the resistance of the material to erosion. In turn, the differences in meander planform are associated with differences in channel cross-section shape, with channels on steeper slopes being wider and shallower. These relationships were further examined by Schumm and Khan (1972), who confirmed that mean velocity and transport capacity of the flow (as indexed by sediment load) increased with valley or flume slope (Fig. 5.32*A, B*). As a result, width increased, mean depth decreased, and width–depth ratio increased (Fig. 5.33*A, B, C*). Sinuosity also changed with flume slope, with discharge and angle of entrance held constant at 4.25 L/sec and 40°, respectively (Fig. 5.33*D*). Below a slope of 0.002, the channel remained straight, but as slope increased above this to 0.008 bank erosion became more intense and rapid and the channel became increasingly sinuous. Above a slope of 0.013, the channel became increasingly unstable and cutoffs created a tendency for braiding to occur. Froude number approached 1.0 as this change occurred. Above 0.017, the channels were completely braided and the channel became straight once more, with a sinuosity of 1. This, of course, is the explanation for the abrupt increase in velocity and sediment load at the pattern change (Fig. 5.32).

Edgar (1973, 1984) extended these investigations by varying discharge as well as slope to determine the effect of stream power (ω) upon channel form. Increasing values of stream power enhanced the erosive ability of the stream, leading to more rapid bank erosion and increased bed-load transport. However, increased discharge caused a decline in sinuosity, probably because of the low cohesive strength of the banks (Fig. 5.34). The relationship between sinuosity and stream power is similar to that presented by Schumm and Khan (1972) between sinuosity and flume slope, with straight channels maintained at low stream power, increasing sinuosity up to some threshold value of stream power, which is different for each discharge, and braiding above that value. As could be expected from the previous discussion, channel width increased, depth decreased, and width–depth ratio increased with slope (Fig. 5.35).

As water flows down a surface with a varying valley slope, it cuts a channel with dimensions appropriate to its discharge and the sediment load it must convey. In order to maintain a constant channel slope either the channel must aggrade or degrade or sinuosity changes can adjust channel slope to changing valley slope.

Not only sinuosity but the channel type can be altered by valley slope. This is consistent with the observations by Schumm and Khan (1972) and Edgar (1973) of thresholds between straight, meandering, and braided channel patterns, and also

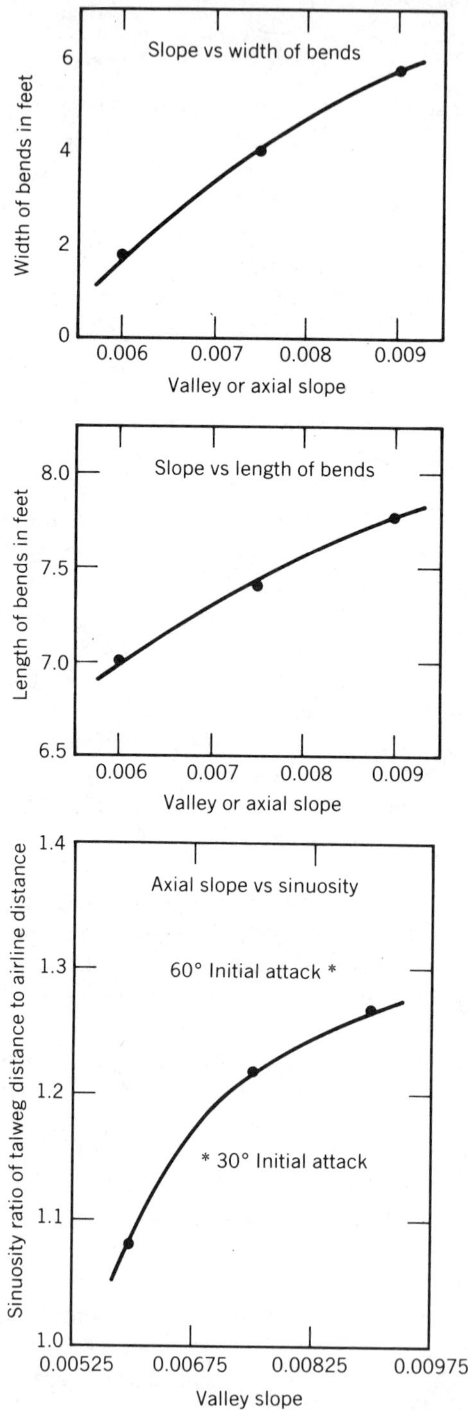

FIGURE 5.31. Effect of valley slope on meander dimensions and sinuosity (from Friedkin, 1945).

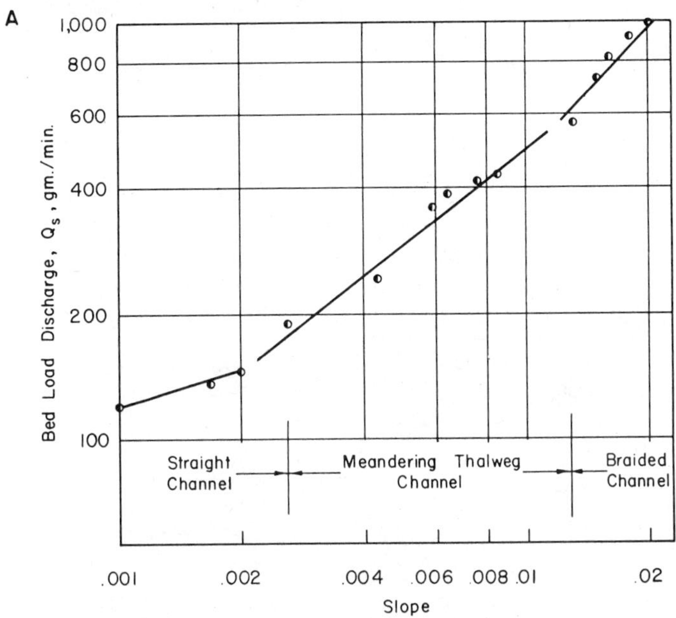

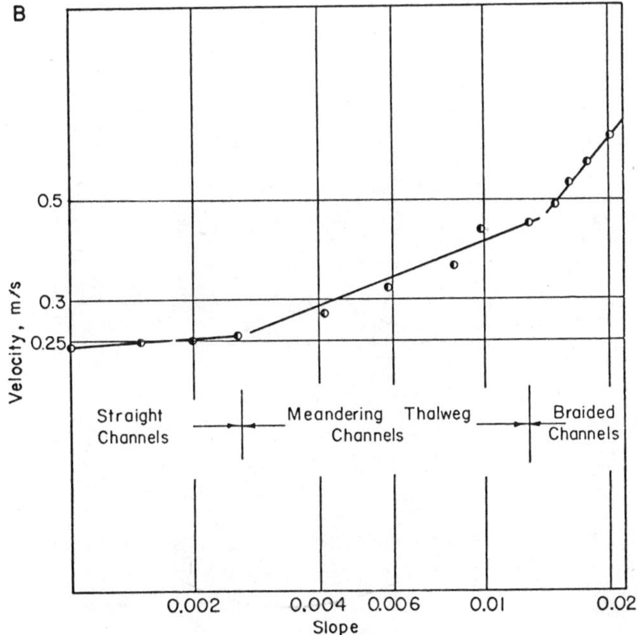

FIGURE 5.32. Effect of valley slope on (*A*) sediment load and (*B*) velocity (from Schumm and Khan, 1972).

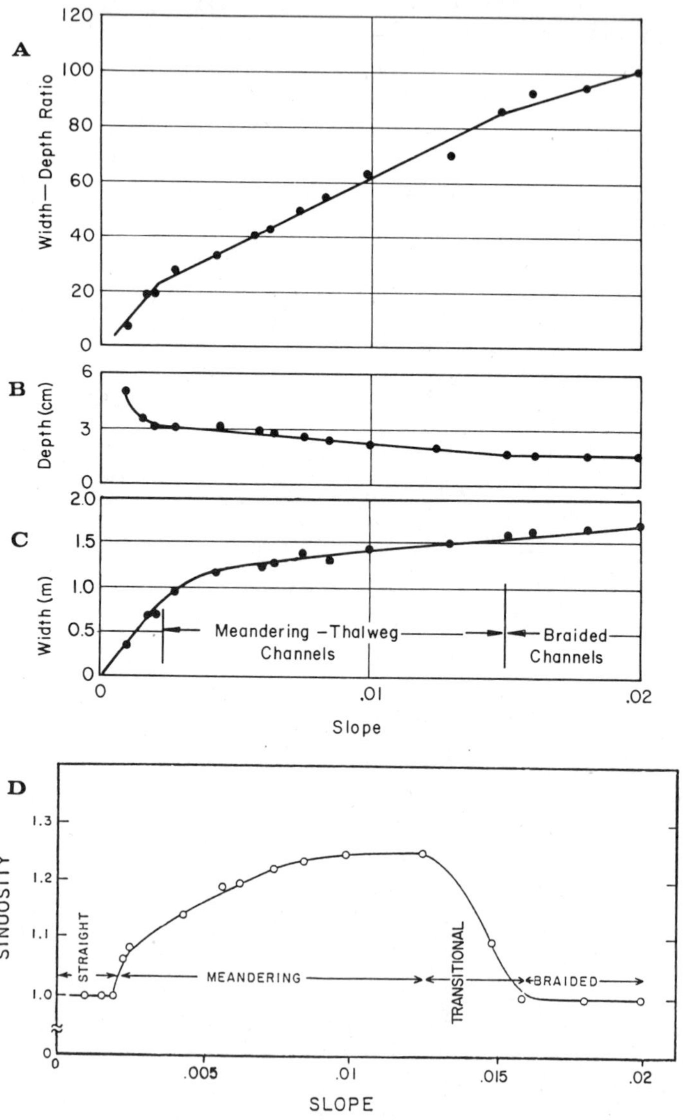

FIGURE 5.33. Effect of valley slope on channel geometry: (*A*) width–depth ratio, (*B*) depth, (*C*) width, and (*D*) sinuosity (from Schumm and Khan, 1971).

with earlier efforts to discriminate between meandering and straight channels by reference to discharge and valley slope. For example, the channels studied by Ackers (1964) and Ackers and Charlton (1970b) were divided into meandering, shoaled (with alternate bars), and straight (Fig. 5.36) based upon valley slope and discharge. In their equations, S is straight-line valley slope, not the slope along the thalweg. As slope increased in Ackers and Charlton's experiments, a critical slope was reached at which bank erosion and meandering began.

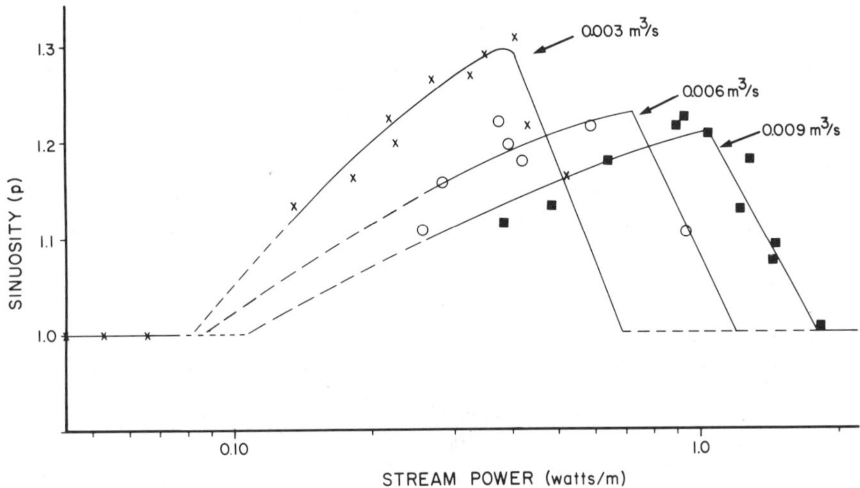

FIGURE 5.34. Effect of three discharge rates (m^3/sec) on the relation between stream power and sinuosity (from Edgar, 1973).

Hydrologic Controls

Discharge

The size and shape of model rivers has repeatedly been shown to be controlled by the volume of water to be carried, other things being equal. Thus, for example, Ackers (1964) found that the best-fit relations describing the morphology of straight channels in medium sand, which carried flows (Q_w) between 0.011 and 0.153 m^3/sec, were

$$A = 0.52\,Q_w^{0.85} \tag{5.9}$$

$$V = 1.92\,Q_w^{0.15} \tag{5.10}$$

$$w = 2.64\,Q_w^{0.42} \tag{5.11}$$

$$d = 0.20\,Q_w^{0.43} \tag{5.12}$$

where A is cross-section area in m^2, V is velocity in m/sec, w is width in m, and d is depth in m. These are comparable to similar equations for natural rivers. However, the dimensions of flume channels are not always easy to define or measure; this is particularly the case with channel width, which is normally used to compute mean velocity and depth. For example, as a meander migrates, relict backwater channels remain that are not infilled by floodplain construction as in natural rivers (Figs. 5.8*C, D*; 5.12; 5.37). Therefore, only a small part of the channel can be regarded as effective. Hickin (1972) circumvented the problem by working with the outer channel only (Fig. 5.37). Comparison of the data collected by Ackers (1964) and Ackers and Charlton (1970a) indicates that meandering channels average twice the width of straight channels, while channels with alternate bars are of intermediate width.

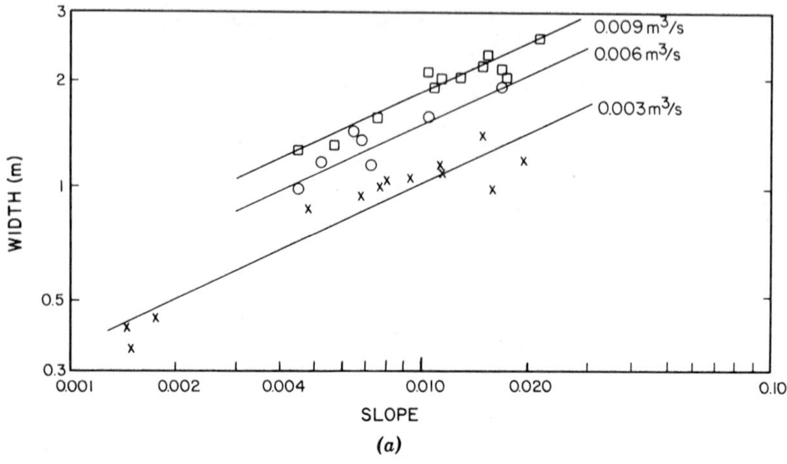

(a)

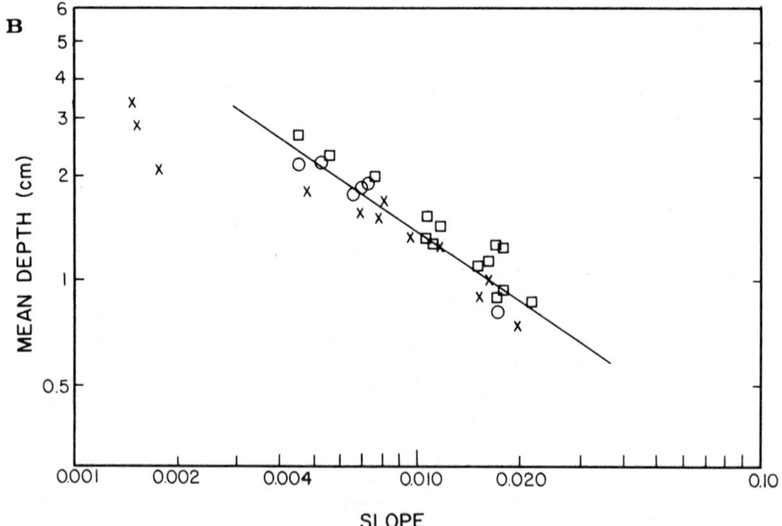

FIGURE 5.35. Relations between valley slope, discharge and (*A*) channel width, (*B*) channel depth, and (*C*) width–depth ratio (from Edgar, 1973).

Planimetric variables are also controlled by discharge. Friedkin (1945) found that meander wavelength, sinuosity, and amplitude increase with discharge, a conclusion echoed by a number of other investigators. The relationship between meander wavelength (*L*) and discharge shows a degree of scatter that is probably caused by differences in sediment size and initial entrance angle. To eliminate the effects of sediment, Ackers and Charlton (1970a) and Shahjahan (1970) resorted to dimensionless plots, as follows:

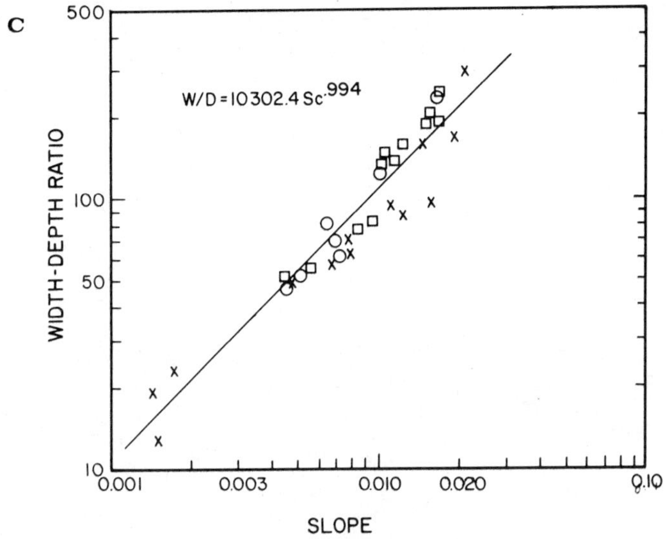

FIG. 5.35. *(Continued)*

$$\frac{L}{d_{50}} = f\left[\frac{Q_w^2}{g d_{50}^5}\right] \tag{5.13}$$

That is, relative wavelength is a function of relative stream size.

Shahjahan (1970) found a series of subparallel relationships of this form for different data sets and suggested that a "relative settling size" of sediment should

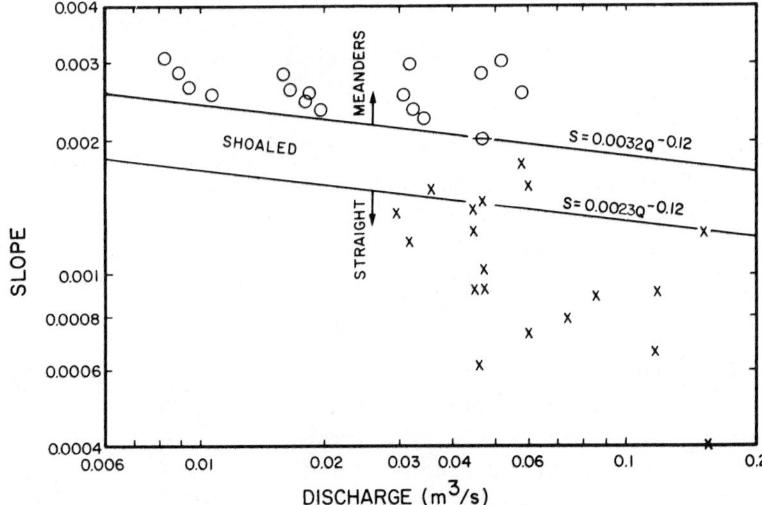

FIGURE 5.36. Discriminant functions for straight, shoaled, and meandering channels (from Ackers and Charlton, 1970b).

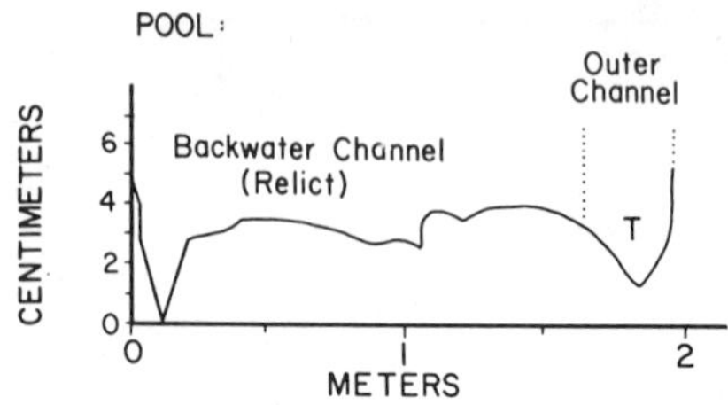

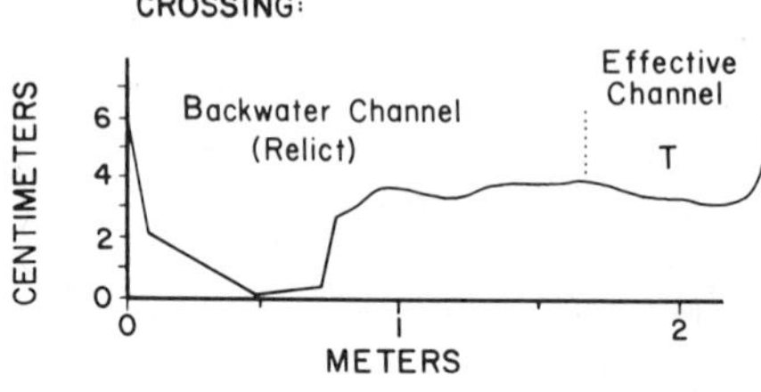

FIGURE 5.37. Typical cross section in sand bed flume channels. The outer, or effective, channel conveys the bulk of water discharge and sediment load (from Zimpfer, 1975).

also be included. He concluded that although earlier investigators had attributed poor relationships between L and Q_w to the difficulty of defining or measuring channel-forming discharge and had used channel width as an index of size, valley slope and sediment load might be responsible for some of the remaining inconsistencies in the plots. For a given discharge, bed-load channels have larger wavelengths than channels transporting finer sediments (Schumm, 1968).

Meander amplitude of model rivers has also been found to be related to discharge, although there is substantial scatter about the best-fit bivariate relationship. This is commonly attributed to the effect of sediment character, and Shahjahan (1970) again obtained an improved fit of relative amplitude upon relative discharge, when the effect of sediment differences was accounted for.

Varying Discharge

During Mosley's (1975c) studies of the effect of tributary angles he varied the flow in the channels, and Fig. 5.38 shows that as the smaller tributary (with discharge Q_1) increases in discharge relative to the larger, bed scour in the confluence becomes more pronounced. When all flow is in one channel, shear and turbulence are at a minimum and water depth in the confluence is equal to the depth in a straight channel. As flow in the smaller tributary is increased, it exerts an increasingly large turning force on flow in the main channel. As a result, shear and turbulence are generated, helicoidal flow cells become established in the confluence, and bed scour increases. The net amount of turning of the flow, generation of turbulence, and hence depth of scour are greatest when the tributary has from half to equal the discharge of the main channel. Tributary widths were varied but they had no clear effect on scour depths.

A perennial problem in river studies has been identification of the dominant discharge responsible for channel formation. Inglis (1949) defined dominant dis-

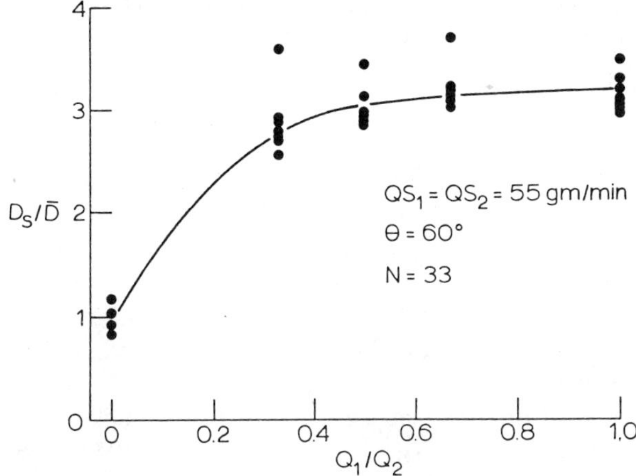

FIGURE 5.38. Relation between relative scour hole depth $D_s/\bar{D}$ in the confluence and tributary discharge ratio Q_1/Q_2. The three outliers were for runs in which flow in the tributaries was not uniform across the channel, but became concentrated near one bank. Discharges per unit width were therefore unusually high and caused increased scour (from Mosley, 1975c).

charge in the following words: "at this discharge, equilibrium is most closely approached and the tendency to change is least. This condition may be regarded as the integrated effect of all varying conditions over a long period of time." He regarded it as a little higher than bankfull discharge. Others have defined dominant discharge as the flow that accomplishes most sediment transport, and which is of intermediate frequency and magnitude (Pickup and Warner, 1976; Wolman and Miller, 1960), but it is a common view that the dominant discharge is most likely equivalent to the flow that just fills the channel, the bankfull discharge. Bankfull discharge tends to occur once every one or two years and to be approximately equal to the most probable annual flood or mean annual flood, with mean recurrence intervals of 1.58 and 2.33 years, respectively. However, there is a lot less regularity than has often been supposed, and Mosley (1981) and Williams (1978) found that bankfull discharge could have average recurrence intervals of from a few months to several years.

Sediment Type

The difference in experience between different experimenters is instructive. For example, Ackers and Charlton (1970b) were apparently able to produce meandering channels that were stable for many days and did not require an entrance bend, whereas Zimpfer (1975) found that, under the experimental conditions used by himself, Edgar (1984), and Schumm and Khan (1971), a regular meandering channel was only a temporary stage on the way to a final, braided pattern. On the other hand, Mosley (1975c), using the same discharge, slope, and operating conditions as Zimpfer but a different sediment in the flume, concluded that a straight, single-thread channel was the equilibrium form. As already noted in preceding sections, sediment character is an important control on channel form, and many qualitative and quantitative observations clarify this influence.

Friedkin (1945) demonstrated the effect of varying experimental material, other factors being equal. A channel flowing in sediment composed of 60% silt and 40% sand was deeper and narrower and meander wavelength and amplitude were smaller than in a channel flowing in 20% silt and 80% sand. Inclusion of a small amount of cement, which was not evenly distributed, in the experimental material similarly affected meander development by limiting channel width, meander wavelength, and amplitude in some bends and creating a channel pattern that was markedly nonuniform in comparison with a channel flowing in homogeneous material.

Ackers (1964) noted a marked effect on channel dimensions by the deposition of an inerodible layer of clay on the banks of the channels he studied. Widths and cross-sectional areas were less than for channels whose banks were able to erode and mean velocity was greater. The data from Mosley's (1975c) experiments with a silt–sand mixture were consistent with those of Ackers for the clay-armored banks. Schumm and Khan (1972) confirmed Ackers's (1964) observation of the effect of clay deposits by inducing a truly meandering channel from a meandering-thalweg channel that had already been formed by a flow of 4.25 L/sec in noncohesive sand. They added kaolinite to the flowing water, to give a concentration of 30,000 ppm, and at the same time reduced the rate of coarse-sediment feed. The thalweg scoured along the length of the channel, so there was no overall change in gradient, and suspended sediment was deposited on the alternate bars. This had the dual effect of reducing the already shallow water depths over the bars and stabilizing them against erosion. Scour along the thalweg lowered the water level, the submerged alternate bars emerged to become point bars, and the channel became narrower, deeper, and more sinuous as a result of the change in character of the sediment load (Figs. 5.39, 5.40). This experimental procedure was followed with four channels, all flowing at 4.25 L/sec but with valley slopes of 0.0026 to 0.0085, and the same decrease in cross-sectional area (Fig. 5.41*A*), increase in width and width–depth ratio (Fig. 5.41*B, C*), decrease of depth and hydraulic radius (Fig. 5.41*C*), and increase in sinuosity was observed in each case. In spite of increased sinuosity (Fig. 5.41*E*), mean velocity also increased (Fig. 5.41*F*) because the decrease of cross-sectional area, the reduction of friction due to the deposition of fines on the channel banks, and the damping effect of suspended sediment on turbulence (Vanoni, 1946) increased efficiency of the narrower, deeper channel. The increased flow velocity caused further erosion of the concave banks, so that the sinuosity of the suspended load channel finally exceeded that of the meandering thalweg in the channels flowing in noncohesive sediment prior to introduction of kaolinite. However, eventually both bed and banks were coated with a veneer of clay, channel changes ceased, and the sand load fed into the channel moved directly to the tailbox.

These qualitative treatments of the influence of sediment load on river morphology are supported by more formal quantitative analysis. Shahjahan's (1970) use of dimensional analysis to include the influence of sediment type on meander geometry has already been mentioned; by grouping the relevant variables in non-

FIGURE 5.39. Channel pattern in CSU flume 1 (*A*) with bedload and (*B*) after addition of suspended load, both with a discharge of 8.5 L/sec, a slope of 0.005, and the same entrance bend.

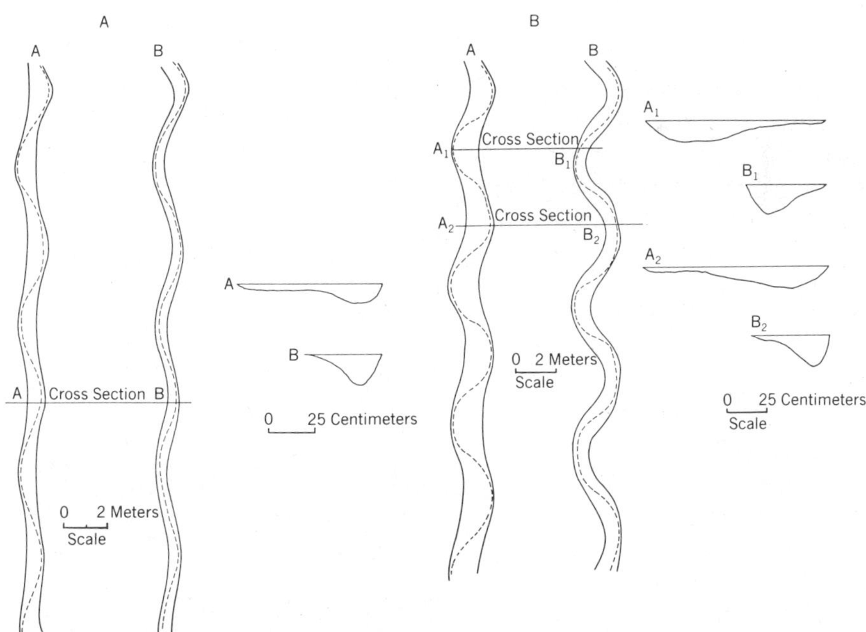

FIGURE 5.40. Channels (*A*) before and (*B*) after introduction of suspended-sediment load. Discharges were 4.25 L/sec (from Schumm and Khan, 1971).

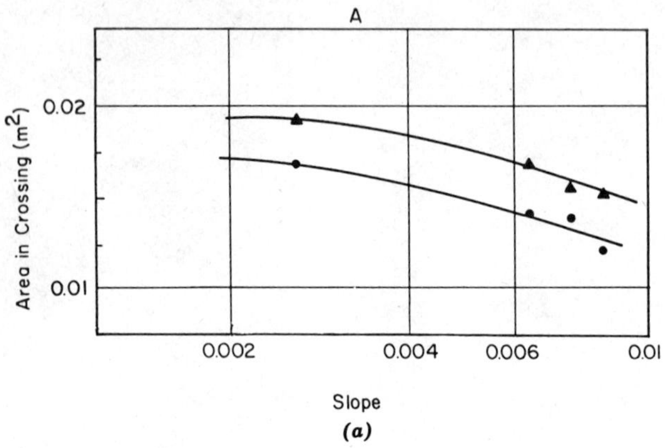

(a)

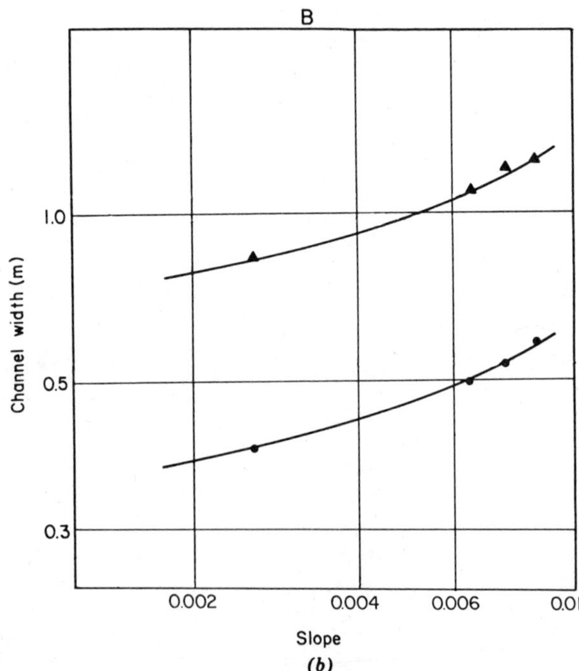

(b)

FIGURE 5.41. Effect of addition of suspended-sediment load on channel's (*A*) area, (*B*) width, (*C*) width–depth ratio, (*D*) hydraulic radius, and (*E*) velocity. In each figure, triangles represent no suspended load and dots indicate a 3% suspended-sediment concentration (from Khan, 1971).

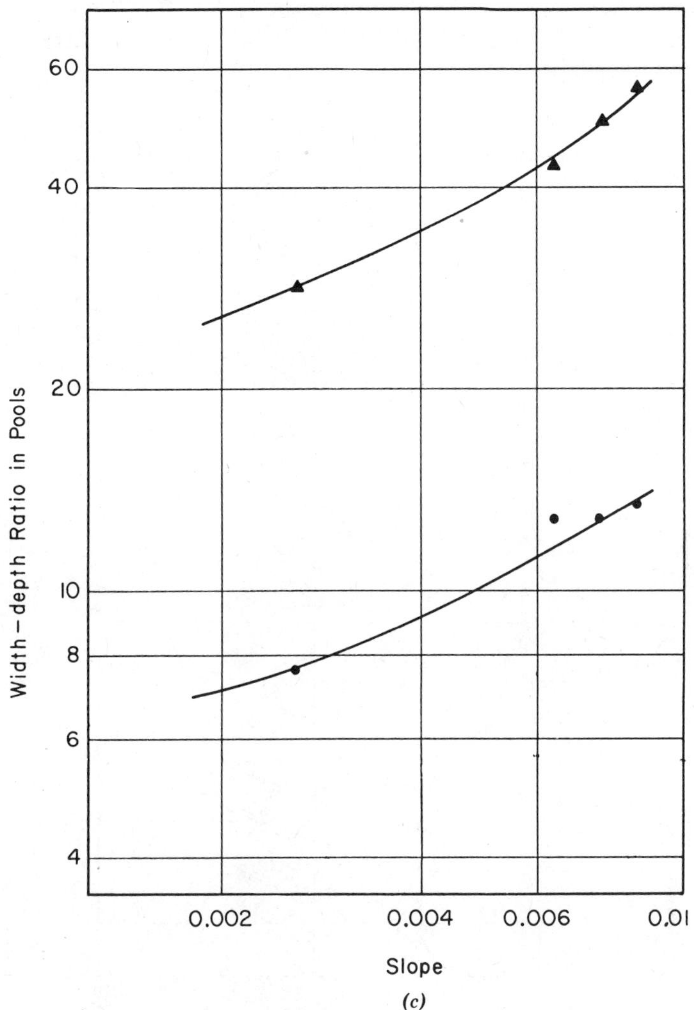

(c)

FIG. 5.41. *(Continued)*

dimensional terms, he was able to show that for a given value of "relative discharge" ($Q_w^{2/5}/d_{50}g^{1/5}$), meander wavelength and channel width increase, but depth decreases with the "relative settling size of sediment" $d_{50}g^{1/3}/^{2/3}$, where d_{50} is median grain size and g is the gravitational constant.

Ackers and Charlton (1970b) pursued a similar approach. They noted that the variables that describe a stable alluvial channel are functions of the imposed flow and force field, sediment properties, and fluid properties, and they examined several different dimensionless groupings of the appropriate variables as predictors of chan-

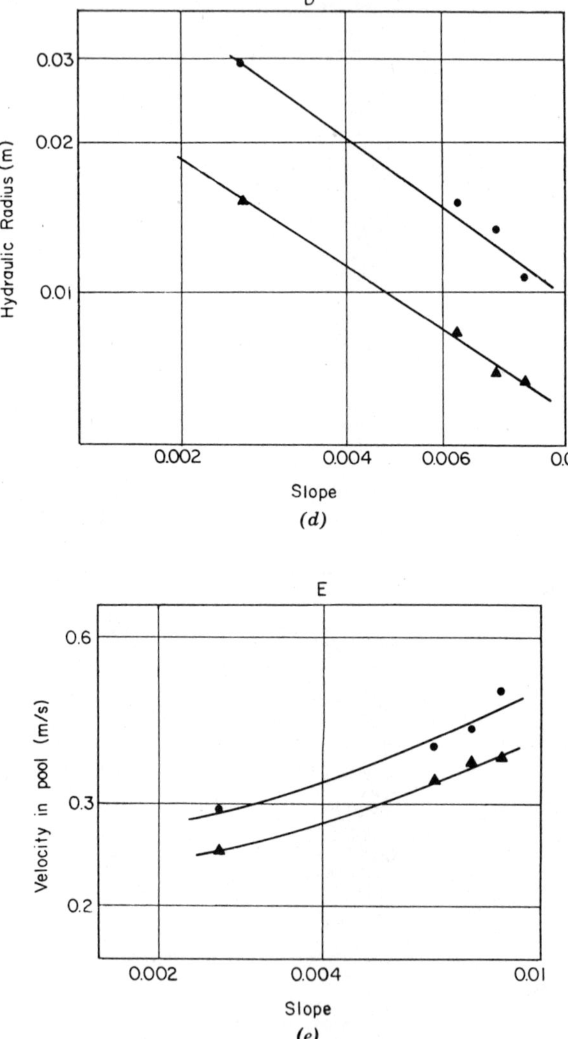

FIG. 5.41. *(Continued)*

nel morphology. There proved to be a good relationship between relative meander wavelength L/d_{50} and relative discharge Q_w/d_{50} given by the best-fit equation

$$\frac{L}{d_{50}} = 123 \left[\frac{Q_w}{d_{50}^2 \sqrt{g d_{50} (S - 1)}} \right]^{0.376} \tag{5.14}$$

where L is meander wavelength in ft, d_{50} is median grain size in mm, and Q_ω is water discharge in ft^3/sec. This implies that for material with a given specific gravity and valley slope (S),

$$L \propto d_{50}^{0.06} Q_{\mathrm{w}}^{0.376} \tag{5.15}$$

However, data sets with different median diameters showed slightly different trends, with exponents of between 0.45 and 0.55. Evidently, d_{50} has a significant, although secondary, effect on meander wavelength.

The controls on meander morphology have been extensively studied by other authors, either using published data for different sediment types or by working with several sediments in a single study. It is widely considered that there is a close relationship between meanders and alternate bars, and information on the conditions necessary for alternate-bar formation might cast light on the reasons why meanders form in some circumstances and not in others. A particularly interesting study that used several sediments was carried out by Chang et al. (1971); they examined the formation of alternate bars in channels composed of sand, plastic pellets, and expanded clay aggregate, with specific gravities of 2.65, 1.05, and 1.80, respectively. Unfortunately, other characteristics of the sediments also varied; median fall diameters were 0.7, 0.18, and 0.48 mm, respectively, while, in addition, the sand was angular and the plastic and clay aggregate were round. The rate of evolution of the channels was strongly controlled by sediment type, with channels in the plastic pellets reaching equilibrium in only a few hours, while the sand channels took from two to three days, and the clay-aggregate channels required an intermediate time. Alternate bars formed in the majority of channels, with discharges between 2.83 and 14.16 L/sec and slopes between 0.00044 and 0.0064. There was no obvious relationship between the wavelength (L) of the alternate bars and discharge or the discharge–slope product ($Q_{\mathrm{w}} S$), but L was markedly influenced by particle differences. Average values of L for plastic pellet, clay aggregate, and sand channel alternate bars were 7.3, 10.7, and 17.1 m, respectively, with standard deviations of 2.6, 4.4, and 4.0 m. Alternate bars failed to form when the width–depth ratio was less than 12 in the fixed-wall flume. The best relationship between the dimensionless wavelength parameter LS/D and other variables was with Froude number, as found by Kinosita (1961).

Jaeggi (1984) carried out a more extensive analysis of the conditions under which alternate bars form, using his own data for flume channels in five different sediments and data from several other authors. He identified both upper and lower limits of channel slope at which alternate bars could form. The upper limit is defined in terms of the relative tendencies for horizontal and vertical deformation (alternate bar and dune formation, respectively) and the lower limit is defined by the beginning of motion of the bed material. Together they define a region in which alternate-bar formation is possible, as controlled by bed-material character, channel slope, and channel shape, both of the latter being a function of the character of bed and bank material.

A good example of the effect of sedimentary materials on channel morphology is Jin's (Jin and Schumm, 1986) study of the effect of resistant outcrops on meander migration. As has been described by many investigators, the downstream shift of the meanders is hindered at the outcrop of resistant materials and the meanders compress themselves upon it (Fig. 5.42), with an increase of sinuosity and meander

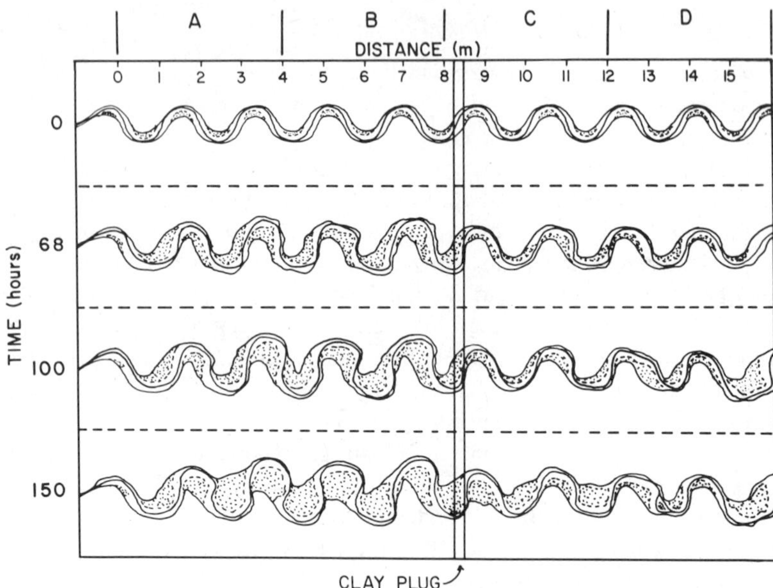

FIGURE 5.42. Effect of a clay outcrop on channel pattern (from Jin and Schumm, 1986). Flow is from left to right.

amplitude upstream (Fig. 5.43). In more erodible sediments the downstream bends would show a more marked shift away from the obstruction and a consequent decrease in sinuosity and increase in wavelength. Jin also constructed a floodplain composed of layered cohesive and noncohesive sediments, which permitted true meander development (see Chapter 8).

Sediment Load

In many flume studies, sediment load (Q_s) or concentration (C_s) has been treated as a dependent variable controlled by discharge, channel slope, and flow depth. Bank erosion is a major source of sediment load in many rivers, and degradation and aggradation of the riverbed may also influence sediment transport rates, so that under natural circumstances sediment load may be a function of channel behavior and morphology. However, in the long term, sediment load is imposed on a channel by conditions upstream in the watershed, and the channel must adjust to convey this imposed load. Design of canal systems able to convey water and sediment without "silting or scouring" has been the aim of much research by waterway engineers.

The close association between channel slope, bank erosion, and sediment load was illustrated by Friedkin (1945), who showed that at high channel slopes, bank erosion rates and sediment loads increased and caused channel widening and an increase in meander amplitude. Raju et al. (1977) carried out similar experiments. Commencing with initial surface slopes of 0.0005–0.0025, they allowed the initial

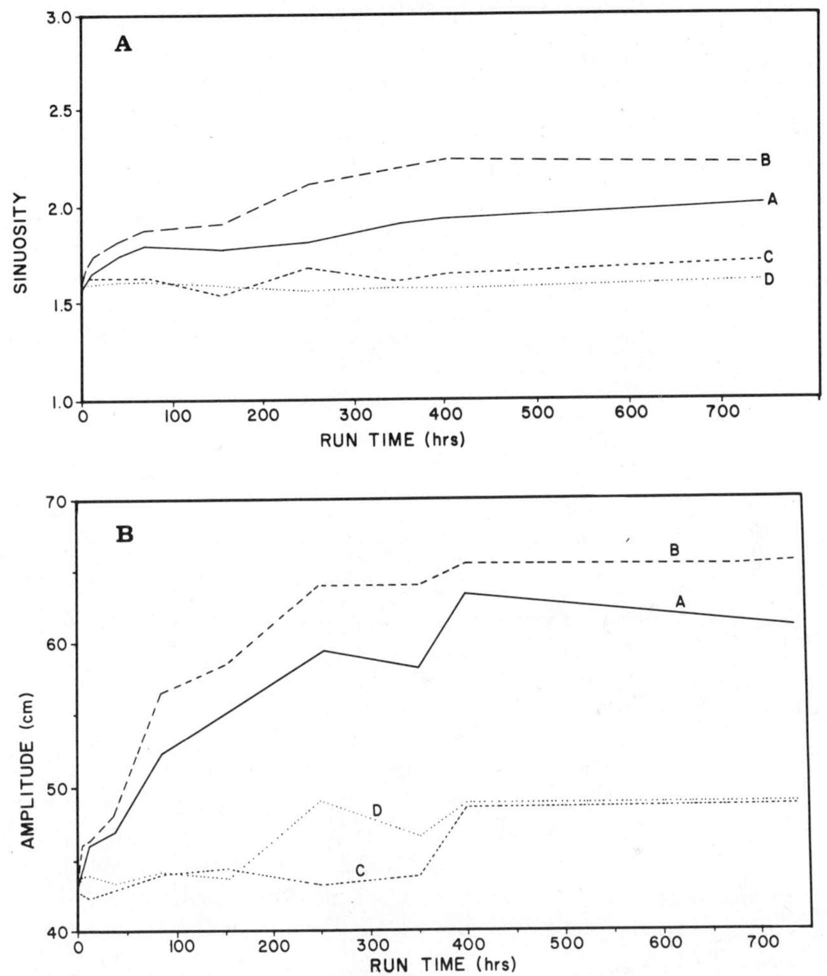

FIGURE 5.43. Effect of obstruction on (*A*) sinuosity and (*B*) meander amplitudes.

straight channels to evolve to a stable condition. They concluded that wetted perimeter was practically independent of C_s, hydraulic radius declined with increasing C_s, and slope increased with increasing C_s. Schumm and Khan (1972) also found that the rate of sediment feed necessary to maintain a stable channel was related to valley slope (Fig. 5.32*A*). They used the same data to suggest that an increase in sediment load Q_s would require a change of the rate of increase of slope as patterns change (Fig. 5.44).

Shahjahan (1970) recorded rather small changes in the dependent morphologic variables over a wide range of sediment concentrations, with other factors (including slope) held constant. Meander wavelength and channel width declined and amplitude and channel depth increased with increasing sediment concentration, but

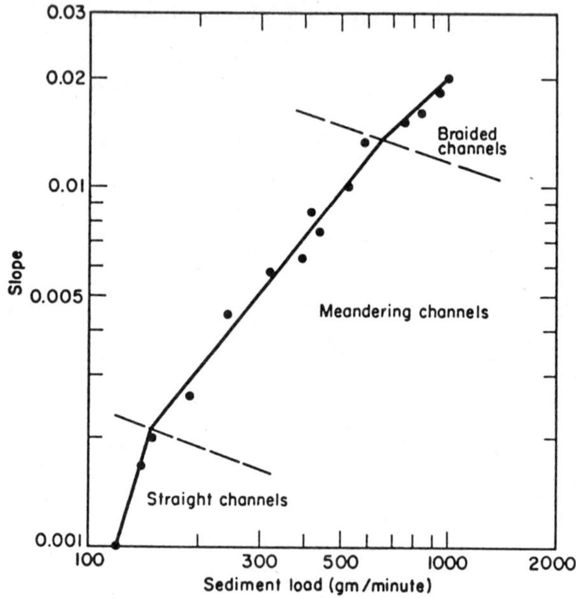

FIGURE 5.44. Effect of sediment load on slope (from Schumm and Khan, 1972).

the small changes indicate that the independent effect of sediment concentration on channel morphology is small. This conclusion was also reached by Ackers and Charlton (1970c) in their dimensional analysis of meander wavelength. Ackers (1964, p. 24–25) indicated that C_s would in theory be included in multivariate relationships with morphologic variables, but that it would be relatively unimportant. The difficulty of deciding whether sediment concentration is controlled by slope or vice versa, and is therefore either a dependent or independent variable, recalls the discussion of the direction of causality at different time scales earlier in the chapter (Table 5.1).

Mosley (1975c) also varied sediment load during his confluence investigation, and Fig. 5.45 shows the effect of varying sediment load, with tributary discharges constant, on relative scour hole depth. As sediment load increases, relative scour depth decreases; but varying relative tributary sediment load (Q_{s1}/Q_{s2}) had no discernible effect upon D_s.

DISCUSSION

The experimental studies reproduced four types of channels, straight, meandering, meander–braided transitions, and braided. The morphology of these were affected to some degree by numerous variables such as initial conditions, water and sediment discharge, type of sediment load, and valley slope. The results have been anticipated and/or supported by field studies. Nevertheless, a major question that arises

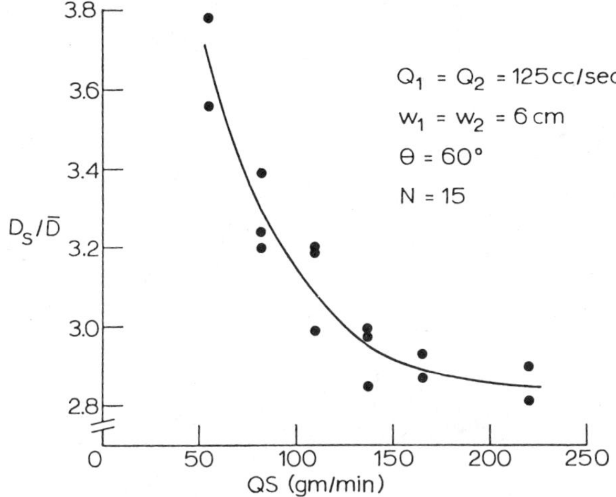

FIGURE 5.45. Relation between relative scour hole depth ($D_s/\bar{D}$) in the confluence and total sediment load (Q_s) through the confluence. Note that confluence angle θ was 60° throughout (from Mosley, 1975c).

from a discussion of experimental studies of alluvial channels relates to the validity of their results. The braided channels provide a good starting point for comparing experimental with natural channels. Qualitatively, the comparisons are remarkably good, both at an overall channel scale (Fig. 5.46) and at the scale of individual bar forms (Fig. 5.20). Quantitatively, too, the model channels were geometrically similar to their natural analogues; the relation between confluence scour depth and total discharge for Mosley's (1975c) experimental channels with a flow of 0.25 L/sec can be extended to confluences in the braided Ohau River carrying discharges of up to 250 m^3/sec (Mosley, 1982a); while Ashmore and Parker (1983; Kjerfue et al., 1979) similarly found close agreement between model and prototype relationships for relative scour depth versus confluence angle.

Straight model channels also show marked similarities with full-scale natural channels and canals in terms of both general morphology—including the presence of bedforms whether ripples, dunes, or alternate bars—and actual dimensions. For example, the equations for mean velocity (V), width (W), and mean depth (D) as a function of discharge (Q_w) for the small channels studied by Ackers (1964) are entirely consistent with those for natural channels [Eqs. (5.9)–(5.12)].

Laboratory meandering channels are not, on the other hand, particularly similar to natural meanders in terms of their general appearance. Their regularity and symmetry are explicable by the uniform sediment of which natural floodplains are composed, but so-called meandering channels in flumes commonly have a very low sinuosity, so that the overall channel is nearly straight, and it is only the thalweg which crosses from side to side of the channel, and reverse (upvalley) curves never occur. The experiments with cohesive materials by Schumm and Khan (1972), Mosley (1975c), and Friedkin (1945) demonstrate the importance of bank

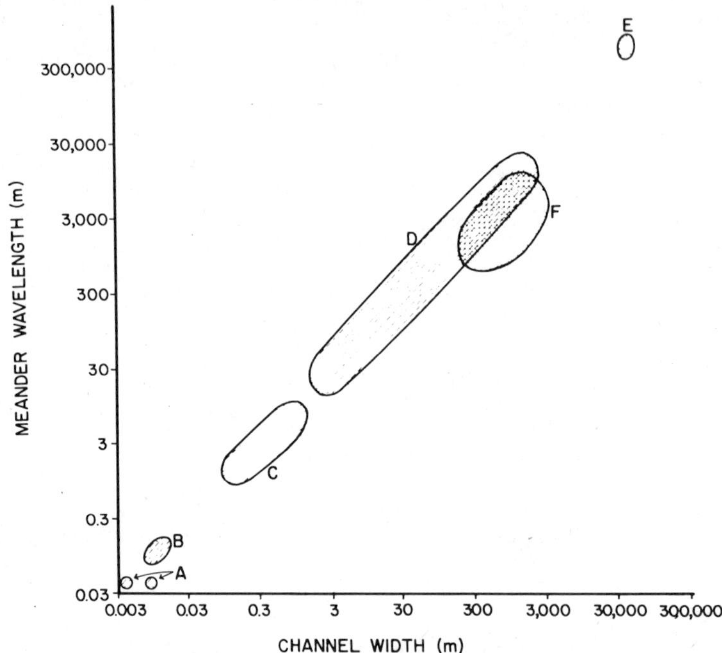

FIGURE 5.46. Relation of meander wavelength of channel width (A) surface tension meanders (Gorycki, 1973); (B) solution channels in limestone; (C) supraglacial meltwater channels; (D) alluvial streams; (E) Gulf Stream meanders; (F) lunar rills (after Davies and Sutherland, 1980).

erodibility, and particularly variations in sediment type (Fig. 5.2), in controlling channel planform. Field investigations that support the experimental results are reported by Richards (1979), Schumm (1960), and Smith and Smith (1984). In each case a dramatic change of sediment type caused morphologic changes that could be expected from the experimental results. Both the field and experimental studies support the classification of Fig. 5.2, which has been used by sedimentologists to explain the diversity of fluvial deposits (Galloway and Hobday, 1983).

When banks are composed of uniformly erodible, noncohesive sediment, the fundamental meandering tendency of the model channels is expressed (Ferguson, 1976), but the dissimilarity in appearance between irregular natural meandering rivers and uniform model channels in homogeneous materials illustrates the major role of sediment character in controlling detailed channel morphology. Nevertheless, the geometric similarity of model meanders to meanders in natural rivers, as demonstrated by their conformity to the same wavelength–channel width relation (Fig. 5.46), is indicative of a fundamental tendency of flowing water to take a winding course, the geometry of which is controlled by discharge. This contrast indicates a major strength of experimental studies of model rivers, that the effect of selected variables upon channel morphology may be highlighted by holding constant other variables or removing them entirely from consideration. Factors such as valley slope and floodplain composition can actually be varied in the laboratory to identify their influence, an impossibility in natural channels. This potential is demonstrated particularly neatly by Friedkin's (1945) experimental design, in which he successively varied each variable while holding all others constant, to give a series of bivariate relationships between the dependent and independent variables. Physical reasoning and dimensional analysis were used by Shahjahan (1970), Quaraishy (1973), and Ackers and Charlton (in their 1970 series of papers) to examine multivariate relationships, that is, simultaneously to analyze the response of channel morphology to variations in several independent variables. While highly effective, the physical significance of the dimensionless combinations of variables that are produced by this approach is not always obvious, and several different combinations may be possible. Indeed, Ackers and Charlton (1970a) concluded that several different equations could emerge, depending upon the precise objectives and procedures of the analysis; and, for all its apparent rigor, dimensional analysis is not yet able to solve all the problems in river engineering.

A particular difficulty encountered by Ackers and Charlton in comparing model results with data from natural channels was the incompleteness and poor quality of the latter. Despite the hundreds of studies that have been carried out, they found that frequently such fundamental variables as discharge were only estimated (e.g., from a regional runoff equation) or unmeasured. Measurement of the character of bed and bank sediment and of sediment discharge is notoriously difficult, but the natural variability of even such apparently well-defined variables as channel width (Fig. 5.37) renders their measurement markedly less simple than is commonly assumed. The situation is not helped by poor definition of variables and measurement procedures by different authors (Ackers and Charlton, 1970c; Richards, 1982). This, of course, implies another strength of experimental studies, that measurements under controlled conditions are more accurate than in the field, and commonly also easier and cheaper. They can, moreover, be confidently and easily replicated (Fig. 5.6).

There seem to be many advantages in utilizing flumes for studying alluvial channel morphology if the experimental design is carefully thought out. Kennedy (1983) expressed the opinion that "modern electronic and optical instrumentation permits conduct of the clever, definitive experiments that should supplant the traditional flume (mud-mill) experiments on steady, uniform flow over sand beds (data from which, however useful they have been and will continue to be, have accumulated until more of these experiments have little value beyond the educational benefit to students who conducted them)." Virtually all the studies mentioned in this chapter are of this "mud-mill" type, and the sophisticated experimental work recommended by Kennedy seems to date to have been restricted to very detailed examination of the structure of turbulent flows, commonly in sediment-free water. One can only look forward to the day when the complexities of non-uniform, unsteady flow through heterogeneous sediments can be studied experimentally with the rigor anticipated by Kennedy.

The full justification for the experimental work comes only when the results can be applied to the field situation, where a fuller understanding of river morphology

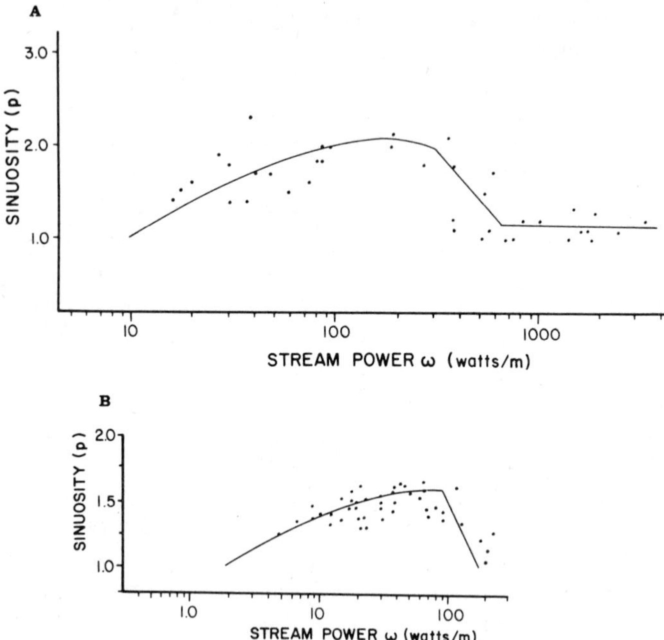

FIGURE 5.47. Relation between sinuosity and stream power for (*A*) Canadian rivers (from Edgar, 1973), and (*B*) Indiana and Illinois streams (from Edgar, 1984).

results. For example, the plot of stream power or valley slope and sinuosity suggests that for a number of rivers in one region thresholds of stream patterns can be identified. Edgar (1973, 1984) was able to demonstrate for a number of rivers (Fig. 5.47) that this relation is valid and that one could identify those subject to change. The same applies to data from the Mississippi River (Fig. 5.48), where the scatter leads to scepticism concerning the validity of applying the curve to these data.

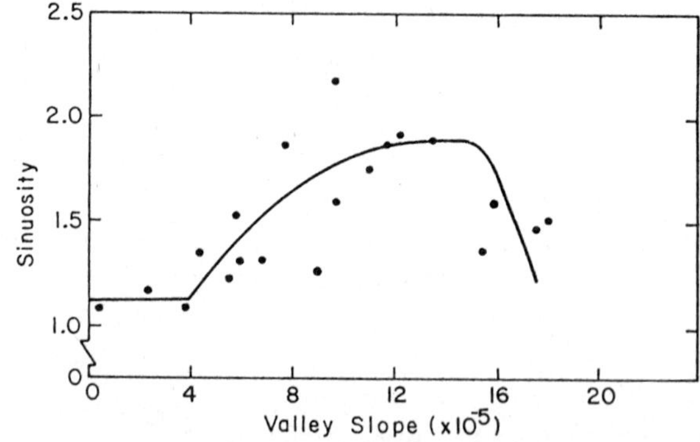

FIGURE 5.48. Relation between valley slope and sinuosity of Mississippi River. Data obtained from 1890 map that precedes much of the man-induced channel change (from Schumm et al., 1972).

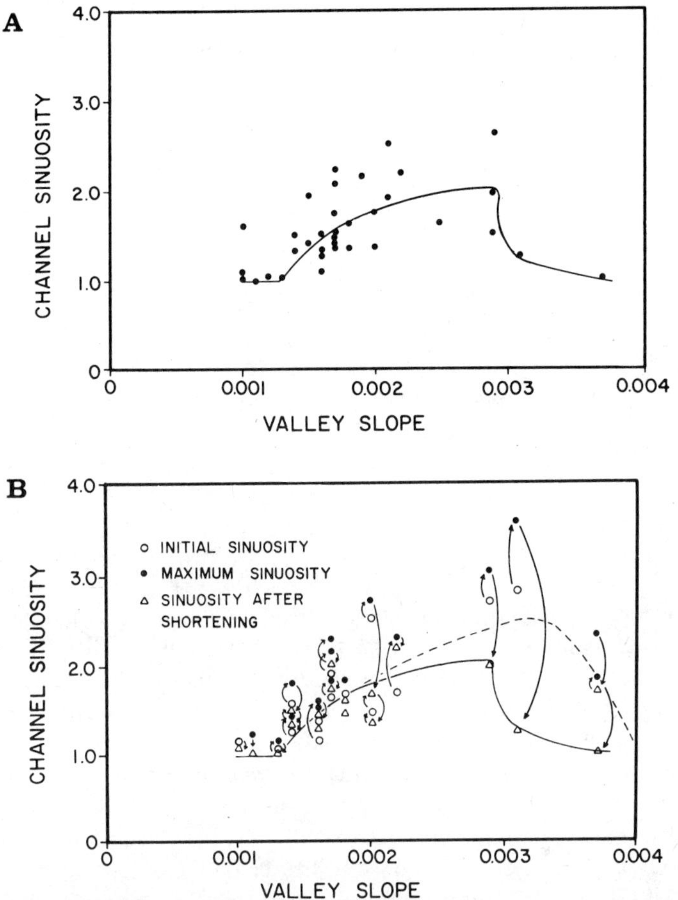

FIGURE 5.49. Relation between valley slope and sinuosity of Powder River between Moorhead and Broadus, Montana (from Martinson, 1983). (*A*) Sinuosity in 1978; (*B*) variation of sinuosity for reaches of Powder River. Many reaches were shortened by natural cutoffs between 1939 and 1978.

However, the approach is supported by Martinson (1983), who showed how the Powder River has changed sinuosity and pattern during the period of record, with the greatest changes occurring within the meander–braided transition zone (Fig. 5.49). The further significance of these relations, and especially their practical significance, lies in the development of such relations for a river or for a group of rivers. This, then, would aid the engineer in river planning and in developing an understanding of the sensitivity of different river reaches.

6 Incised Channels

Incision of alluvial fills by steep-sided, generally ephemeral-stream channels has received much attention from land managers, geomorphologists, and geologists because of its implications for land use and for understanding the recent history of the landscape. Not only does the extension of gullies destroy agricultural land and undermine structures, but farm access is affected, water tables may be lowered, and sediments eroded from the gully may have an impact on water quality and reduce reservoir and channel capacities downstream.

Incised channels range in size from small rills, most commonly observed on road cuts, to major arroyos or gullies that may be on the order of 10 to 15 m deep (Table 6.1). Clearly, they are major sediment producers. For example, the infamous Rio Puerco arroyo in New Mexico (Fig. 6.1*A*), although comprising only 20% of the Rio Grande drainage basin above Elephant Butte Reservoir and producing only 8% of the runoff, delivers almost 50% of the sediment that reaches the reservoir (Patton, 1973), and channelized streams in northern Mississippi (Fig. 6-1*B*) are the source of large amounts of sediment that cause aggradation in the Mississippi (Schumm et al., 1984).

An incised channel is commonly characterized by one or more abrupt breaks in its longitudinal profile, particularly at its headward end. Such a break in the profile is known as a nickpoint, from the German term *knickpunkt* (see Chapter 4). The abrupt change in elevation at the head of an incised channel will be referred to as a primary nickpoint or a headcut (Fig. 6.2). Such a change elsewhere within an incising channel will be referred to as a secondary nickpoint, and a steeper reach of channel, which represents a headward-migrating zone of incision, will be referred to as a nickzone (Fig. 6.2*B*). In addition, a small nickpoint frequently precedes the primary nickpoint and is referred to as a precursor nickpoint.

Several approaches to the study of incised channels have been adopted. First,

TABLE 6.1 Classification of Incised Channels Based on Size[a]

1. Rills
2. Gully
   a. Valley-side gully
   b. Valley-floor gully
      i. Discontinuous gully
      ii. Tributary gully
3. Entrenched stream
   a. Ephemeral or intermittent (natural arroyo)
   b. Perennial or intermittent (man-induced, channelized stream)
4. Drainage system rejuvenation

[a] From Schumm et al. (1984).

studies of alluvial deposits have enabled establishment of the sequence of past periods of trenching and filling (Antevs, 1952; Bailey, 1935); second, historical evidence of long-term changes in incised-channel morphology has been examined (Bryan, 1928; Cooke and Reeves, 1976; and, third (Fig. 6.3), direct observations of incised-channel evolution have been made (Heede, 1974; Ireland et al., 1939; Piest et al., 1975, Schumm et al., 1984). In addition, incised-channel dimensions and growth rates have been related statistically to watershed characteristics (Beer and Johnson, 1965; Seginer, 1966), and there have been some initial attempts at mechanistic treatments of limited aspects of incised-channel behavior (Bradford et al., 1973; Pickup, 1975).

Despite all the effort that has been invested, however, Cooke and Reeves (1976) observed that "a reader of the prodigious arroyo literature may be justifiably perplexed by the shifting currents of conflicting arguments, the discharge of unsubstantiated assertions, the pools of controversy, and the shoals of abandoned hypotheses." This observation referred to the particularly difficult topic of arroyo initiation, but there is more agreement on the qualitative aspects of later phases of gully evolution. Nevertheless, there are several major methodological difficulties with the study of incised channels. The first is that they are not equilibrium land-

FIGURE 6.1. Aerial photograph of incised channels. (*A*) Rio Puerco, New Mexico, has incised deeply and straightened. (*B*) Oaklimiter Creek, Mississippi. Channel has incised following channelization. Former sinuous channel remains on floodplain to right.

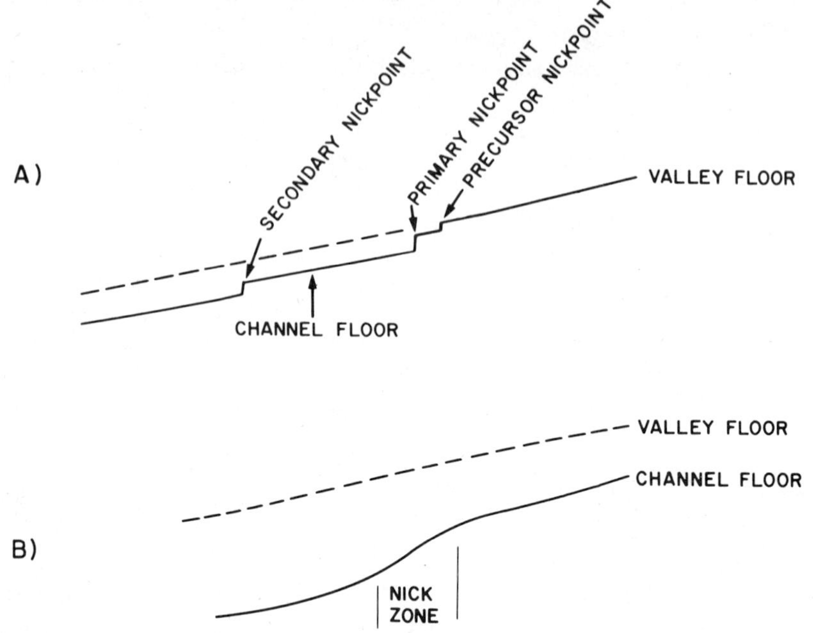

FIGURE 6.2. Sketch showing nickpoint types. (*A*) Primary, secondary, and precursor nickpoints. (*B*) Nick zone.

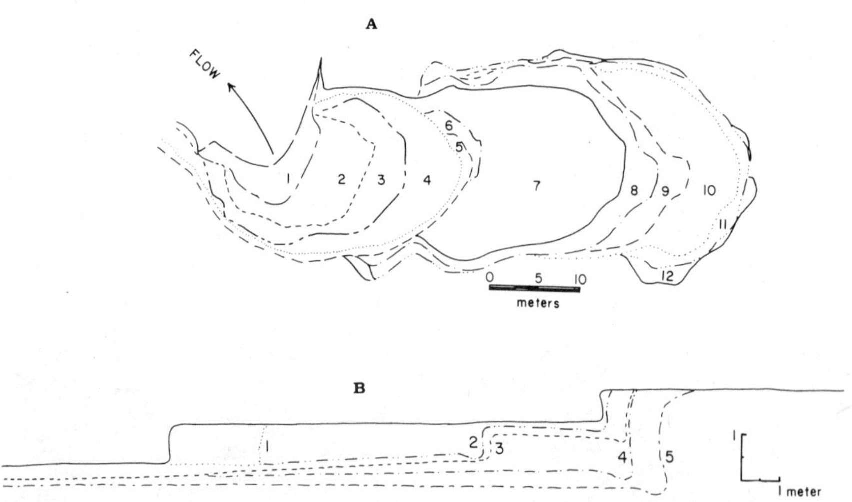

FIGURE 6.3. Gully evolution studies using direct observation. (*A*) Evolution of a gully head near Treynor, Iowa, 1965–1973. Each line represents the position of the gully head after a period of times as follows: (1) 5 months, (2) 2 months (3) 20 months, (4) 3 months, (5) 8 months, (6) 10.5 months, (7) 1 month, (8) 30 months, (9) 11.5 months, (10) 12 months, (11) 5 months, (12) 12 months (from Piest et al., 1975). (*B*) Migration of headcuts in response to storms in a gully in Manitou Experimental Forest, Colorado (from Heede, 1974). Each line represents the position of the headcut after a period of time as follows: (1) 22 months, (2) 1 month, (3) 18 days, (4) 23.3 months, (5) 24 days.

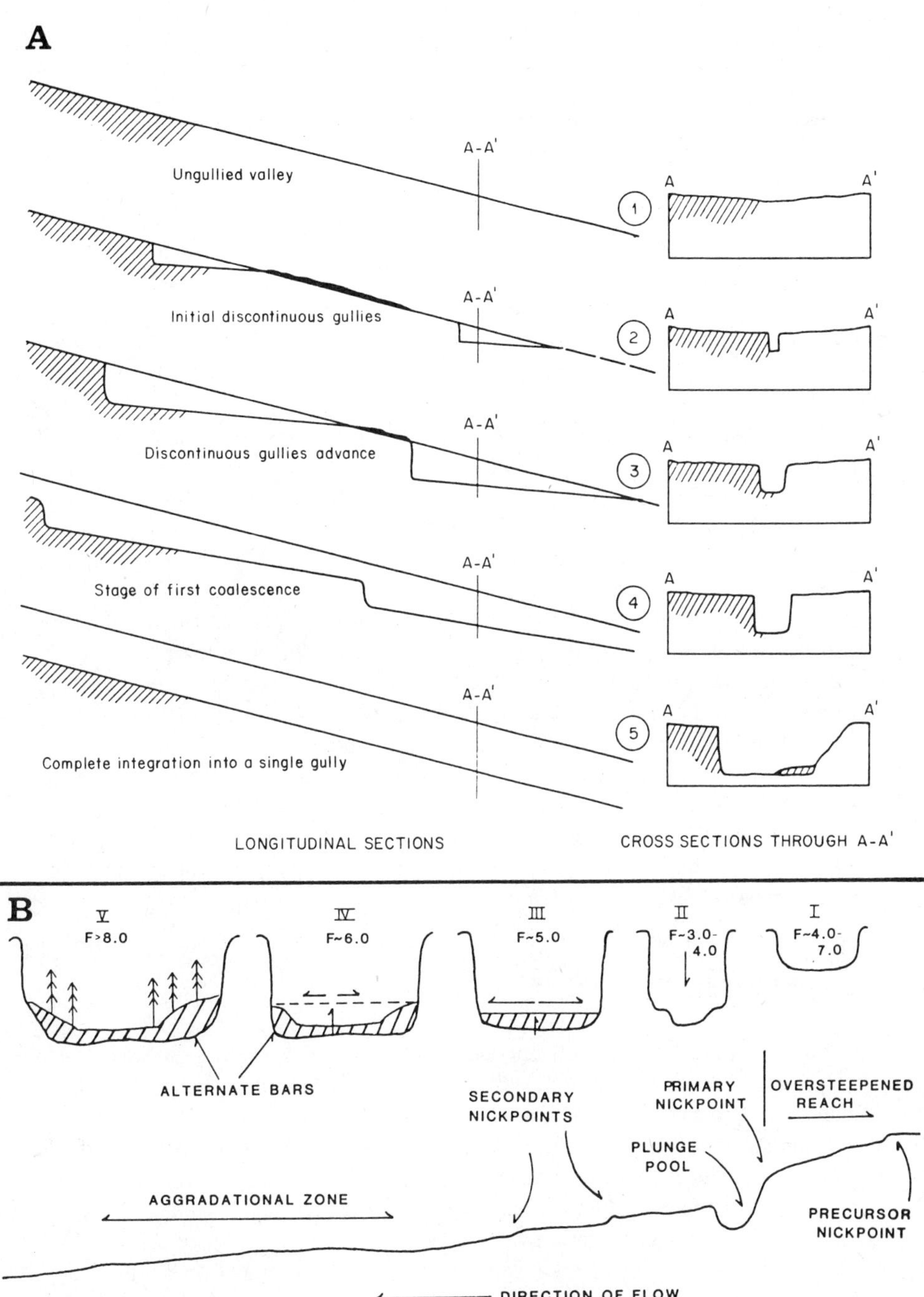

FIGURE 6.4. Stages of incised channel evolution. (*A*) Discontinuous gully evolution (from Leopold and Miller, 1956). (*B*) Schematic longitudinal profile and cross sections of an incised channel (from Schumm et al., 1984). *F* is width–depth ratio and arrows indicate type of channel behavior (incision, deposition, lateral erosion).

forms but are actively evolving. Rates of change are never constant but are normally highly variable or episodic; moreover, different stages of channel evolution may have different characteristic forms (Fig. 6.4). Shape cannot be related to some "dominant discharge" like that of an equilibrium alluvial river, because form is related not only to hydraulic factors but also to factors such as the amount of baselevel lowering. Finally, incised channels (gullies) evolve on a time scale of several years or decades (i.e., the human time scale). The history and manner of initiation are both important controls on present form and behavior and are in themselves a prime area of research interest. However, historical studies must always have a degree of uncertainty about them because of the difficulty of obtaining detailed evidence of past events.

Physical modeling of incised-channel evolution in the laboratory appears to offer several advantages over field studies, and it has been used by a number of investigators to study gully development and evolution (Begin, 1979; Brush and Wolman, 1960; Holland and Pickup, 1970; Leopold et al., 1964; Meyer, 1984; Mizutani, 1985). As with physical modeling of drainage basins (Chapters 2 and 4), the major advantages are that all aspects of the channel environment may be controlled, so that the effect of varying just one factor may be isolated; detailed and continuous observations and measurements can be made, time scales can be accelerated, and simplification is possible by elimination of complications such as storms or human influences. Although the studies referred to above have not explicitly attempted to maintain formal similarity relations with natural incised channels, it seems that a model incised channel that forms in cohesive material in the laboratory is not grossly different in process, mode of evolution, or material from natural channels that have incised into valley fills. Therefore, extension of results from model to natural incised channels presents fewer problems of extrapolation and interpretation than does that for complete drainage networks (Chapter 2).

EXPERIMENTAL STUDIES

An extensive study of incised-channel development by nickpoint retreat was carried out by Begin (1979) and Meyer (1986) in the REF and in a 20-m-long tilting flume (Table 6.2). The sediments used were cohesive, sandy clays (Fig. 6.5), with median grain sizes of 0.09 mm (flume) and 0.25 mm (REF). They were placed and shaped to give surfaces with a longitudinal slope of 0.01 and a shallow V-shaped cross section. In the 20-m flume, clear water was circulated at a constant discharge; whereas in the REF, precipitation was applied to a 15 × 7-m source area, and the drainage network that developed supplied water and sediment to a 15-m-long, 2-m-wide valley in which the experimental channels entrenched (Fig. 6.6). The experimental setup was very similar to that of Harvey (Fig. 4.8).

Each experiment consisted of two or three runs (3A, 4, and 5 in the REF, and 11 and 12 in the flume; Table 6.2). Water was applied to the initial surface without baselevel lowering and a channel was allowed to develop, the course of which was

TABLE 6.2 Hydraulic Data—Mean Values for Runs[a]

Run	REF			Flume	
	3A	4	5	11	12
Length of run (min)	120	720	1800	1320	1020
Discharge (cm^3/sec)	678	733	740	749	749
Average velocity (cm/sec)	31.8	30.9	31.1	24.2	25.3
Slope (cm/cm)	0.0123	0.0142	0.0172	0.0150	0.0170
Froude number	1.22	0.80	0.76	0.71	0.80
Amount of baselevel lowering (cm)	3.5	7.6	7.5	7.2	7.5
Average diffusion coefficient, k_a(cm^2/min)	7000	7000	1800	500	1000

[a]From Begin (1979).

guided by the shallow V of the initial surface. When the channel had ceased evolving, the run was stopped and baselevel was lowered. The channel was then allowed to develop until no further significant change of shape was observed. Baselevel was then lowered again, the final channel in the preceding run providing the initial channel for the new run, and degradation was renewed. Detailed measurements of headcut locations, channel cross sections, and sediment loads were made during all runs.

Nickpoint Retreat

In response to baselevel lowering, a nickpoint migrated upstream (Fig. 6.7), sometimes evolving to a nickzone until it was finally eliminated. The initial planform of the nickpoints in all runs was straight, but they soon attained an arcuate, concave-

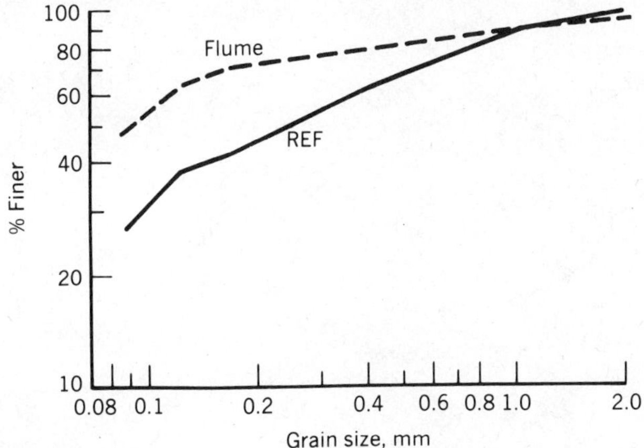

FIGURE 6.5. Grain size distribution of sediments used in incised-channel experiments (from Begin, 1979).

FIGURE 6.6. The experimental channel in the REF during run 4 (from Begin, 1979).

FIGURE 6.7. Headcut formed after lowering of baselevel, run 3A (from Begin, 1979).

downstream form. Their exact shape reflected microtopography and flow patterns immediately upstream; flat valley or channel floors gave rise to straight nickpoints with water discharge evenly distributed across the lip. Nonflat valley or channel floors increased depth and shear stress in the center of the channel, which in turn increased the rate of nickpoint retreat and created an arcuate form. In the extreme case, topographic influence was indicated by the development of multiple nickpoints where the initial channel branched around a valley floor irregularly. Nickpoint profiles varied from vertical headcuts with a basal plunge pool (Fig. 6.8*A*), through weakly developed nickpoints with a markedly steepened reach upstream (Fig. 6.8*B*), to a nickzone consisting simply of a steepened reach (Fig. 6.8*C*). In both Figs. 6.8*A* and 6.8*B* the primary nickpoint (headcut) is preceded by a steeper reach and a small precursor nickpoint. During some runs, secondary nickpoints about 1 cm high also appeared repeatedly, downstream of the main headcut (Fig. 6.2).

In runs 3A and 4, the main headcut soon lost its identity, above 500 cm in run 3A and 750 cm in run 4, and subsequent degradation upstream was by a series of small nickpoints and scour holes. In general, nickpoint migration rate declined with time; this decline was uniform in runs 3A, 4, and 5 and is approximated by a power curve (Fig. 6.9). Of special interest are the secondary nickpoints that developed as a result of sediment storage in the channel. They migrated rapidly

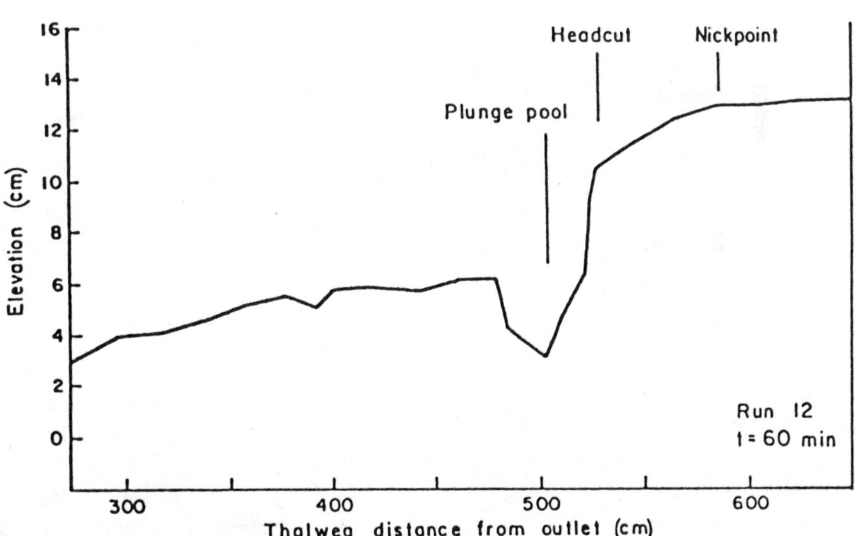

FIGURE 6.8. Nickpoint types. (*A*) Vertical headcut, run 12; (*B*) headcut with marked oversteepened reach, run 5; (*C*) an oversteepened reach, nickzone, downstream of a nickpoint, run 12 (from Begin, 1979).

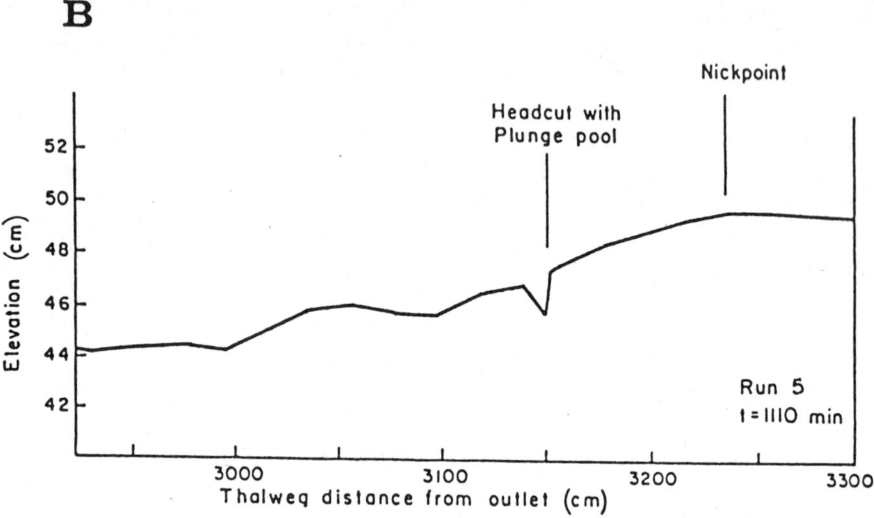

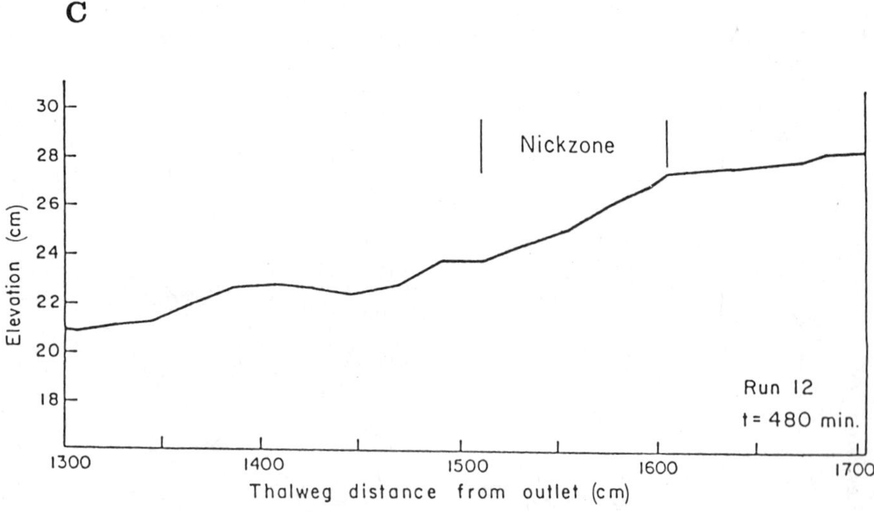

FIGURE 6.8. *(Continued)*

upstream and disappeared after 1–3 m. At about 300 min three secondary nickpoints were active in the channel of run 4 (Fig. 6.9) and many formed during run 5 (Fig. 6.10). They were the result of the development of valley fans that produced convex-up segments of channel in which sediment was stored. Renewed incision produced the secondary nickpoints and eroded the stored sediment (Chapter 4).

Gardner (1983) observed nickpoint development and destruction in a flume that was filled with cohesive sediments (70% sand, 19% silt, 11% kaolinite). Baselevel

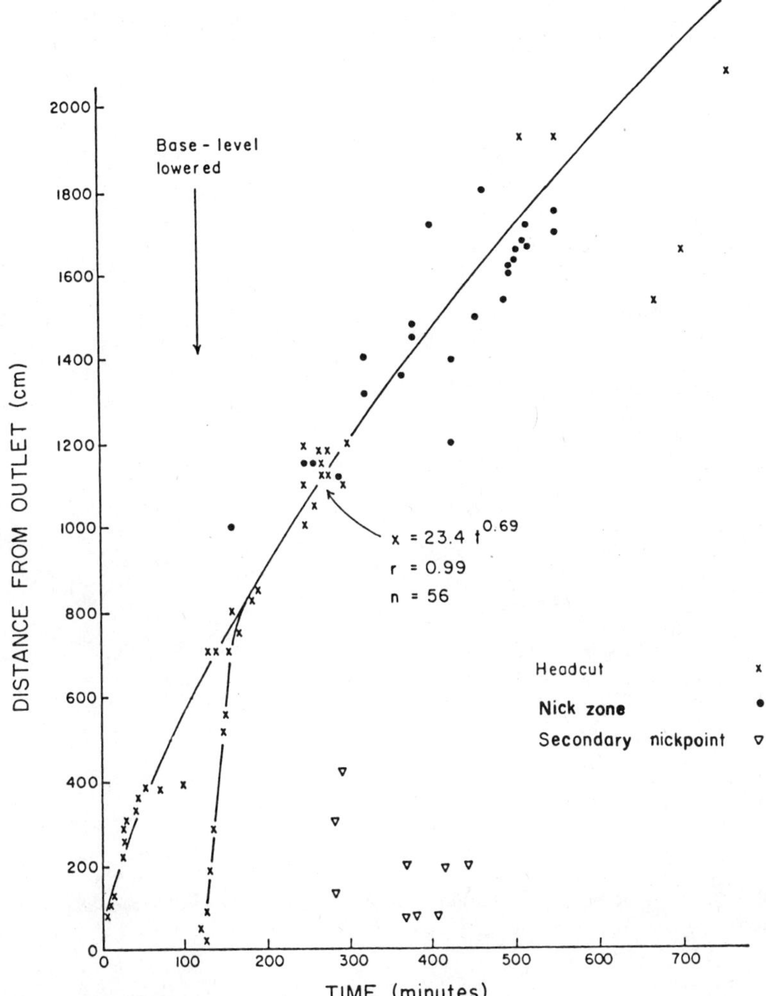

FIGURE 6.9. Rate of nickpoint migration, runs 3A and 4. Note occurrence of secondary nickpoints (from Begin, 1979).

lowering produced a vertical nickpoint, and this abrupt change in the longitudinal profile caused the water surface profile to steepen toward the lip of the nickpoint. This drawdown zone extended about 7 cm upstream from the nickpoint, and average flow velocity in this reach increased from 29 to 60 cm/sec. The high shear forces caused erosion of the lip and face of the nickpoint, which lowered it, and the face of the nickpoint rotated backward. Gardner concluded that a nickpoint can only be preserved and retreat parallel to itself if a resistant layer protects the upper part of the nickpoint.

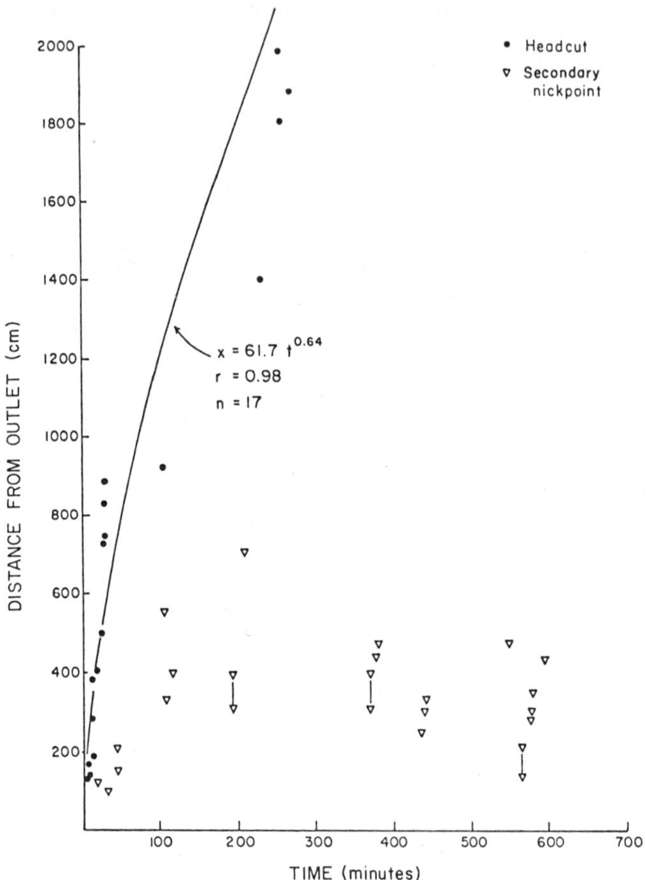

FIGURE 6.10. Rate of nickpoint migration, run 5. Note occurrence of secondary nickpoints (from Begin, 1979).

Evolution of Longitudinal Profiles

In response to baselevel lowering, the experimental channels degraded (Fig. 6.11). To initiate run 3A ($t = 0$), baselevel was lowered by 3.5 cm, after the channel had been developing for 42 hr. This created a vertical nickpoint that migrated upstream 3 m in 30 min. However, at a distance of 6 m from the outlet, the channel was degrading without nickpoint activity. Note on Fig. 6.11 the steeper reach on the initial profile ($t = 0$) above 6 m. The erosion of this steeper reach is analogous to erosion of steep alluvial reaches and valley fans, where discontinuous gullies form (Chapter 4). The primary nickpoint merged with this steeper profile and disappeared. At the end of this run of 120 min, the channel was steeper, but it was as irregular as the initial profile.

The next run was begun after 120 min with an additional 7.6 cm of baselevel lowering. A vertical nickpoint formed and migrated upstream. After 45 min it

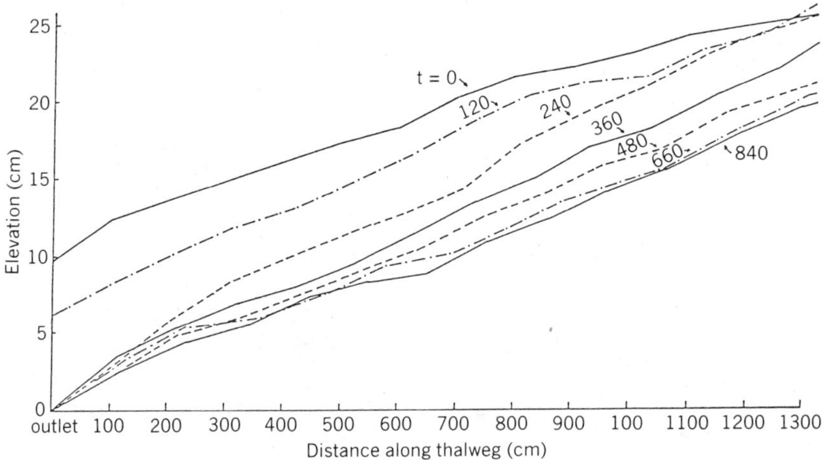

FIGURE 6.11. Profile development during runs 3A and 4. Baselevel was lowered twice, at t = 0 and at t = 120 min. Numbers denote run time in minutes (from Begin, 1979).

reached 7.5 m, where it also lost its identity in the steeper reach. Degradation occurred on the steeper reach, as a series of small nickpoints migrated upstream between 10 and 13 m. At the end of this run (240-min profile) the profile had steepened as a result of baselevel lowering and contained two broad convexities.

The profiles (Fig. 6.11) show downstream steepening as a result of the two lowerings of baselevel, and degradation is greatest at a profile convexity. For example, compare the profiles at 240 and 360 min and the profiles at 660 and 840 min. Aggradation occurred at the downstream end of the 660-min profile, and the four convexities on that profile may also be the result of local sediment storage. During the final 180 min of this experiment, these convexities had not been removed completely, and the final profile at 840 min was still irregular.

Profile irregularities are obviously important in determining channel behavior during incision. The development of secondary nickpoints was a dominant feature of the run (Fig. 6.9). They formed on channel convexities (Figs. 6.11, 6.12) caused by local accumulations of sediment. These small headcuts, about 2.5 cm high, migrated rapidly and disappeared after moving between 1 and 3 m.

The experiment was continued by lowering baselevel again (Fig. 6.12), the initial profile (t = 0) being the same as the 840-min profile in the preceding run (Fig. 6.11). At the end of this series of runs, after 1800 min aggradation was general along the profile and final profiles showed fewer irregularities than earlier profiles. Two convexities in the lower 5 m of the final profile were due to sediment deposition. As in the previous experiment, degradation was variable in space and time and depended to a large extent on the profile irregularities that existed on the previous profile of the sequence (Begin, 1979).

A third series of experiments on the effect of baselevel lowering was carried out in a large flume. The sediments contained 20% more silt and clay than the previous

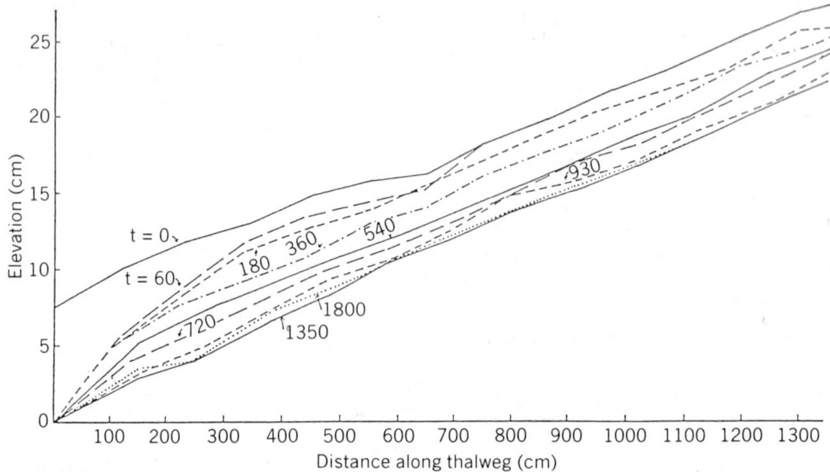

FIGURES 6.12. Profile development during run 5. Time 0 (t_0) is 56.00 hr, which produced final profile of run 4. Numbers denote run time in minutes (from Begin, 1979).

experiments in the REF, and they also contained about 4% more sediment that was coarser than sand size (Fig. 6.5). The increased cohesion resulted in maintenance of a nickzone following baselevel lowering (Fig. 6.13), and the coarser fraction of the sediment formed a channel armor that effectively prevented additional degradation. The result was a steeper final profile.

Begin (1979) used an index of relative degradation, the ratio of degradation (Δy_o) to total baselevel lowering (Y_0), to show how degradation changed with time. As expected, relative degradation decreased upstream (Fig. 6.14), but eventually even the most upstream reach (13 m from baselevel) approached a relative

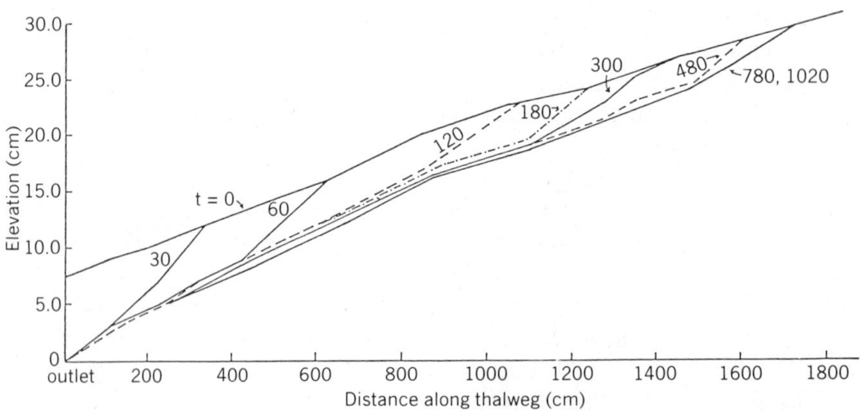

FIGURE 6.13. Profile development, run 11. Numbers denote time in minutes (from Begin, 1979).

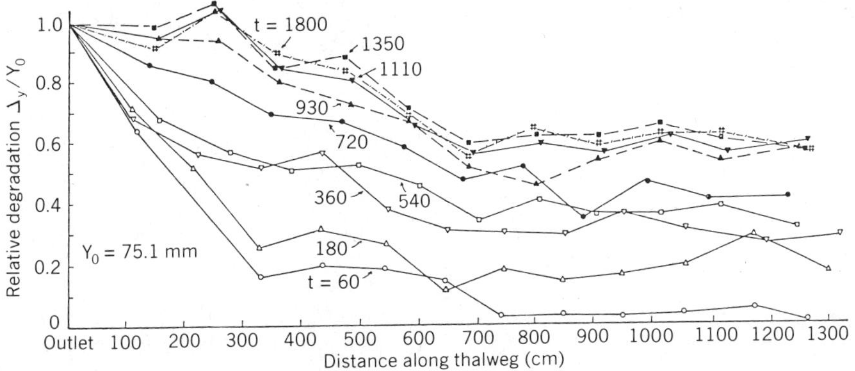

FIGURE 6.14. Change in relative degradation ($\Delta y/y_0$) during run 5 (from Begin, 1979).

degradation value of 0.7. Although little change was observed after 1800 min of run time, additional slow degradation was possible. The effect of baselevel lowering migrated upstream and decreased with time (Fig. 6.14), and the rate of incision declined in an upstream direction (Fig. 6.15).

Cross-Section Evolution

As the long profile of an incised channel evolves to a new condition of stability following baselevel lowering, channel cross sections adjust to the incision. Mass failure of the banks depends on the depth of incision and the strength of bank materials. Almost immediately after the passage of a major nickpoint, the walls of

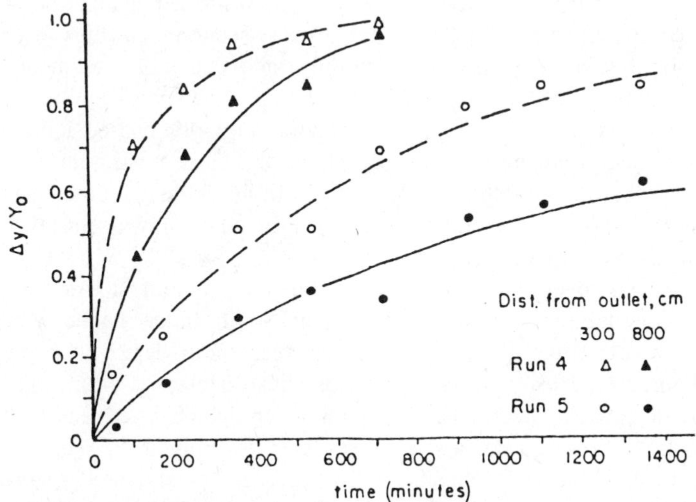

FIGURE 6.15. Relative degradation ($\Delta y/y_0$) at selected locations (distance from outlet), runs 4 and 5 (from Begin, 1979).

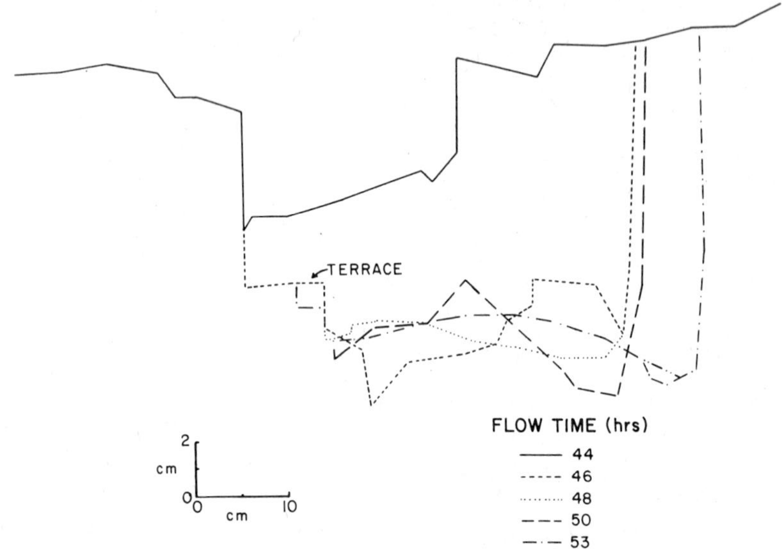

FIGURE 6.16. Cross-section evolution, about 1 m from outlet after baselevel lowering at 44 hr (run 4) (from Meyer, 1986).

an incised channel begin to fail. This is especially the case when flow is directed against the base of the bank, as a result of flow deflection by bar growth. For example, the banks were relatively stable during incision to a depth of 4 cm during the 44 hr of runs preceding run 4 (Fig. 6.16; Table 6.2), but additional incision during the next 2 hr of run 4 (46 hr) led to rapid channel widening.

Meyer (1986) observed a close relation between incised-channel behavior and sediment load, as illustrated by the change in cross-section area (Figs. 6.16, 6.17). Nickpoint migration and bank failure usually produced more sediment than the channel downstream was capable of transporting. The increased sediment load produced by incision, bank failure, and nickpoint migration caused additional widening of the incised channel (Fig. 6.17, 44–46 hr). Once maximum incision was achieved, the thalweg shifted laterally (Fig. 6.16; 46–53 hr). As the banks failed, sediment was deposited as alternate bars, which forced the flow against the opposite bank and triggered further bank failure.

Following baselevel lowering at the start of run 4, sediment load increased as a result of headcut migration and incised-channel width, inner-channel width (width of flowing water) (Fig. 6.18), and inner-channel width–depth ratio (Fig. 6.19) increased abruptly. However, as the sediment load decreased with time, incised-channel width became generally constant and inner-channel width and width–depth ratio decreased. Nevertheless, a second baselevel lowering at the start of run 5 (56 hr) caused bank failure to become increasingly important and incised-channel width, inner-channel width, and inner-channel width–depth ratio increased dramatically (Figs. 6.18, 6.19). As the inner channel became more sinuous during alternate bar

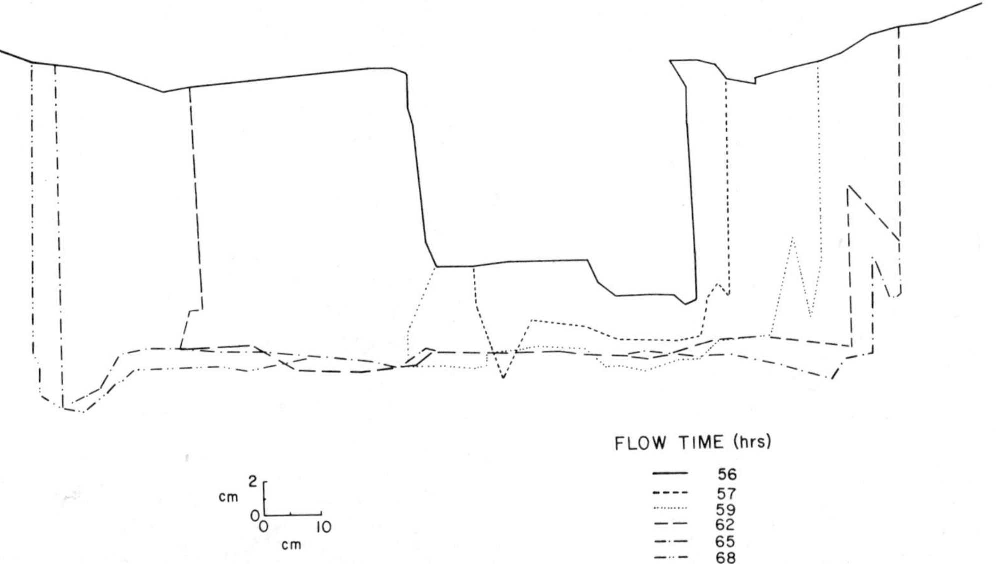

FIGURE 6.17. Evolution of cross section at about 2 m from outlet, run 5 (from Meyer, 1986).

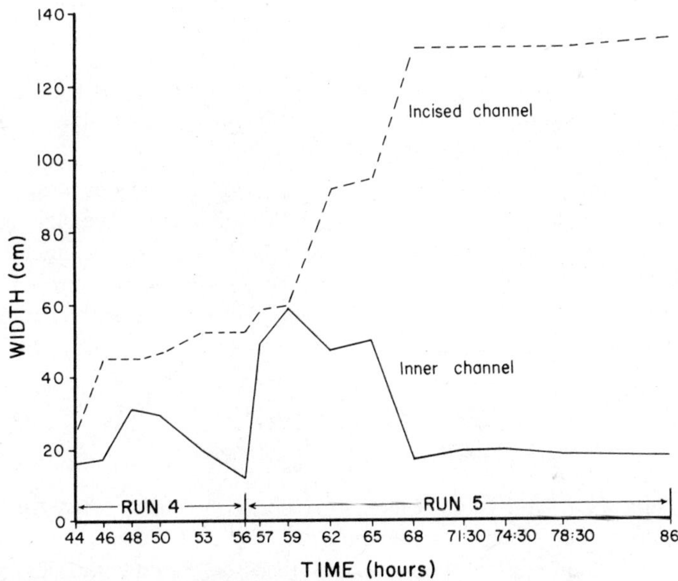

FIGURE 6.18. Changes of inner-channel and incised-channel width about 1 m from outlet during runs 4 and 5 (Table 6.2). Baselevel was lowered before each run (from Meyer, 1986).

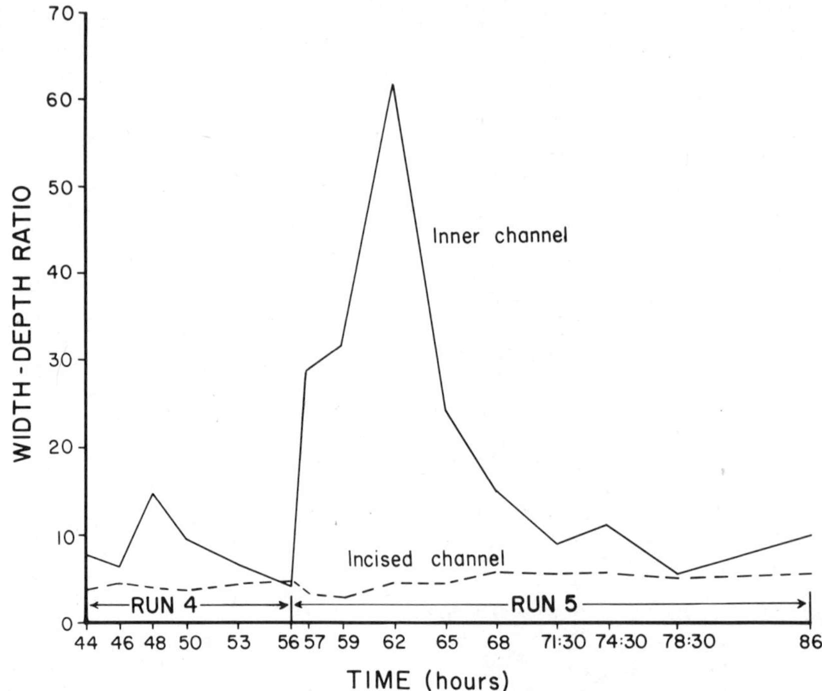

FIGURE 6.19. Changes of width–depth ratio of inner channel and incised channel about 1 m from outlet during runs 4 and 5. Baselevel is lowered before each run (from Meyer, 1986).

construction, bank erosion increased (Fig. 6.17). During this time, the width and width–depth ratio of the inner channel decreased and the incised channel widened (Figs. 6.18, 6.19). Again, when sediment loads decreased, widening ceased (Fig. 6.18) and width–depth ratio stabilized (Fig. 6.19).

These runs demonstrated that incised-channel widening was at least partly dependent on sediment supply. Although nickpoint migration produced large amounts of sediment that overloaded the channel, once the nickpoints became stable, the bars responsible for forcing the flow against the channel walls were eroded and channel widening ceased. In order to study the effect of sediment delivered to the channel independent of nickpoint migration, sediment was added at the head of the flume during run 14, without baselevel lowering. Inner-channel width and incised-channel width increased as a result of the sediment influx.

Sediment Yields

Sediment yield is an expression of nickpoint recession, channel bed degradation, and bank failure. The trend of sediment yield at the flume outlet was therefore closely related to the trends of these erosion processes. Sediment discharge (Q_s in g/sec) during the period that an incised channel was responding to baselevel low-

by bank collapse, and flushing of the channel with formation of secondary nick-points.

Terraces

The development of multiple terraces following incision was attributed in Chapter 4 to the influx of sediment to a reach from rejuvenated tributaries. However, terraces also formed in incised channels that were cutting into both alluvium and simulated bedrock. For example, a well-developed set of terraces formed during Gardner's (1973) experiments (Fig. 6.22) and, in fact, terraces formed in reaches of incision during most of the baselevel-lowering experiments. In each case, rapid incision produced more sediment than could be transported through the downstream reach. It was deposited and then incised.

A different technique was used by Sugitani (1984) to produce incision. After establishing a graded condition in a 5.5-m flume, he tilted the flume but did not increase the sediment fed into the channel. As a result, the channel scoured in the upper part of the flume. The scour started near the inlet and progressed downstream, forming a flight of discontinuous terraces.

When in the field some or all of a flight of terraces cannot be explained by climatic fluctuations or tectonics, they are termed degradation terraces by New Zealand geomorphologists (J. Soons, personal communication). The experiments suggest that they may be formed as a result of natural variation of discharge and sediment load or as a result of overloading of the channel, which results in episodic erosion (Wildman, 1981; Womack and Schumm, 1977). For example, Bull and Knuepfer (1983) identified 12 post-Wisconsin terraces in the Charwell River valley, New Zealand, that they attribute to this mechanism. Ritter (1982) and Brakenridge (1984) and Young and Nanson (1982) have used this explanation for degradation terraces in Alaska, Tennessee, and New South Wales.

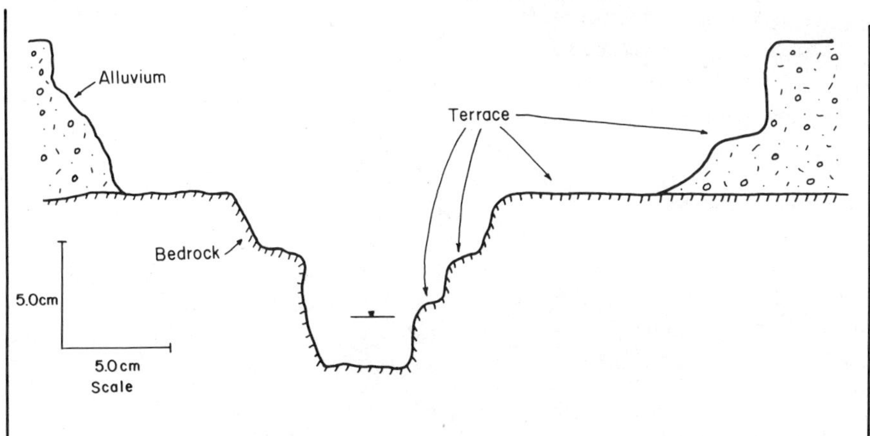

FIGURE 6.22. Development of four terraces in alluvium and simulated bedrock following one baselevel lowering (from Gardner, 1983).

Development of a "Diffusion Model" of Incised-Channel Evolution

Begin's (1979) experimental work was carried out in conjunction with a theoretical analysis of incised-channel development. The experimental data were used to calibrate and test his model and to suggest refinements such as those necessitated by the occurrence of bed armoring and bank collapse. Formal scaling was not carried out, but the mathematical model that was developed fitted both the experimental data and limited data from natural channels.

For the two modes of upstream nickpoint migration observed in the experiments, migration of a vertical headcut and migration of an initially vertical or steep nickpoint that flattens and is finally eliminated, equations of motion may be written that explain the form of the migration curves of Figs. 6.9 and 6.10 and the longitudinal profiles of Figs. 6.11 and 6.12. This development by Begin (1979; Begin et al., 1980a, b; 1981) is based on continuity for the channel sediments, coupled with an assumption concerning sediment motion, which results in an equation that is analogous to the equation of diffusion of heat or chemical substances.

In an alluvial channel the two-dimensional equation of sediment continuity is

$$\frac{\partial y}{\partial t} = \frac{1}{\gamma_s}\frac{\partial q_s}{\partial x} + B \tag{6.2}$$

in which y is bed elevation, x is distance along the channel, t is time, γ_s is bulk weight of sediment, q_s is sediment discharge by weight per unit width, and B is lateral sediment influx.

Many sediment transport equations have the form (Gessler, 1971)

$$Q_s = C_1(\tau - \tau_c)^P, \tag{6.3a}$$

where C_1 and P are empirical constants and τ and τ_c are the bottom shear stress and critical shear stress for sediment motion. In some transport equations, P takes the value $\frac{3}{2}$ (Meyer-Peter and Muller, 1948), so that if $\tau >> \tau_c$, Eq. (6.3a) reduces to

$$Q_s = C_1\tau^{3/2} \tag{6.3b}$$

Mean flow velocity V is given by

$$V = \sqrt{\frac{8}{f}\, gRS_e} \tag{6.4}$$

where f is the Darcy–Weisbach friction factor, g is the acceleration due to gravity, R is hydraulic radius, and S_e is energy slope. Following Gessler (1971), it is noted that for wide channels, $V = Q_w/R$, where Q_w is the water discharge per unit width, and substituting this value in Eq. (6.4) and rearranging leads to

$$R = Q_w^2\left[\frac{f}{8gS_e}\right]^{1/3} \tag{6.5a}$$

Since $\tau = \gamma_w R S_e$ (where γ_w is water density), the substitution of R from Eq. (6.5a) yields

$$\tau = \left[\gamma_w\left(Q_w^2 \frac{f}{8g}\right)^{1/3}\right] S_e^{2/3} = C S_e^{2/3} \qquad (6.5b)$$

and substituting this value of τ in Eq. (6.3b) yields the following expression for the sediment discharge:

$$Q_s = (C_1 C_2^{3/2}) S_e = K S_e \qquad (6.6a)$$

with

$$K = C_1 Q_w \sqrt{\gamma_w^3 f/8g} \qquad (6.6b)$$

That is, the sediment discharge is linearly proportional to the energy slope if Q_w is assumed to be constant. If the energy slope S_e can be approximated by the bed slope:

$$S_e = S_b = \frac{dy}{dx}$$

then substitution in Eq. (6.7a) yields

$$Q_s = K \frac{dy}{dx} \qquad (6.7a)$$

from which

$$\frac{\partial Q_s}{\partial x} = K \frac{\partial^2 y}{\partial x^2} \qquad (6.7b)$$

And substituting Eq. (6.7b) into Eq. (6.2) yields

$$\frac{\partial y}{\partial t} = \frac{K}{\gamma_s} \frac{\partial^2 y}{\partial x^2} + B \qquad (6.8)$$

defining:

$$k = \frac{K}{\gamma_s} \qquad (6.9)$$

Eq. (6.8) finally becomes

$$\frac{\partial y}{\partial t} = k \frac{\partial^2 y}{\partial x^2} + B \qquad (6.10)$$

Equation (6.10) is a version of the heat or diffusion equation, so the k becomes a "diffusion coefficient" (Table 6.2) with dimensions L^2/T (Begin et al., 1981). The equation was developed using different lines of reasoning by other authors, and it has been solved for the elevation $y(x, t)$ for various initial and boundary conditions.

Equations for Profile Degradation in Response to Baselevel Lowering
In Begin's experiment (Begin et al., 1981), the boundary condition, at time $t < 0$, is a channel profile at equilibrium, which is described by the linear equation

$$Y_{(x,0)} = ax + Y_0 \tag{6.11}$$

in which a is initial bed slope and Y_0 is a constant. The initial conditions are set at time $t = 0$, by instantaneously lowering baselevel by an amount Y_0. It is also assumed that

1. the channel is entrenching into a homogeneous, isotropic material,
2. the process can be described by a two-dimensional solution, and
3. sediment is produced by the banks at a rate which changes with distance (e.g, Fig. 6.26*B*).

$$B = B_0 e^{-bx} \tag{6.12}$$

in which b is a constant and B_0 is the volume of sediment produced by the banks at $x = 0$, per unit channel width, per unit length, per unit time. Under these conditions and assumptions, the solution of Eq. (6.10) for the case when bank erosion is significant (Carslaw and Jaeger, 1959) is

$$y_{(x,t)} = ax + \left[Y_0 + \frac{B_0}{kb^2}\right] \operatorname{erf} \frac{x}{2\sqrt{kt}} + \frac{B_0}{kb^2} D \tag{6.13a}$$

in which

$$D = 1 - \exp(-bx) + \frac{1}{2}\exp(kb^2t - bx)\operatorname{erfc}\left[\frac{b}{\sqrt{kt}} - \frac{x}{2\sqrt{kt}}\right]$$
$$- \frac{1}{2}\exp(kb^2t + bx)\operatorname{erfc}\left[\frac{b}{\sqrt{kt}} + \frac{x}{2\sqrt{kt}}\right] \tag{6.13b}$$

and in general: erf(z) is the error function for the argument z and erfc(z) = 1 − erf(z). Values of the error function are available in mathematical tables. If bank erosion can be neglected, $B_0 = 0$, and Eq. (6.13a) reduces to

$$Y_{(x,t)} = ax + Y_o \operatorname{erf} \frac{X}{2\sqrt{kt}} \tag{6.13c}$$

Defining

$$\Delta y = y_{(x,0)} - y_{(x,t)} \tag{6.14}$$

and combining Eqs. (6.11), (6.13b), and (6.14) gives a solution for relative degradation $\Delta y/Y_0$ at different x and t:

$$\frac{\Delta y}{Y_0} = \left(1 - \operatorname{erf}\left[\frac{(x)}{2\sqrt{kt}}\right] - \frac{B_0}{Y_0 kb^2}\left[D - \operatorname{erf}\frac{x}{2\sqrt{kt}}\right]\right. \tag{6.15}$$

The second term on the right is the amount by which relative degradation is reduced due to the introduction of bank sediments.

Equations of Nickpoint Motion

The equation of motion for a nonvertical nickpoint that flattens or rotates as it migrates is derived from the "diffusion" model described above (Begin et al., 1980a). Definition of the "point of disturbance" to be dealt with is slightly more complicated than that of an obvious vertical headcut; one might consider (1) a "disturbance" of specified Δy (great enough to undermine bridge abutments, for example, or the smallest detectable change in elevation or (2) the location of the maximum rate of degradation.

For case 1, Eq. (6.15) may be used to provide a solution for the migration curve. A simple example in which sediment input from bank erosion is neglected will demonstrate the application. Assuming a value of k of 1000 cm^2/min (see below for determination of k), a disturbance of magnitude $\Delta y/Y_0 = 0.1$ will propagate upstream according to

$$0.1 = 1 - \operatorname{erf}(x/2\sqrt{1000t}) \tag{6.16}$$

from which $\operatorname{erf}(x/2\sqrt{1000}\ \sqrt{t}) = 0.9$. Looking up the value of the argument for which the value of the error function is 0.9 (Abramowitz and Stegun, 1964), one finds that

$$x/2\sqrt{1000}\ \sqrt{t} = 1.16$$

$$x = 73.36\sqrt{t} \tag{6.17}$$

The results of sample calculations of x for given values of $\Delta y/Y_0$ following this procedure are given in Fig. 6.23.

For case 2, the migration curve for the locus of maximum degradation rate follows from Eq. (6.13c). If sediment supply from bank caving is negligible, the simpler Eq. (6.13c) may for convenience be used. The rate of degradation is given by

$$-\frac{\partial y}{\partial t} = \frac{Y_0 x}{2\sqrt{\pi k t^3}} \exp\left[\frac{-x^2}{4kt}\right] \tag{6.18}$$

At any time t, $(-\partial y/\partial t)$ is a maximum where

$$\frac{\partial}{\partial x}\left(\frac{-\partial y}{\partial t}\right) = 0$$

Differentiating Eq. (6.18) with respect to x:

$$\frac{\partial}{x}\left(\frac{-\partial y}{\partial t}\right) = \frac{y_0}{2\sqrt{\pi k t^3}} \exp\left[\frac{-x^2}{4kt}\right] - \left[\frac{-2x}{4kt}\right] \exp\left[\frac{-x^2}{4kt}\right]\left[\frac{Y_0 x}{2\sqrt{\pi k t^3}}\right]$$

$$= \frac{Y_0}{2\sqrt{\pi k t^3}} \exp\left[\frac{-x^2}{4kt}\right]\left[1 - \frac{2x^2}{4kt}\right]$$

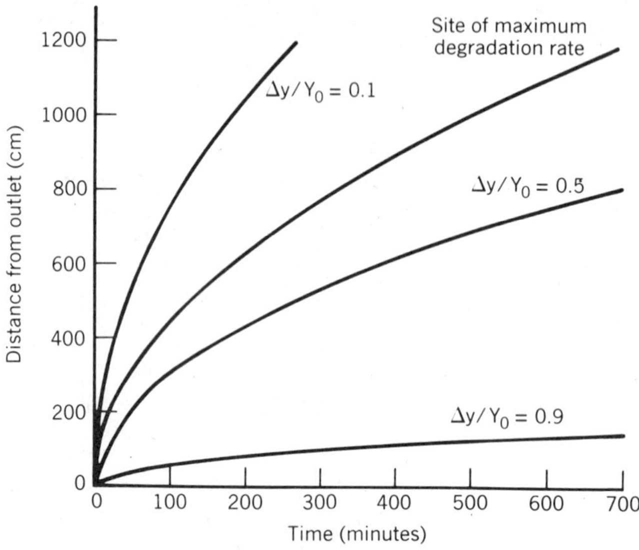

FIGURE 6.23. Migration curves for the propagation of disturbances of different magnitudes $\Delta y/Y_0$. Note the slower rate of propagation of disturbances of greater magnitudes (Begin, 1979).

From this it may be seen that $(\partial/x)(-\partial y/\partial t) = 0$ only when $1 - (2x^2/4kt) = 0$, so that

$$x \text{ (maximum degradation)} = \sqrt{2k}\,\sqrt{t} \qquad (6.19)$$

Equation (6.19) describes the distance x from the outlet at any time t of the locus of maximum degradation; the rate of maximum degradation decreases with time and distance [Eq. (6.18)].

Equations of Headcut Motion

To derive the equation of motion of a vertical headcut (Begin et al., 1980a), we assume a headcut of height h in a bed of width w, a channel slope (S) and sediment bulk density γ_s (Fig. 6.24). Upstream from the headcut, the sediment discharge per unit width is $q_s(\text{u})$ and downstream it is $q_s(\text{d})$. If $q_s(\text{d}) > q_s(\text{u})$, a sediment deficiency exists, which is associated with upstream headcut migration. Assuming that the headcut does not change form or height:

$$q_s(\text{d}) - q_s(\text{u}) = \gamma_s h\left[\frac{dx}{dt}\right] \qquad (6.20)$$

and if $q_s(\text{u})$ is small in comparison with $q_s(\text{d})$ (that is, most erosion in the channel is accomplished by headcut erosion), then

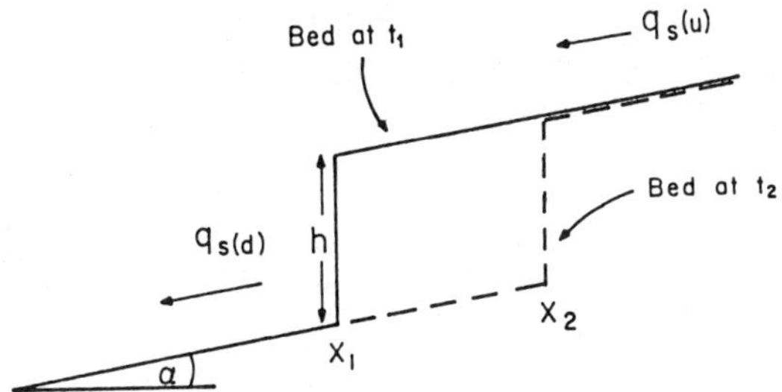

FIGURE 6.24. Definition sketch for equation of motion of a nickpoint (from Begin, 1979).

$$\frac{ds}{dt} \doteq \frac{q_s(\mathrm{d})}{\gamma_s h} \tag{6.21}$$

If q_s(d) and h remain constant, the rate of headcut migration remains constant, as is probably the case for a headcut with a plunge pool from which all sediment is flushed. However, according to the "diffusion" theory of degradation in response to baselevel lowering, the slope of the channel should decrease with time and with distance from the outlet and its changing values may be found by differentiating Eq. (6.12) with respect to distance:

$$\frac{dy}{dx} = \frac{Y_0}{\sqrt{\pi k t}} \exp\left[\frac{-x^2}{4kt}\right] + S \tag{6.22}$$

Theoretical longitudinal profiles of a channel responding to a baselevel lowering Y_0 of 10 cm with values of diffusion coefficient $k = 1000\ \mathrm{cm^2/min}$ and initial slope $a = 0.01$ are shown in Fig. 6.25.

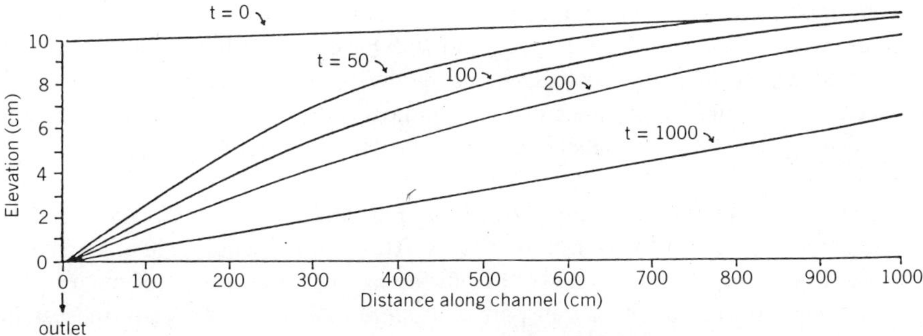

FIGURE 6.25. Theoretical development of longitudinal profiles of an alluvial channel in response to baselevel lowering by 10 cm. Numbers denote time in minutes from baselevel lowering. The diffusion coefficient of the process is $k = 1000\ \mathrm{cm^2/min}$ and the initial bed slope is 0.01 (from Begin, 1979).

Since $q_s = \gamma_s k(dy/dx)$ [Eq. (6.6) and (6.9)], sediment discharge per unit width at time t and distance x may also be calculated:

$$Q_s = \gamma_s k \frac{dy}{dx} = \gamma_s k \left[\frac{Y_0}{\sqrt{\pi k t}} \exp \frac{[-x^2]}{4kt} + S \right] \tag{6.23}$$

Sediment discharge at a point thus decreases with time and distance from the baselevel; substituting Eq. (6.23) into Eq. (6.21) gives the propagation velocity of a headcut (Begin et al., 1980a):

$$\frac{dx}{dt} = \frac{\frac{Y_0}{h} \sqrt{k/\pi}}{\sqrt{t}} \exp\left(\frac{-x^2}{4kt}\right) + bk \tag{6.24}$$

Comparison of the Model with Profile Degradation

If the solution of Eq. (6.10), given by Eq. (6.15), describes degradation in the experiments, then degradation is governed by the coefficients k, B_0, and b, which must be evaluated. Values of B_0 and b under the experimental conditions were readily estimated from measurements of bank erosion (Fig. 6.26). Bank erosion by slumping occurred primarily in the lower reaches, where banks were highest.

Estimation of k is more difficult; as a first approximation, Begin (1979) assumed zero lateral sediment influx. For each measurement point along the channels the value of $(1 - \Delta y/Y_0)$ was calculated. According to Eq. (6.12), this is the value of the error function erf(z) corresponding to the argument $z = x/2\sqrt{kt}$; the argument z may be read from tables of the error function, and k calculated for the values of x and t at which the measurements of Δy were made. A mean value k_a for each run was calculated and adjusted for mean discharge per unit width by

$$k_q = k_a \frac{q_x}{\bar{q}} \tag{6.25}$$

where q_x is discharge per unit width at a measurement point and $\bar{q}$ is mean discharge per unit width. Based on this average k_q and with the estimates of B_0 and b, expected values of $\Delta y/Y_0$ were then calculated using the reverse procedure. The plot of expected versus observed values of $\Delta y/Y_0$ indicates that the model of degradation provided by Eq. (6.15) is satisfactory, despite the simplifying assumptions (Fig. 6.27).

The results for all runs, summarized in Fig. 6.28, show the scatter that is characteristic in sediment transport studies. However, it appears that the basic model [Eq. (6.16)], which is derived from the law of conservation of matter plus the assumption that sediment transport is proportional to bed slope, is successful in describing degradation due to baselevel lowering.

The model predicts that in a homogeneous material not subject to armoring, baselevel lowering ultimately should result in degradation by the same amount along the whole channel, as observed in many natural gullies (Leopold and Miller, 1956; Fig. 6.4*A*). However, this would be the case only if degradation could take

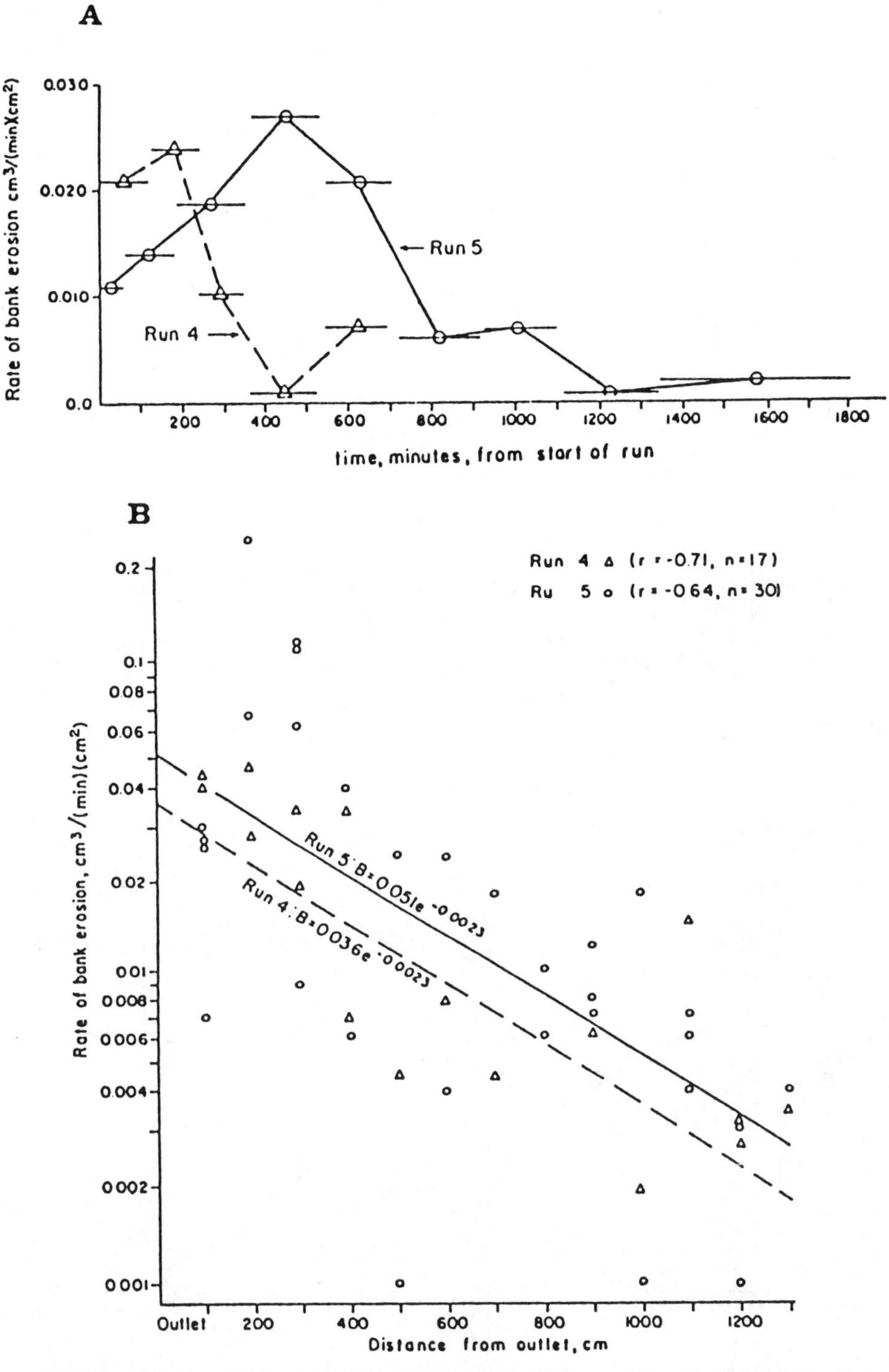

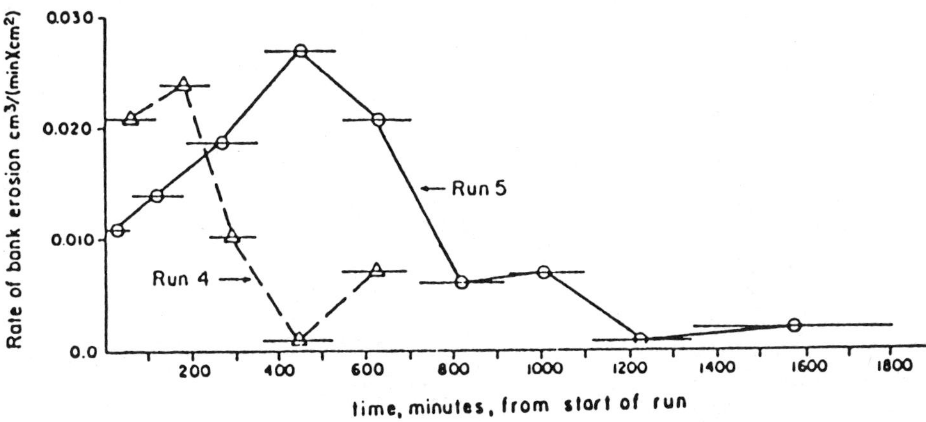

FIGURE 6.26. (*A*) Bank erosion rate as a function of time in runs 4 and 5. (*B*) Bank erosion rate as a function of distance from the outlet in runs 4 and 5, and estimation of the parameters *b* and *B*.

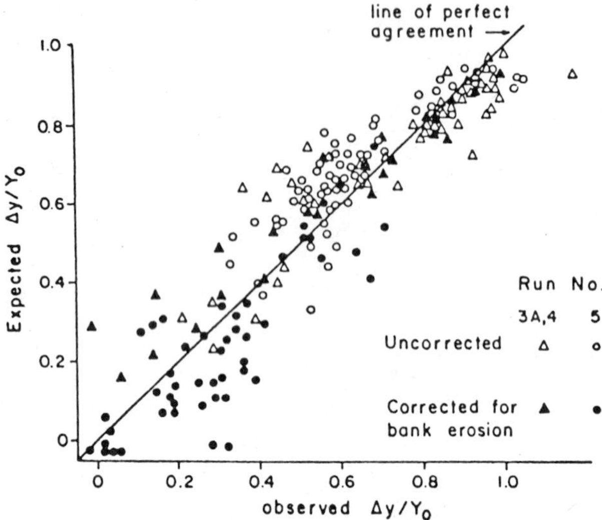

FIGURE 6.27. Observed versus expected values of $\Delta y/Y_0$, runs 4 and 5. For the earlier periods of run 4 (t = 0–240 min) and run 5 (t = 0–540 min) the expected values were obtained from Eq. (6.15), with B_0 and b taken from Fig. 6.26*B*. For the later periods of each run, points are not corrected for bank erosion, assuming $B_0 = 0$ in Eq. (6.15) (from Begin et al., 1981).

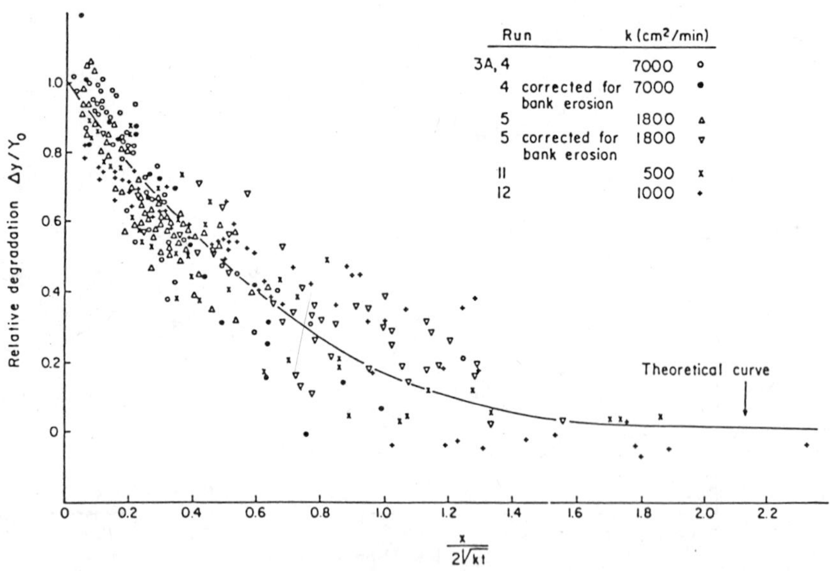

FIGURE 6.28 Plot of all experimental results on a nondimensional degradation diagram that shows good agreement between theory and results (from Begin et al., 1981).

place at the watershed divide. Because this is not possible, degradation decreases to zero in the uppermost reaches of streams. The main impact is near the mouth and soon after baselevel lowering. The rate of degradation at any location reaches a peak and then slowly decreases, and the peak rate of degradation decreases upstream. The degradation near the outlet may lead to bank instability, slumping, and channel widening; temporary storage of this sediment will prevent further bed degradation until it is flushed out. Bank slumping is thus integral to the overall degradation process, since its rate is partially controlled by gully depth (Δy), and in turn it controls degradation rate (via the parameters B_0 and b). It may be dealt with in a separate, linked submodel that deals with the soil mechanical aspects of the degradation process (Bradford et al., 1973; Thorne et al., 1981). This suggests, therefore, that the basic two-dimensional model may be extensible to three dimensions. For example, downstream meander migration may be regarded as causing sediment production at a constant average rate B. An analytical solution for degradation in this case is given by Carslaw and Jaeger (1959, p. 79):

$$\frac{\Delta y}{Y_0} = \left[1 - \operatorname{erf}\frac{x}{2\sqrt{kt}}\right]\left[1 + \frac{Bx^2}{2Y_0k}\right] - Bt \operatorname{erf}\frac{x}{2\sqrt{kt}} - \frac{Bx}{k}\sqrt{\frac{kt}{\pi}}\exp\left(\frac{-x^2}{4kt}\right) \tag{6.26}$$

Comparison of the Model with Nickpoint Migration

The "diffusion" degradation model was also tested by fitting Eq. (6.24) to Begin's (1979) data on headcut migration. The height h was not measured and was therefore approximated by $Y_0/2$; values of k were chosen by trial and error.

With estimated k values of 800 and 1800 cm^2/min for runs 4 and 5, respectively, a good fit of the observations by the theory was obtained (Fig. 6.29*A*). For runs 11 and 12, values of k of 500 and 1000 cm^2/min gave fair agreement for the first part of the migration curves, before the effects of bed armoring became dominant (Fig. 6.29*B*). Since armoring progressively limited degradation and sediment discharge thereafter, headcut migration slowed and eventually stopped in runs 11 and 12 and the model could not be fitted.

Parker (1977) monitored headcut migration in the main channel and three tributaries in the REF drainage basin, although difficulties were experienced because nickpoints were easily washed out, some became multiple nickpoints, and others coalesced. In Experiment 1, the rate of nickpoint migration declined as a power function with time (Fig. 4.2). The parallelism of the retreat curves for the three tributaries in Fig. 4.2 indicates a similar rate of retreat up each tributary, which declined in the headward direction as stream order, basin area, and discharge declined (Fig. 4.4).

The data for headcut retreat are insufficiently detailed to permit analysis using Begin's (1979) model, but extension of the drainage network onto the initial surface, where extension was initiated by baselevel lowering, is amenable to simple analysis. The upstream limit of an expanding channel may be objectively defined in terms of some measurable depth Δy_{tip}, and Eq. (6.15) (with $B_0 = 0$) provides an appropriate model for the distance x_{tip} upstream that a "disturbance" of mag-

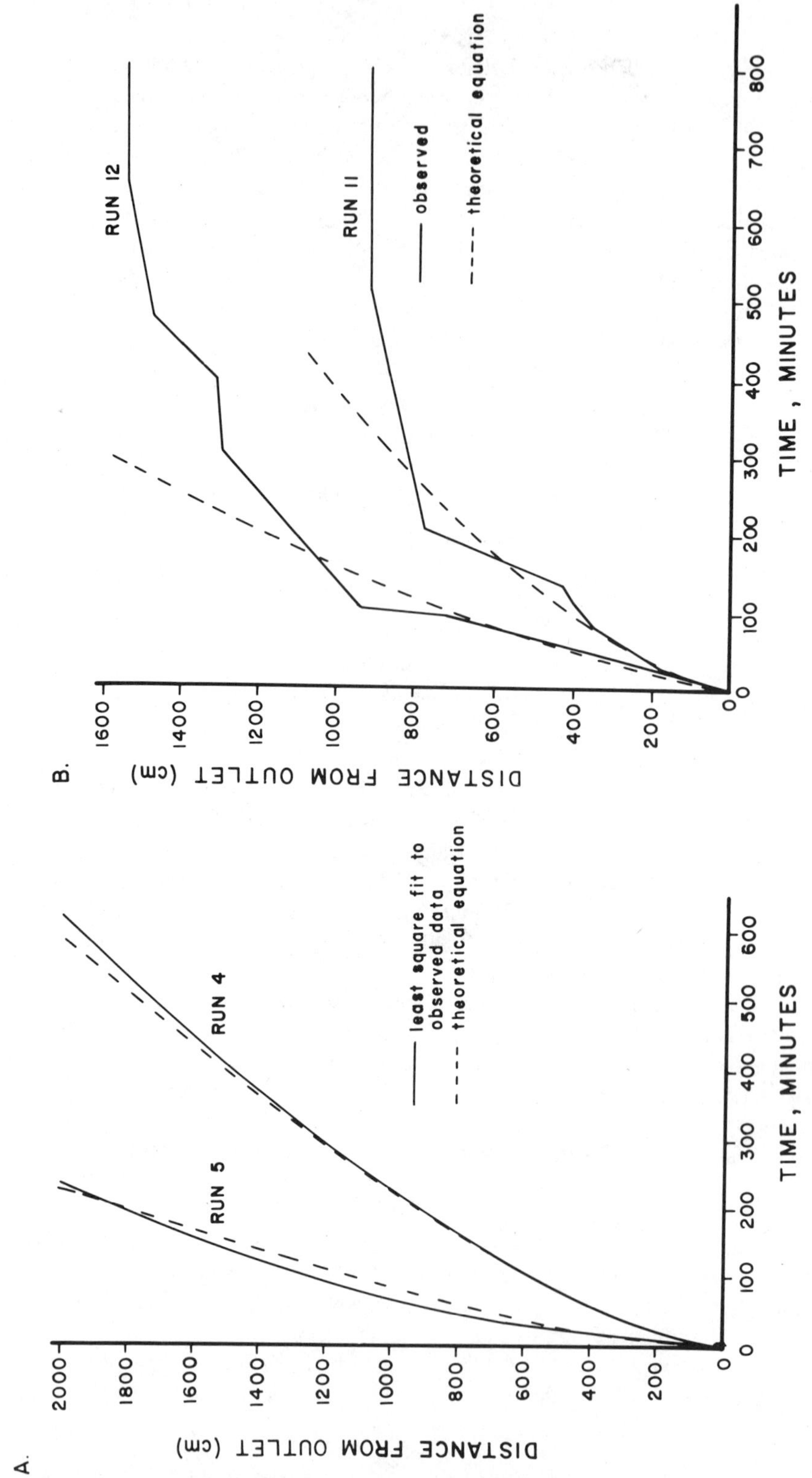

A.

DISTANCE FROM OUTLET (cm)

2000
1800
1600
1400
1200
1000
800
600
400
200
0

RUN 5
RUN 4

—— least square fit to observed data
- - - - theoretical equation

0 100 200 300 400 500 600

TIME, MINUTES

B.

DISTANCE FROM OUTLET (cm)

1600
1400
1200
1000
800
600
400
200
0

RUN 12
RUN 11

—— observed
- - - - theoretical equation

0 100 200 300 400 500 600 700 800

TIME, MINUTES

nitude $\Delta y_{tip}/Y_0$ will travel. Neglecting the effects of bank erosion, which was not significant in the headwater tributaries of the REF drainage networks, the distance x_{tip} traveled in time t is

$$x_{tip} = 2\,C_{tip}\sqrt{kt} \tag{6.27}$$

where C_{tip} is taken from the relationship

$$\operatorname{erf}(C_{tip}) = 1 - \frac{\Delta y_{tip}}{Y_0} \tag{6.28}$$

Parker (1977) attributed the decline with time of channel extension rate in the REF to declining drainage area and discharge as the channel tip approaches the drainage divide (Chapter 2). However, the "diffusion" model also predicts a decline with constant discharge. Parker also noted that the geometric mean length of first-order streams in Experiment 1 (Chapter 2) did not change significantly during network extension, which was initiated by baselevel lowering. He suggested that this was because a first-order channel grows to some limit and then bifurcates, thus being eliminated from the class of first-order channels. However, if S-type first-order channels are considered, geometric mean length tended to increase in Network 1–5 of Parker's Experiment 1 from 30.9 to 46.7 cm. Since its drainage area and discharge is declining as a channel extends, the actual mechanism for bifurcation is not clear. Begin (1979) suggested that, since extension of the channel tip implies deepening, the depth of the channel would eventually exceed some minimum value, Δy_{fork}, necessary to cause incision of a subbasin. If the area of this subbasin (probably delimited by the initial flowline network) exceeds the constant of channel maintenance, then exceeding the critical depth of Δy_{fork} will enable nickpoint retreat and incision to commence.

DISCUSSION

The experimental studies provide a detailed record of channel incision following baselevel lowering and essentially confirm the conclusions reached from field observations of arroyos, channelized streams, and gullies elsewhere (Schumm et al., 1984). The adjustment of a channel to a baselevel lowering in all cases is found to be complex and episodic, with periods of incision followed by sediment storage and renewed incision. This is particularly apparent when the incision affects an entire drainage system as simulated during experiments in the REF (Chapter 2), when increasing amounts of sediment were delivered downstream as tributary channels were rejuvenated by nickpoint migration.

FIGURE 6.29 Theoretical and observed curves of nickpoint migration for (*A*) runs 4 and 5 and (*B*) runs 11 and 12 (from Begin, 1979). Theoretical curves are based on Eq. (6.24) with the following parameters:

Run	4	5	11	12
Y_0 (cm)	9.5	7.5	7.2	7.5
b (cm/cm)	0.015	0.015	0.010	0.015
k (cm^2/min)	800	1800	500	1000
h (cm)	$y_0/2$	$y_0/2$	$y_0/2$	$y_0/2$

The most significant contribution of Meyer's (1986) experiments is the conclusion that incised-channel change is closely related to sediment supply. When incision was shallow and bank erosion small, the channel stabilized quickly. However, when incision was deep and banks were unstable, erosion was promoted by bar construction and meander shift, with the result that stability was long delayed. Once the nickpoint stabilized, widening ceased and point bars were eroded. When sediment was artificially fed into the channel, point bars began to grow again and the channel widening was renewed. Meyer noted field situations where coarse sediment is deposited in alternate bars and arroyo widening is significant—for example, in some reaches of San Pedro, San Simon, and Santa Cruz arroyos in Arizona.

Incised-channel widening was greatest during the experiments when sediment loads were highest, and the major enlargement of the REF channel in the lower part of the drainage system was probably the result of a similar process (Fig. 2.20). This suggests that incised-channel and valley widening can be retarded if sediment contributions can be reduced. A reduction in sediment loads will reduce valley widening, and when the channel incises into the valley floor, underfitness will develop without a change of hydrologic conditions (Paine, 1985). These observations add a new dimension to the controversy over the significance of underfit streams (Dury, 1958).

A related study by Yoxall (1969) supports Meyer's (1986) conclusions. Usually baselevel is lowered abruptly during experimentation, but Yoxall used a continuously falling baselevel to investigate the effect of rate of baselevel change on lateral erosion and found that maximum lateral erosion occurred at a baselevel lowering rate of 1 cm in 9.6 hr. Rates slower than that produced less erosion owing to less tractive force on the gentler slopes; whereas at more rapid rates of baselevel lowering, vertical incision was increasingly dominant. However, at some point continued incision should cause bank or valley-wall failure, and the added sediment should cause lateral erosion and a slowing of the incision rate.

The cross sections were observed to go through phases of degradation, aggradation, and stability in the vertical direction and undercutting, accretion, and stability in the lateral direction. In spite of this apparent complexity, the studies of Barnard (1977), Daniels (1960), and Schumm et al. (1984), as well as the previously discussed gully investigations, reveal a characteristic sequence of incised-channel evolution (Fig. 6.4). The rate and extent of channel incision and adjustment varies from place to place, in response to differences in sediment loads, flow regime, and the character of the alluvium, but comparison of eroding, stable, and stabilizing channels in an area should provide clues that will enable an investigator to identify the optimum situations for instituting channel-stabilization measures.

Despite complexity and irregularity of response to baselevel lowering, Begin's (1979) observations on longitudinal profiles (Fig. 6.11), nickpoint migration (Figs. 6.9, 6.10), and sediment discharge (Fig. 6.20) during channel incision indicate that all show a fundamental, systematic pattern of change that can be described by diffusion-type equations (Carslaw and Jaeger, 1959).

7 Valleys and Channels in Bedrock

The origin of meandering bedrock valleys has intrigued and frustrated geomorphologists for decades. A major difficulty is the inability actually to observe and measure the processes and rates of erosion in bedrock channels or to monitor long-term changes in their position. Even studies of meandering valleys, which are presently evolving in response to active stream erosion, have severe limitations because of the slow rate of operation of the relevant processes (Blank, 1970; Cole, 1930; Lewin and Brindle, 1977; Milne, 1979). Tinkler (1971) suggested the active valley meanders in south-central Texas are evolving in response to flood flows with a recurrence interval of from 10 to 50 years. Therefore, the practical difficulty of studying such features is apparent.

Early debate concentrated on the historical implications of meandering valleys. Davis (1893; Tarr, 1924) considered that a two-cycle origin would account for the incised pattern of the Osage River in the Ozark Plateau, suggesting that the Osage was once freely meandering in alluvium on a peneplain and that uplift and rejuvenation had caused vertical incision and superimposition of the river pattern. He concluded therefore that the modern pattern has been inherited with little or no modification from a pattern established under very different conditions. Winslow (1893), however, argued that the modern pattern was established during incision, by simultaneous vertical and lateral erosion. According to him, the river originally had a relatively straight course, but irregularities were enlarged by lateral cutting on concave banks to produce a winding valley with steep valley sides at the outsides of the bends and more gentle ''slip-off'' slopes on the inside of the bends.

Other early work concentrated more on the environmental factors that would influence the processes of formation of meandering valleys. Rich (1914) considered that the relative importance of three general processes—incision in the vertical direction, lateral cutting into the valley walls, and downstream ''sweep'' of a meander train—would control the type of resulting valley. He recognized three broad valley types: the entrenched-meander valley is produced when vertical incision is dominant; the ingrown-meander valley is produced when lateral erosion is dominant; and the open valley, which has relatively straight, smoothly trimmed valley walls, results when downvalley sweep is dominant (Fig. 7.1). Rich thought that rate of uplift was the most important control on valley type through its influence on the relative efficacy of incision, lateral erosion, and sweep. Moore (1926b) extended Rich's views by pointing out that the relative importance of incision and lateral erosion may also be controlled by lithology and sediment load. He suggested that in the Colorado Plateau the insignificant amount of lateral cutting in the entrenched meanders of the San Juan River appears to be due to "underloading" of the river. On the other hand, ingrown meanders are characteristic of rivers heavily loaded with coarse sediment.

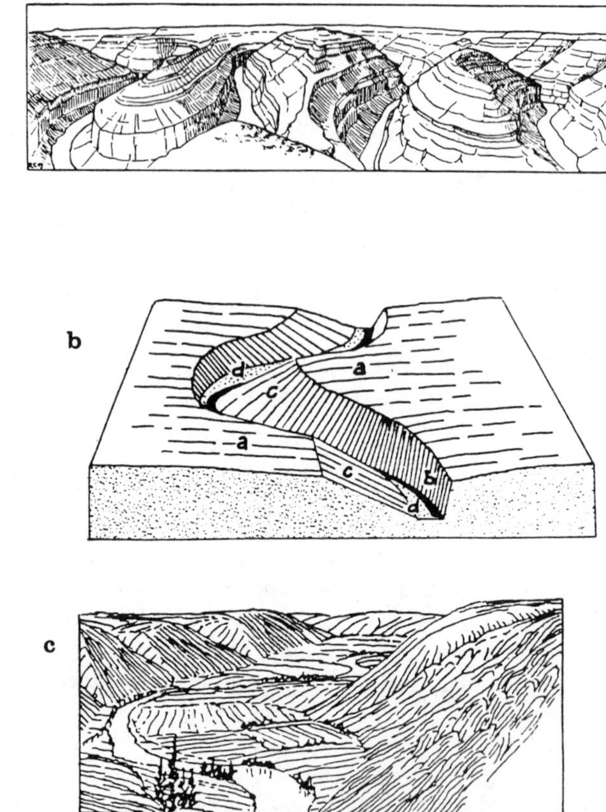

FIGURE 7.1. Examples of entrenched, ingrown, and open valleys. (*a*) Entrenched meanders of the Goosenecks, San Juan River, Colorado Plateau (from Moore, 1926a). (*b*) Idealized sketch of an ingrown meandering valley (from King and Schumm, 1980). (*c*) The open valley of the River Tardes, Central Plateau of France (from Davis, 1912).

One of the main lines of inquiry into the form and origin of valley meanders has used extensive subsurface exploration to determine whether the bedrock floor of a meandering valley, infilled with alluvial deposits, has a form similar to that of the riffles and pools of a meandering stream in alluvium (Dury, 1954, 1970). Such studies have confirmed the existence of a bedrock surface configuration in many valleys that is homologous to that of an alluvial channel bed, with deep "pools" at the outside of valley meander bends and shallower symmetrical "crossings" at the inflection point between bends (Fig. 7.2). Palmquist (1975) found that the probability of the active channel being located at any given point across the valley floor in 105 meandering and 100 straight valley reaches tended to be proportional to the depth to bedrock at that point (Fig. 7.3). Noting also that the depth of scour by an alluvial river is approximately twice the "normal" bank height, he suggested that the asymmetric cross-profile of the bedrock surface in a valley meander might simply be related to the proportion of time that the river is located, and is scouring, at each point on the cross section, so long as the depth of the valley fill is less than twice the bankfull channel depth. To the extent that this hypothesis is true, Dury's (1964b) theory that valley meanders are cut by large ancestral channels that completely filled the valley bottom must be questioned.

Dury (1976) has summarily rejected Palmquist's hypothesis, but without stating his reasons, and it might be noted that some of Dury's own work provides evidence that the modern streams, which he regards as "underfit" and incapable of modifying the valleys in which they flow, are in fact scouring the bedrock surface at

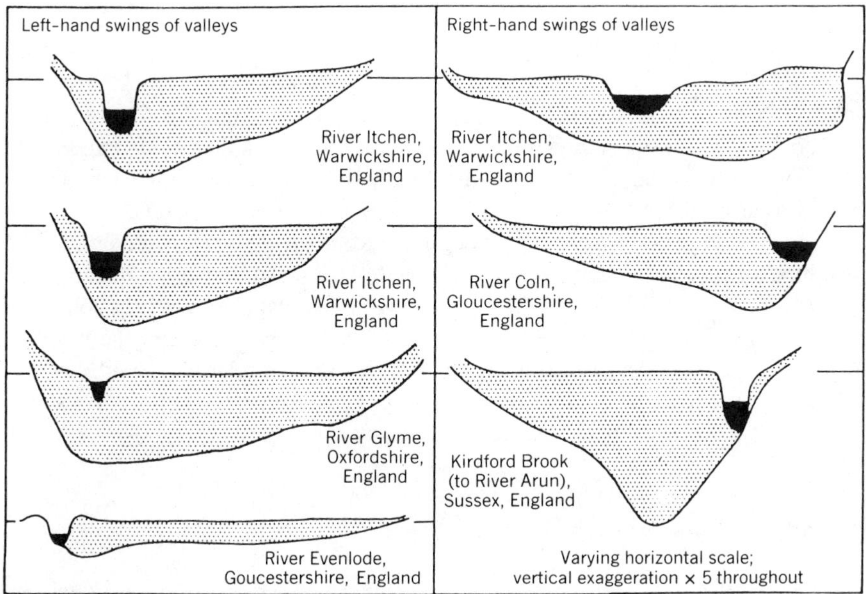

FIGURE 7.2. Asymmetry of the bedrock surface across valley bends of selected English rivers, looking downstream (from Dury, 1964b).

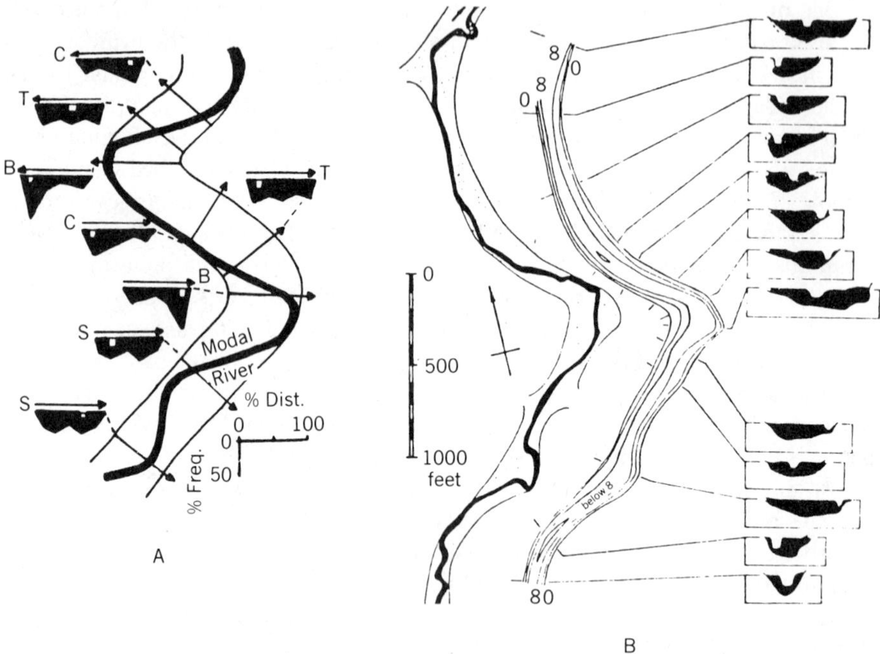

FIGURE 7.3. (*a*) The most common position of a stream within its valley and the resulting transverse profiles as predicted by the preferred position model of Palmquist (1975). (*b*) Transverse bedrock profiles and a winding valley containing an underfit stream, the Itchen River (from Dury, 1954).

present (Dury, 1954). Nevertheless, there is additional information on the form of the bedrock surface in valleys that cannot easily be reconciled with Palmquist's hypothesis. It is commonly found that in dam-site excavations, for example, the cross-profile of the bedrock surface includes a narrow inner channel, or "gut" (Fig. 7.4). Such features have been recognized for many decades; Matthews (1917) described long, linear "deeps" cut into the bedrock floor of the Susquehanna River, revealed during a survey for the McCalls Ferry power plant; Bretz (1924) observed a similar inner channel at The Dalles on the Columbia River, where the river has incised 35 m below sea level 305 km from the sea. Similarly, a field study of Tertiary gravel-filled valleys ancestral to the Yuba River in California revealed inner channels that contained rich gold deposits.

EXPERIMENTAL STUDIES

The form of the bedrock surface and the planimetric form of valleys are thus seen to be of scientific and practical interest, but study is hampered by inaccessibility

FIGURE 7.4. Inner channels. (*a*), (*b*) An inner bedrock channel exposed in the excavation for Prineville Dam, Oregon (photographs courtesy of U.S. Bureau of Reclamation). (*c*) An inner bedrock channel in the Buller River, New Zealand.

beneath valley fills, the infrequency of formative events, and in many cases the fossil nature of the landforms. An experimental approach would thus appear to be potentially fruitful, and a series of laboratory studies of the morphology of streams flowing in simulated bedrock have been carried out by Gardner (1973) and Shepherd (1972b). Specific objectives of these studies were to document changes in channel morphology during the incision of a stream from alluvium into bedrock, investigate scour patterns on the floor of an incised channel, and investigate the factors that control lateral and vertical incision of a stream into bedrock.

Equipment and Procedure

A tilting flume 18.3 m long, 1.22 m wide, and 0.76 m deep was used for both studies. Discharges of between 6 and 7 L/sec were supplied either by direct feed from the water supply or from a recirculating pump, and a vibrating electrical sand feeder delivered the desired sediment load to the channels. A point gauge on a movable carriage mounted on brass rails could measure any point in the flume to 0.03 m horizontally and 0.003 m vertically.

An experimental material had to be developed that would simulate the behavior of bedrock. The work of Allen (1971b) and Partheniades (1965) suggested that some mixture of sand, silt, and clay would be appropriate, and a series of preliminary studies examined the performance of different combinations of these materials. A 14:1 mixture of commercially available "fine sand," which included 19% silt–clay, and kaolinite was finally selected (Fig. 7.5); two 6.1-m^3 batches of sand were each mixed with 522 kg of kaolinite and water in a cement truck, and the mix was poured as a slurry into the flume. The material was allowed to dry by evaporation for two weeks. The final water content varied between 18 and 22%. Two types of sand were fed into the channel to provide sediment load; these sands had median diameters of 0.286 mm and 0.7 mm and graphical standard deviations of 1.59 and 2.22, respectively (Fig. 7.5). In some runs, the bedrock surface was covered by a fine sand with 20% silt–clay and 1% small pebbles to simulate an alluvial cover.

The experimental procedure was generally as follows: (1) shape bedrock surface and cover it with alluvium, if necessary; (2) excavate initial channel; (3) adjust flume slope; (4) initiate flow, adjust discharge, and adjust water level in flume tailbox; (5) introduce sand load at head of flume; (6) record all pertinent data and photograph evolving channel when appropriate; (7) after desired run time, measure

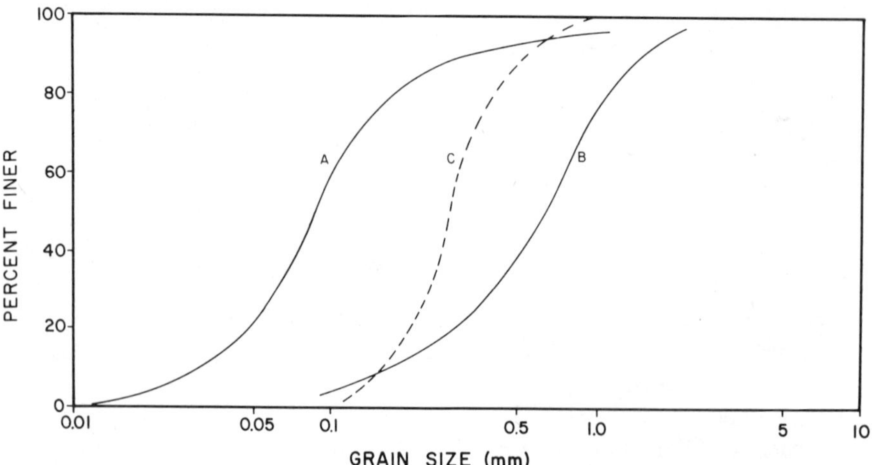

FIGURE 7.5. Grain-size distributions of (*A*) simulated bedrock, (*B*) Bernardo Sand, and (*C*) BF sand (from Shepherd, 1972).

transverse and longitudinal water profiles, terminate flow and sediment feed, record all morphometric data, and make a photographic record. Shepherd's (1972a, b; Shepherd and Schumm, 1974) study involved five experiments and Gardner's (1973), two (Table 7.1).

A certain degree of trial and error was necessary because no previous studies of this type had been attempted. Thus, for example, in Gardner's first experiment, the initial flume slope and discharge were selected on the basis of earlier work, but several adjustments of slope were necessary before a slope was finally identified that allowed the desired meander pattern to form. Similarly, several methods of inducing incision—lowering baselevel, increasing slope, increasing discharge, and introducing sediment load—were tested during Shepherd's Experiments 1 and 2 before a satisfactory result was obtained. The failures were often as instructive as the successes, because they suggested new analogies with nature and new lines of inquiry that could be pursued in succeeding experiments.

Straight-Channel Incision

Cross Section

The first four experiments performed by Shepherd (1972b) involved scour in a straight "bedrock" channel. Although baselevel was lowered and slope was increased twice to 0.018, only minor scour occurred during the first 200 hr of Experiments 1 and 2 (Table 7.1). However, with the introduction of a sediment load of 25 g/min after 214 hr, bed erosion increased significantly. Erosional bedforms included potholes, erosional ripples, and longitudinal grooves similar to those observed by Allen (1971a), and reminiscent of the features described in the Columbia and Susquehanna rivers by Bretz (1924) and Matthews (1917). The sequence of erosional events in the straight channel is illustrated in Figs. 7.6 and 7.7. After a few hours, faint longitudinal lineations, potholes, and transverse erosional ripples had appeared (Fig. 7.7*A*). The lineations progressively enlarged into prominent longitudinal grooves, which became the dominant erosional form, although in some places potholes were at first more prominent (Fig. 7.6*b*, *c*, and *d*; 7.7*B*). The grooves then coalesced and their number decreased (Figs. 7.6*e*, *f*, *g*; 7.7C, D, E). Finally, a single narrow, deep inner channel conveyed the entire flow (Figs. 7.6*h*, 7.7*f*). These features developed at Froude numbers less than 1.0 (mean for 291 hr of run was 0.77; Shepherd, 1972b). Similar features formed during all of Shepherd's experiments.

The development of the erosional lineations, grooves, and inner channel may have been in response to the establishment of secondary circulation cells or longitudinal vortices in the channel, with the erosional ripples and potholes being associated with rollers and flow separation (Allen, 1971a; Leliavsky, 1955). The linear features varied in shape because of local variations in turbulence and shear stress, and they grew in size because of the lack of sediment deposition where flow had an upward component. However, the significant increase in rate of evolution of the bed forms after sediment was introduced at 214 hr indicates the great importance of corrasion by a solid load in the flowing water, in comparison to purely hydraulic stresses.

Table 7.1 Summary of Incised-Channel Experiments by Shepherd (1–5) and Gardner (6, 7).

Experiment Number	Total Run Time (hr)	Discharge Range (L/sec)	Slope	Sediment Supply Rate (g/min)	Initial Channel			Comments
					Pattern	Width (m)	Material	
1	82.25	1.44–3.68	0.00485	0	Straight, trapezoidal	0.16	Bedrock	Baselevel lowered to initiate incision.
2	291.25	3.94–5.69	0.00484–0.0175	0–90	Straight, trapezoidal	0.27	Bedrock	Slope increased after 56 and 109 hr to initiate incision. Zero sediment supply to 214 hr, then progressively increased to 90 g/min.
3	97.0	2.83	0.0173–0.01	180–60	Straight, rectangular	0.40	Bedrock	Slope decreased after 65 hr. Sediment supply progressively reduced to 60 g/min by 42 hr. Initial channel had 0.22-m layer of sand on bottom.
4	111.5	1.42–7.09	0.0016–0.012	20–60	Straight, rectangular	0.72	Alluvium	Straight entrance to 72.5 hr; angled entrance thereafter.

5	107.0	2.83	0.009–0.0167	25–50	Sinuous, semielliptical	0.15	Bedrock	
6	1588.0	0.6	0.008–0.001	10 g/min, for 1375–1444 hr only	Straight, semicircular	0.09	Alluvium	Initial bend at head of flume. Planar bedrock surface beneath alluvial cover. Baselevel lowered by 0.02 m at 197 hr, and by 0.05 m to bedrock contact at 321 hr. Baselevel lowered 0.05 m beneath bedrock at 364 hr. Flood discharge introduced between 868 and 916 hr.
7	642.0	0.6	0.005	0	Straight, semicircular	0.09	Alluvium	Initial bend at head of flume. Anticlinal bedrock surface beneath alluvial cover. Baselevel lowered 0.08 m at 210 hr; 0.09 m to bedrock surface at 306 hr; 0.08 m below bedrock surface at 443 hr.

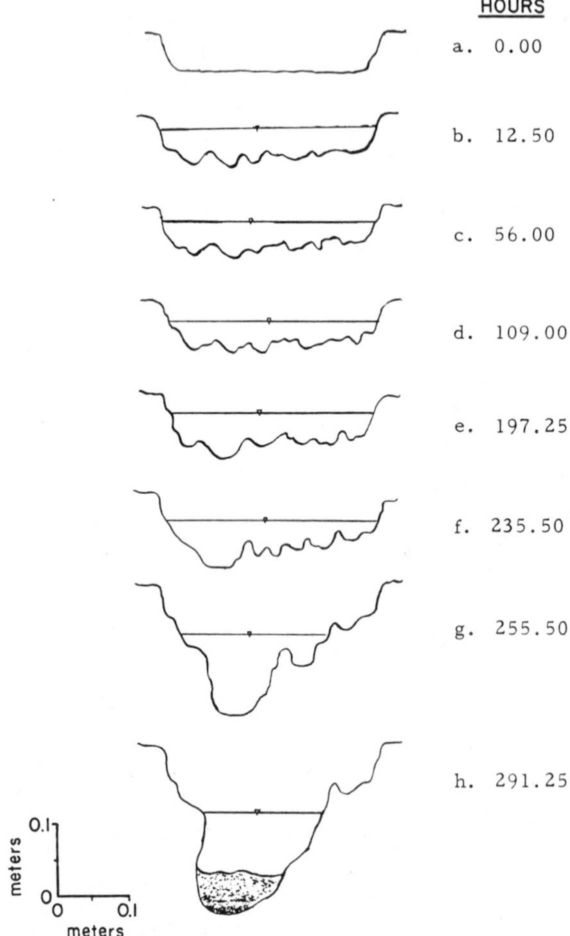

FIGURE 7.6. Series of transverse profiles measured in Experiment 2, 10.4 m upstream from baselevel (from Shepherd, 1972b).

Figure 7.8 shows a series of cross sections that developed during a 97-hr run of Shepherd's Experiment 3. The narrow inner channel widens in a downstream direction, and becomes shallower, as baselevel is approached. Shepherd's analysis of the data from this run shows that top width, area, wetted perimeter, and width–depth ratio increase in a downstream direction. Velocity and Froude number remained relatively constant and slope, stream power, and shear stress decreased in a downstream direction. Note the irregularity of these cross sections. Deposition of sediment tended to cause undercutting of the "bedrock" walls (9.14 m, 7.01m) in a manner similar to the widening of incised channels as sediment loads increased (Figs. 6.17, 6.18).

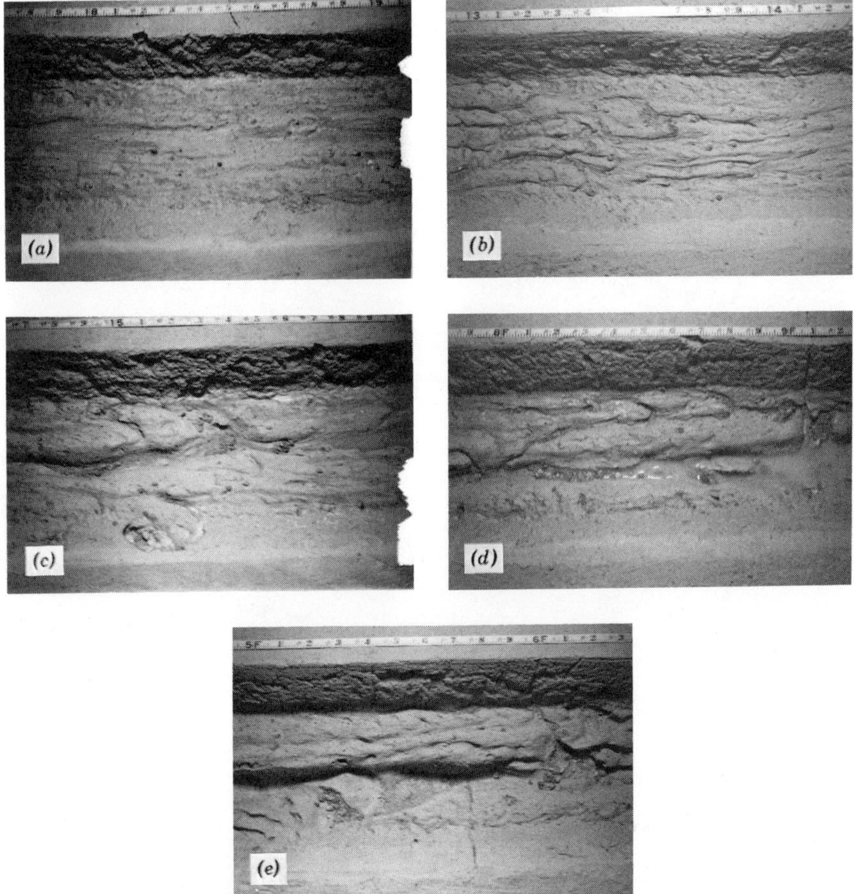

FIGURE 7.7. Scour features observed in Experiment 2. Scale is in tenths of feet and flow was from right to left (from Shepherd, 1972b). (*a*) Early stages of development of transverse ripples, potholes, and grooves. (*b*) Longitudinal grooves becoming dominant; potholes also present. (*c*) Grooves enlarging and number of grooves decreasing. (*d*) A large groove becoming dominant. (*e*) Final development of an inner channel. The reduced gradient now permits sand deposition in the inner channel.

Longitudinal Profiles

During incision, the "bedrock" was eroded below baselevel (Fig. 7.9) and the profile of the bedrock inner channel was not smoothly sloping in a downstream direction. The deep erosion was not simply a result of local turbulence resulting from scour at the baselevel. When the sand bed was removed by the flow, "bedrock" incision began. An inner channel eventually conveyed all of the flow. At 64.75 hr (Fig. 7.10) no sediment had been deposited in the upper 10 m of the

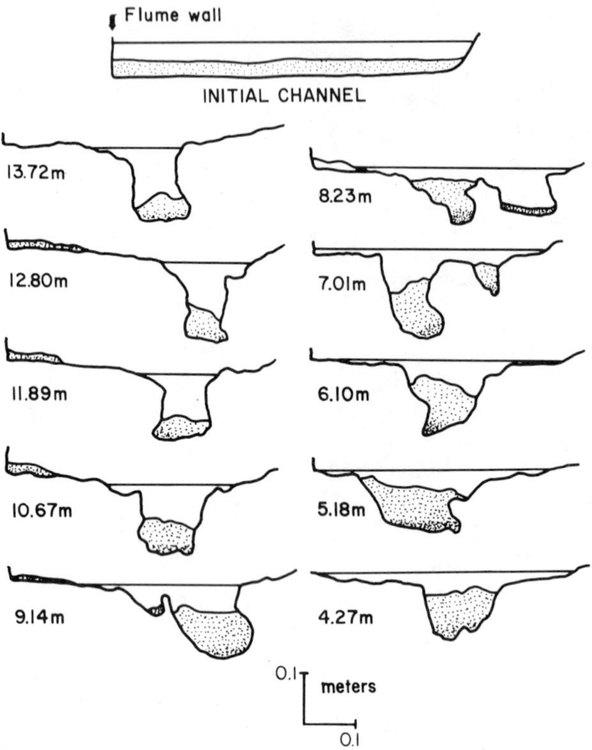

FIGURE 7.8. Transverse profiles after 97 hr in Experiment 3 following aggradation, at specified distances upstream from baselevel (from Shepherd, 1972b).

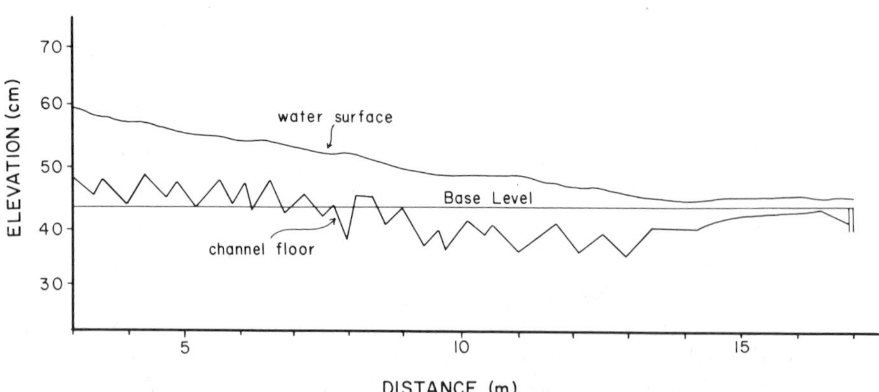

FIGURE 7.9. Water and bed surface profiles of channel 3 at 64.75 hr of elapsed run time. Profiles drawn by computer using straight-line point connections (from Shepherd, 1972b).

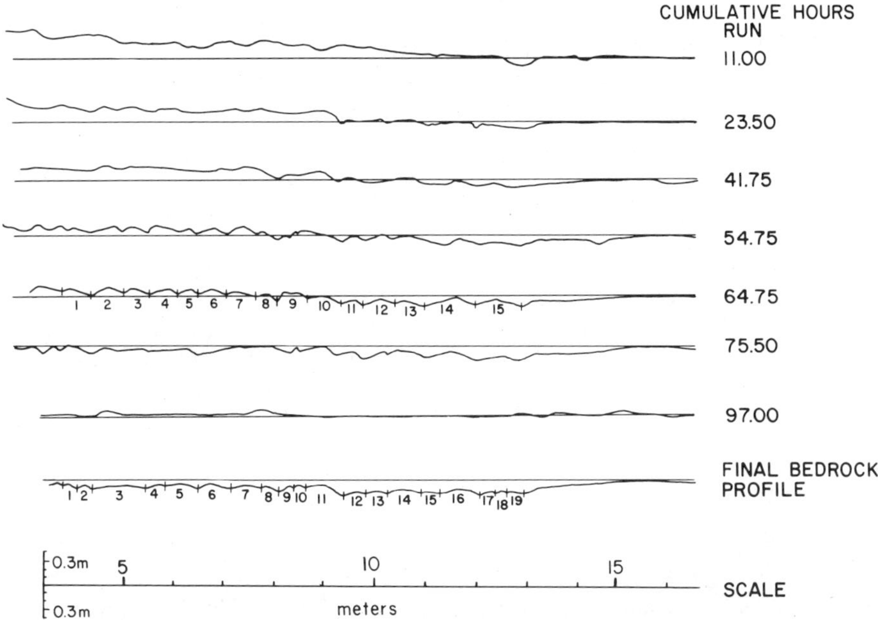

FIGURE 7.10. Longitudinal profiles along the streambed of channel 3 (from Shepherd, 1972b).

channel, but the backwater effect of the baselevel caused deposition in the lower 2 m and the presence of sand in the channel prevented bedrock scour. Shepherd noted that the channel was markedly narrower at the points of maximum scour and wider at the crests between scour holes, so that flow cross-sectional area and water surface slope were conserved.

After 65 hr a reduction of flume slope permitted deposition and burial of the inner channel (Table 7.1; Fig. 7.10). Statistical tests of the regularity of spacing of the scour holes (Shepherd, 1972b) were applied to the actively incising bedrock profile at 64.75 hr in Experiment 3 and the final, inactive profile that was cleared of its alluvial fill (Fig. 7.10). The tests indicated that the actively incising channel had a weakly regular spacing of scour lows, but that in the final channel (Fig. 7.10) the scour holes were randomly spaced. It is inferred that the weakly regular pattern of active scour holes, generated in response to the hydraulic processes that seem to be responsible for the development of regularly spaced alternate bars (Chapter 5), was altered by local scour as the channel aggraded, and a lesser regularity of the final bedrock profile resulted. Keller and Melhorn (1978) report a regularity of bedrock pools, with a good relation between bedrock pool spacing and channel width in natural bedrock channels.

During Shepherd's experiments, bedrock features developed that are analogous to those observed in natural rivers (Bretz, 1924; Matthews, 1917) and glacial spillways (Kehew, 1982). Bretz (1924, p. 146) suggested that requirements for the "Dalles-type" channel are (1) large discharge and (2) high gradient of the stream, and (3) close and vertical jointing of the rock. The experimental studies did not demonstrate any marked response of the model streambed to variations in discharge and slope, and the simulated bedrock was isotropic. However, the need for the presence of a sediment load that provided the flowing water with a ''tool'' with which to corrade its bed was indicated. Bretz's (1924, p. 146) description of the expected evolution of the Dalles-type channel is remarkably consistent with the experimental observations:

> All Dalles-type channels are early stages in the development of a single channel. Trenching of a central main channel eventually results and with this trenching the high water channels are gradually abandoned. By this trenching the gradient is lowered, plucking decreases, and the holes in the rock floor disappear by simple abrasion of the higher parts and perhaps also by filling of the lower. The hole 115 feet below sea level at the head of Five Mile Rapids eventually will be filled in large part.

Incised Meanders

Both Shepherd and Gardner attempted to develop incised meanders in the simulated bedrock. Initially a layer of sand was placed over the bedrock surface so that alluvial meanders would develop as in Khan's experiments (Fig. 5.8). These meanders would then be incised by an increase of flume slope or by a lowering of baselevel. However, as the relation between sinuosity and valley slope indicates (Fig. 5.33*D*), an increase of slope can destroy alluvial meanders and a braided channel will form. This was the result of Shepherd's (1972) attempt to produce entrenched meanders, which suggests that incision into bedrock of an originally meandering alluvial stream is an unlikely result of regional slope increase. Lateral erosion and destruction of the initial meander pattern should be the result, and the hypothesis of superimposition of symmetrical meanders from an uplifted peneplain or alluvial plain into bedrock is difficult to accept.

To pursue the question of meander incision into bedrock, Shepherd excavated a channel in bedrock with a semielliptical shape and sinuous pattern (Fig. 7.11). At each bend, erosion was initially greatest at the inside of the bend, but when a sediment load was introduced, the location of greatest erosion shifted to the outside of the bend and the concave bank was undercut as deposition occurred on the inside of the bends and the flow moved to the outside (Fig. 7.12). However, at the crossings between bends the channel incised almost vertically.

This behavior is explained by the experimental studies of Hooke (1975), Ippen et al. (1962), and Yen (1970). In their experiments bed shear stress was found to be at maximum on the inside of a bend, which accounts for the scour on the inside of the bend before significant sediment was transported through the channel. Sediment transport per unit width of channel is also greatest on the inside of a bend (Hooke, 1975). This phenomenon is familiar to engineers, who avoid siting an irrigation offtake on the inside of a bend because it will also extract a greater sediment load than an offtake sited elsewhere. However, with high sediment trans-

FIGURE 7.11. Scour on inside of bend during incision of sinuous channel.

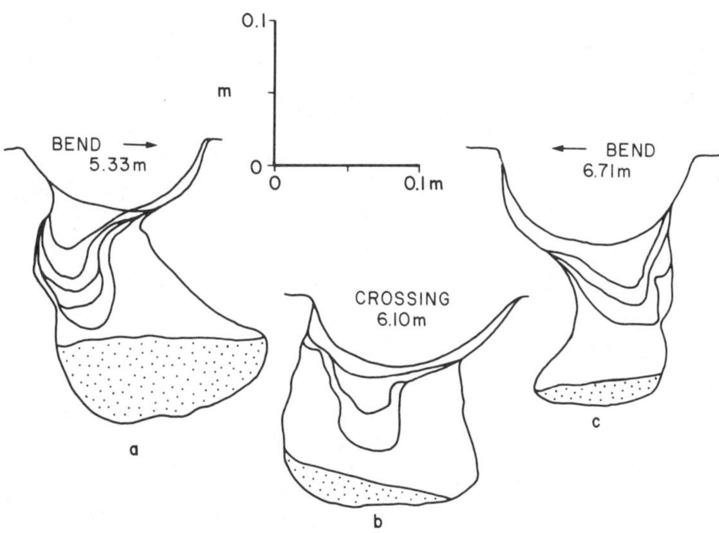

FIGURE 7.12. Series of cross-profiles through two bends and the intervening crossing as incision occurred during Experiment 5 (Table 7.1). Distances upstream from baselevel are given. Arrows point to outside of bends (from Shepherd, 1972b).

port and deposition of a point bar, maximum scour is shifted to the outside of the bend (Fig. 7.12).

These experimental studies lend support to Moore's (1926a) suggestion that

> the effectiveness of sideward cutting in proportion to downward cutting seems to be controlled by the relative loading of the stream, especially with coarse material, and the strength of the rocks. . . . With added load there is less tendency to cut downward and more to corrode sideward and forward.

It appears that the nature of incising meanders will therefore depend at least partially on the amount of load supplied to and entrained by the incising stream.

Gardner (1973, 1975) continued the meandering-channel experiments (Table 7.1). He recognized that incision of meanders would not occur by steepening of slope, so, during Experiments 6 and 7 he caused channel scour by lowering baselevel. He fashioned a planar "bedrock" surface, which was covered to a depth of 0.08 m with sand, and a straight channel with an initial bend was excavated.

After water was introduced into the channel, a sinuous thalweg progressively developed in a downstream direction as a result of bank erosion and point bar deposition. During the first 250 hr of run, kaolinite was added to the water; this was deposited on and stabilized the point bars, enhancing sinuosity (Fig. 7.13; see Chapter 5). It was necessary to make small adjustments to flume slope during the first several hundred hours, over the range 0.005–0.008, to permit the development of a meandering thalweg. At 0.005, minor bank erosion permitted only slow meander development, whereas at 0.008 meanders tended to form, cut off, and braid. The final slope of 0.007 gave the desired pattern.

A stable meandering-thalweg pattern, with a sinuosity, amplitude, and wavelength of about 1.25, 0.5 m, and 1.7 m, respectively, was attained by 197 hr, when baselevel was lowered to induce changes in the alluvial meandering channel (Fig. 7.14*a*). This baselevel lowering resulted in (1) vertical corrasion, (2) increased channel gradient and increased transport capacity, and (3) decreased width–depth ratio. In addition, as channel gradient steepened in response to baselevel lowering and progressive downstream displacement of the meander bends, downvalley sweep became important.

When baselevel was lowered below the bedrock contact (364 hr), incision and migration of the meanders down the dip-slope surface of the bedrock occurred. Evidently, the stream encountered less resistance to erosion downvalley through the alluvium than to incision vertically into bedrock. The resulting increased sediment load caused the channel to aggrade and develop a straight, braided pattern in the lower part of the flume. Incision into bedrock worked headward along the braided channel, creating a straight, incised channel in the lower 5 m with a sinuosity of 1.10 (Fig. 7.14*b*). At this stage in the experiment, it seemed that a decrease in flume slope might inhibit dip-slope migration and downvalley sweep, and slope was decreased to 0.001 at 436 hr. The immediate response to the slope reduction was a decrease in water velocity and greatly reduced bank erosion. Dip-slope migration of meanders was no longer possible on the nearly horizontal bed-

FIGURE 7.13. Stable alluvial meanders in Experiment 6, looking downstream (from Gardner, 1973).

rock surface, and with the reduced bank erosion, braiding ceased and a stable meander pattern developed; nevertheless, the stream could still incise even though lateral migration was greatly reduced.

Flood discharges of up to 2.8 L/sec were passed down the channel for periods of 30–36 sec during 868–916 hr in order to determine the effect of flood flows on the pattern. Meander amplitude, radius of curvature, and wavelength were all increased slightly (Fig. 7.14*c*), but there were greater effects on vertical incision than on lateral migration. For example, the upstream point of incision had migrated only 0.76 m upstream in 400 hr during normal discharge, but it moved headward 3.96 m in only 45 hr during the flood discharge, at a rate 45 times greater.

Bedrock was exposed in pools near the apices of meander bends during the flood discharge. Once the channel had eroded into the bedrock, it was effectively anchored at that point and downvalley sweep was inhibited. Thus meanders became deformed before complete incision had occurred, because the sections of stream

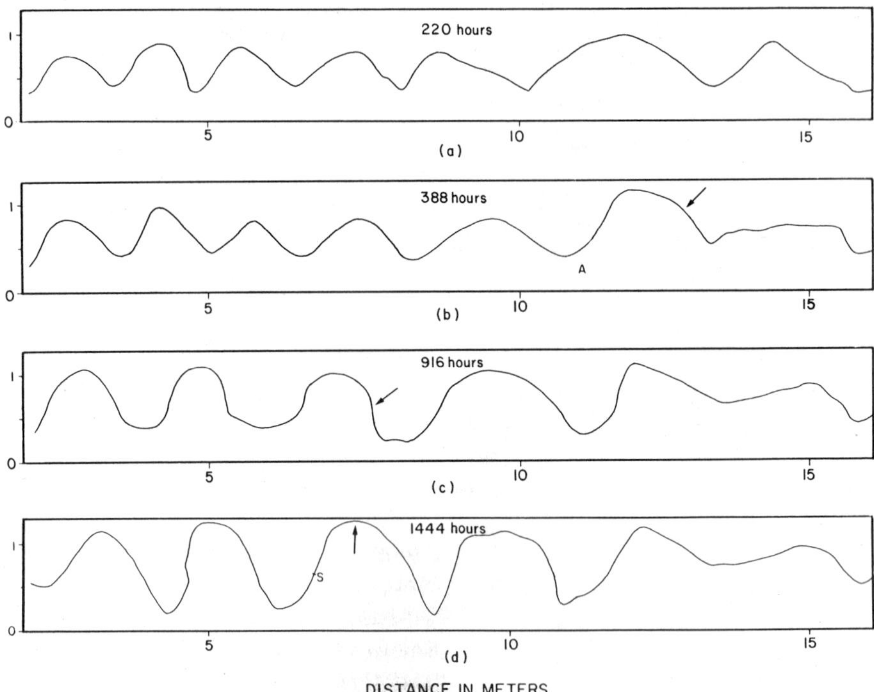

FIGURE 7.14. Plan view of thalwegs in Experiment 6. Arrows indicate upstream extent of bedrock channel at given time. Flow is from left to right. Note scale distortion (from Gardner, 1973).

still flowing in alluvium could shift position but the bend apices were fixed in bedrock (Fig. 7.14*d*).

At 1375 hr, sediment was fed to the channel at point S on Fig. 7.14*d*, which was 0.76 m upstream from the upstream limit of incision into bedrock. Aggradation resulted, which not only stopped vertical incision but also covered exposed bedrock for a distance of 1.5 m below the former upstream limit. The sand was deposited on the inside of the next meander bend downstream, displacing the thalweg to the outside of the bend. This caused an increase in amplitude and wavelength, but the next bend downstream was already anchored in bedrock and could not move. The meander pattern therefore became distinctly deformed, and after the sediment feeder was stopped at 1444 hr the channel reincised in a deformed pattern.

It follows, then, that increased sediment load inhibits vertical corrasion and permits greater lateral migration, as observed by Shepherd (Fig. 7.8) and Meyer (Chapter 6). In circumstances where some bends are laterally constrained and others are free to shift, this will encourage deformation of a preexisting meander pattern and creation of asymmetrical or irregular ingrown meanders.

Although the experimental bedrock was intended to be isotropic, there was a variation in rock resistance at point A on Figure 7.14*b*. At point A the bedrock was locally softer, so the stream widened, which reduced its velocity, depth, and transport capacity. Deposition on the inside of the bend was associated with a progressive shift of the current to the outside, because the stream was apparently more able to erode the bedrock than to transport the coarse sand and pebbles on the point bar. As the channel was still incising in response to a drop in baselevel, incision was both lateral and vertical and an ingrown meander formed with a diagnostic slip-off and undercut outer bank. Thus variations in rock resistance may also tend to produce ingrown meanders or, if gravel or boulder bars form, weak bedrock may be eroded more readily than the remobilization of the coarse alluvium.

In Gardner's Experiment 7 the behavior of a river incising into an asymmetrical anticline from an alluvial cover was examined. The bedrock surface was shaped in Fig. 7.15 and covered with sand to give an alluvial cover having a planar surface and variable depth (Table 7.1). Flume slope and discharge were held constant during the experiment at 0.005 and 0.6 L/sec, respectively, and no sediment load was introduced. These conditions insured that braiding did not occur, although meander development was slow. A stable meandering-thalweg pattern, with amplitude, wavelength, and sinuosity of 0.52 m, 1.7 m, and 1.45, respectively, developed by 70 hr (Fig. 7.16*a*). Adjustments were subsequently made as noted on Table 7.1.

The development of the stream pattern and longitudinal profiles during the experiment are shown in Figs. 7.16 and 7.17. The evolution of the alluvial channel was essentially the same as that in the previous experiment, and the response to baselevel lowering at 210 hr was also similar. Vertical corrasion and increased channel slope initially caused entrainment of all sediment, but downvalley meander sweep eventually overloaded the stream, causing aggradation and deterioration of the meander pattern. In addition, however, meander sweep became so active that two meander cutoffs occurred in the next 24 hr, increasing meander wavelength in the lower flume to 2.4 m. Amplitude increased rapidly to its former value after the cutoffs (Fig. 7.16*A*).

When baselevel was lowered to the bedrock surface, bedrock first appeared in the channel bed at the crest of the anticline and permanent incision rapidly occurred over a distance of 3 m up- and downstream as the channel bed was lowered (Fig.

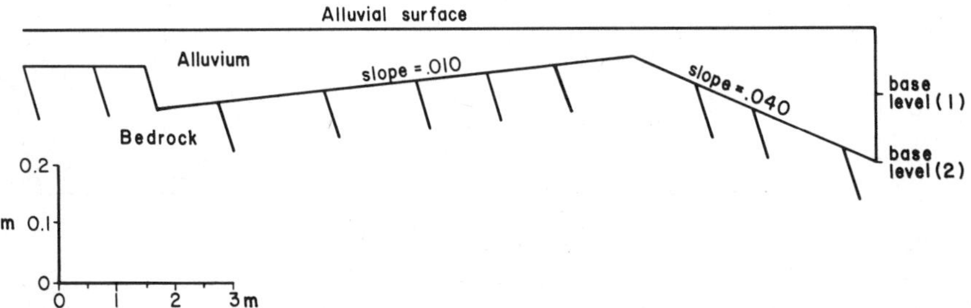

FIGURE 7.15. Longitudinal profile of bedrock surface used in Experiment 7 (from Gardner, 1973).

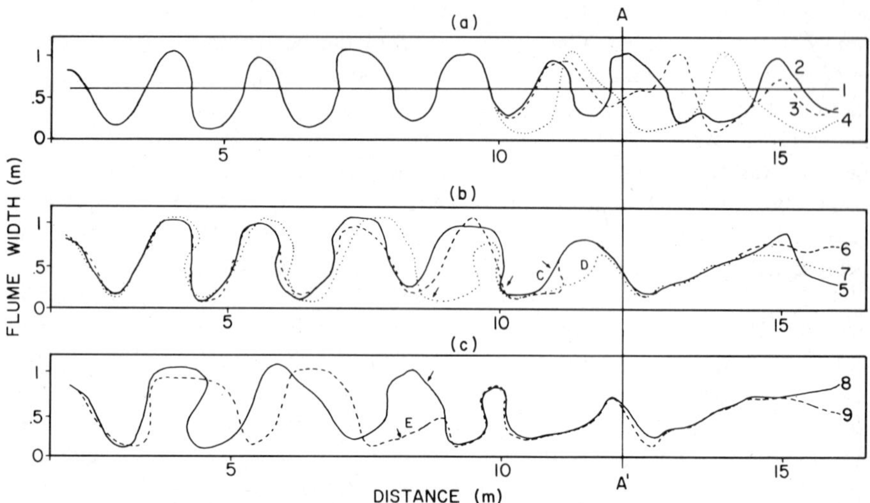

FIGURE 7.16. Plan view of thalwegs in Experiment 7; flow is from left to right. Arrows indicate upstream extent of bedrock channel. Note scale distortion. Vertical line A–A′ indicates crest of bedrock profile (Fig. 7.19) (from Gardner, 1975). Numbers indicate hours of run time for each pattern as follows: (1) 0 hr, (2) 93 hr, (3) 234 hr, (4) 302 hr, (5) 329 hr, (6) 378 hr, (7) 450 hr, (8) 546 hr, (9) 642 hr.

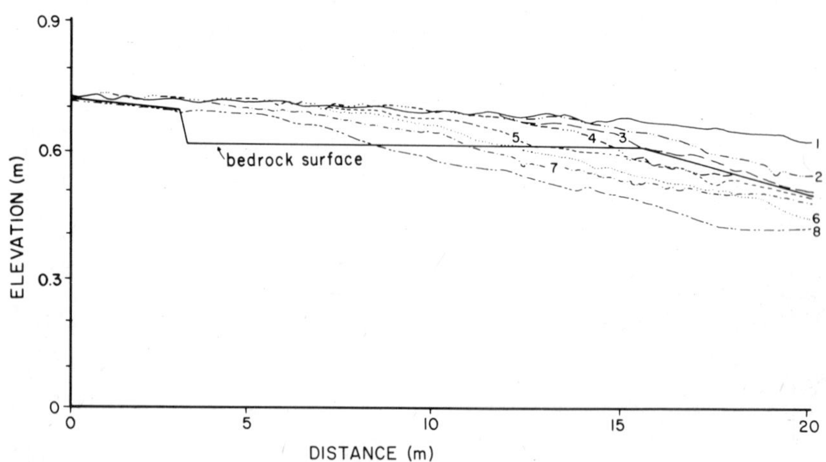

FIGURE 7.17. Longitudinal profiles in Experiment 7 (Gardner, 1975). Time from start of run is as follows: (1) 93 hr, (2) 239 hr, (3) 307 hr, (4) 329 hr, (5) 398 hr, (6) 450 hr, (7) 546 hr, (8) 642 hr.

7.17). Thus meanders were incised before they could be deformed by downvalley sweep. Downstream, dip-slope migration of the meanders along the bedrock surface dominated, producing an essentially straight, braided channel, as in the previous experiment. When baselevel was lowered at 433 hr, incision worked headward up the braided reach, giving an almost straight incised channel downstream from the anticline, with a sinuosity of 1.08 (Fig. 7.16*b*).

Above the crest of the anticline, where the bedrock slope was opposed to that of the alluvial surface, incision was significantly different from that at and below the crest of the anticline. In the thicker alluvium, meander sweep occurred in response to the slight gradient increase associated with incision at the anticline crest (the local baselevel). As a meander migrated downvalley in response to the increased gradient, the downstream limb encountered the bedrock surface, incised, and became fixed in position, while the upstream limb continued to migrate downvalley. The upstream limb also eventually encountered the highest point on bedrock surface as it migrated downvalley, and incised and became fixed. As a result, meanders were compressed and deformed as they migrated downvalley, and as incision worked headward, a deformed meander pattern was incised into the bedrock with a sinuosity of 1.45 (Fig. 7.18). This process was aided by the fact that, because sweep was restricted by bedrock, bank erosion was significantly reduced and sediment load was lower than where sweep or dip-slope migration were unrestricted, so the channel did not aggrade or braid.

Farther upstream, where the alluvium was deeper (Fig. 7.17), meander sweep was unrestricted by the bedrock floor and rejuvenation produced large quantities of sediment, which caused aggradation of the previously incised channel. Deformation of the meander pattern continued and an extensive point bar formed on the upstream side of the bend at location C (Fig. 7.16*b*). The channel aggraded and water flowed over the projecting bedrock spur to create a cutoff at location D. A similar cutoff occurred at location E by 642 hr (Fig. 7.16*c*). Such cutoffs could only occur where aggradation in the incised channel was sufficient to allow water to take a more direct path over a bedrock spur and then erode down through it.

DISCUSSION

A number of observations made during the experiments have application to natural channels flowing in bedrock and allow testing of hypotheses about the processes of and controls upon channel incision into bedrock. There was a remarkable similarity between the scour patterns observed in the experimental bedrock and in natural bedrock channels. The linear grooves and potholes in the model channels appear to be analogous to those observed in The Dalles of the Columbia River and in many other rivers flowing in bedrock, and even to the fossil forms found in the Channelled Scabland of the Columbia Plateau (Baker, 1978).

Theoretical and experimental studies by Ashida and Sawai (1976) show that shear distribution on the floor of a channel is irregular, which leads to the devel-

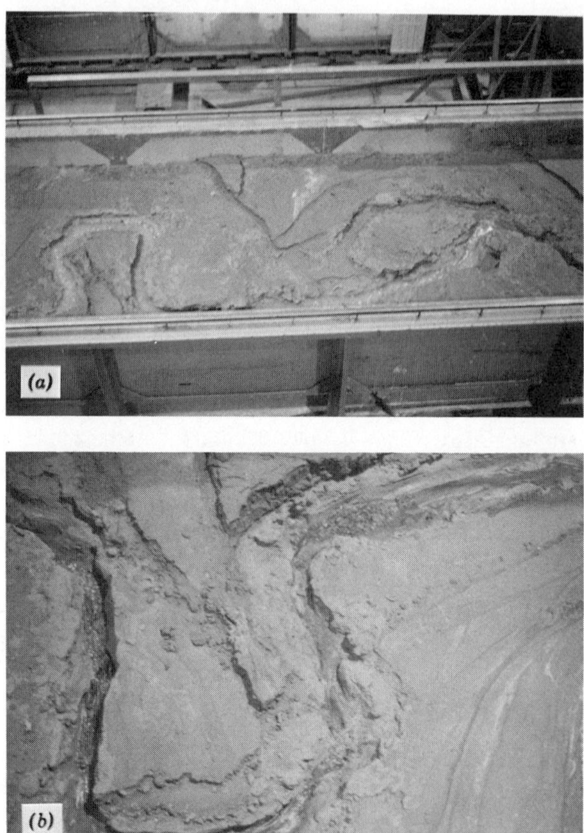

FIGURE 7.18. Deformed meanders in Experiment 7 after 450 hr (from Gardner, 1973). (*a*) Between 9 and 13 m (Fig. 7.16); (*b*) at 10 m (Fig. 7.16).

opment of parallel troughts and ridges (Figs. 7.6, 7.7). Such assemblages of linear features appear to be an intermediate erosional stage, and eventually a single inner channel becomes dominant, as Bretz (1924) suggested. The experimental observations prompted a search for natural examples, which proved quite fruitful. Excavations at dam sites have revealed inner bedrock channels (Fig. 7.4), which have commonly been interpreted as low flow channels.

Not all bedrock valleys have inner channels, but recognition of their possible presence is important for engineering purposes and possibly for geologic exploration. An inner channel on a bedrock floor may be the site of concentration of heavy minerals by hydraulic sorting, and determination of the location and trend of an inner channel underlying Holocene valley deposits may facilitate exploration for gold, diamond, or other placer deposits. For example, in South Africa, alluvial diamonds are found in narrow and discontinuous channels incised into a bedrock surface that is covered by a thin veneer of recent alluvium and soil. The diamonds are associated with coarse fluvial gravels in the scour channels, which may be modified portions of an inner channel system like that in the experiments (Schumm, 1977).

A weakly regular sequence of scour holes formed in actively incising channels (Figs. 7.9, 7.10); the channel was narrow at the deepest points of the scour holes and wide at the crests in between. The regularity of scour was altered when aggradation occurred, as local scour during aggradation eroded many of the bedrock highs, leaving irregularly spaced scour holes. The floors of many of the bedrock scour holes were incised below baselevel. On the floor of a valley and elsewhere, these deep scours could breach the contact between two lithologic units of different permeability, causing leakage of groundwater into or out of the valley alluvium, and it could create a situation permitting later migration of ore-bearing solutions from one location to another.

There seems no doubt that some rivers, like the San Juan River in southeast Utah, have incised vertically for hundreds of meters while maintaining a preexisting sinuous pattern, whereas others have migrated laterally during incision, creating slip-off slopes and steep or undercut outer banks. The experimental work provides evidence in support of Rich's (1914) view that the final form of an incised bedrock valley depends on the relative importance of vertical incision, lateral corrasion, and downvalley meander sweep, and it indicates the conditions under which each may be dominant (Table 7.2).

During the experiments, an alluvial meandering pattern was rapidly destroyed after channel slope was increased, either by tilting of the flume or locally, by baselevel lowering. A straight braided pattern developed, and downcutting into

Table 7.2 Relationship of Bedrock Channel Pattern to Various Controls[a]

	Controls			
Pattern	Bedrock Slope	Orogenic Movement	Thickness of Alluvial Cover	Sediment Load
Straight incised pattern	Slope in direction of stream flow	Movement that increases gradient	Thick	High
Superposed meander pattern	Horizontal bedrock surface	Vertical uplift or baselevel lowering	Thin	Low
Deformed incised meander pattern	Opposite to direction of stream flow	Movement that decreases gradient	Uniform thinning downstream	Moderately high
Ingrown meanders	Variations in resistance to erosion of bank material			Coarse

[a] From Gardener (1975).

bedrock created a straight, incised channel. In Shepherd's Experiment 5 (Table 7.1), when the initial condition was a meandering channel cut into bedrock, there was also an initial tendency for channel straightening as corrasion by the sediment load and incision were concentrated at the inside of bends (Table 7.2). However, as incision reduced slope, stream competence was reduced, deposition occurred at the inside of bends, and lateral migration started to occur simultaneously with incision. This observation is consistent with Moore's (1926a) view that the relative importance of lateral migration and vertical incision is controlled by variations in the stream's ability to entrain the load supplied to it from upstream. Moore indicated that either an excessive load of fine sediment or a smaller sediment load composed of coarse material, cobbles, and boulders could exceed the competence of a stream, and the experiments confirmed that both could be responsible for inducing lateral erosion in bedrock meanders. The rate of lateral migration during the experiments was also seen to be modified by rock resistance variations, leading to deformed ingrown meanders such as those along lines of bedrock weakness described by Strahler (1946) in Conodoguinet Creek, Pennsylvania, and by Cole (1930) in Coy Glen, New York.

Vertical entrenchment of meanders from an overlying alluvial cover was observed in two circumstances (Table 7.2). First, entrenchment occurred when flume slope was reduced to near horizontal, and a lowering of baselevel permitted headward retreat of a nickpoint along a winding channel. The lack of significant downvalley slope inhibited lateral migration or downvalley sweep of the alluvial meanders, so that they were inherited unmodified by the bedrock channel. Second, downcutting in response to a baselevel lowering caused meanders to entrench, without modification, into the crest of an anticlinal bedrock surface covered to a shallow depth with alluvium.

Superposition onto an anticline (Fig. 7.16) caused different modes of incision in different places along the course of a channel meandering in an overlying alluvial cover. Where the alluvial cover was shallow over the crest of the anticline, rapid downcutting permitted incision of the bends without distortion. Downstream, downdip migration of the bends in thick alluvium created a straight braided channel, and an incised straight channel resulted (Table 7.2). Upstream of the crest of the anticline, downvalley sweep of the meanders occurred, but the downstream limbs of the incising meanders were successively locked into place as they encountered the rising bedrock surface, and a distorted meander pattern was incised into bedrock. Farther upstream, the pools in the bend apices were the first places at which bedrock was exposed in the incising channel during flood discharges. Their positions became fixed, but continued migration of the meander limbs deformed the meander pattern, so that distorted incised meanders resulted.

These observations are closely analogous to conditions in the Colorado Plateau, and the experimental results are highly relevant to an understanding of the drainage system in that area (Gardner, 1975). Where the Green and Colorado rivers flow against the structural dip of the bedrock into the northern extension of the Monument Upwarp, incised meanders are well developed in bedrock (Fig. 7.19, points A′, B) and their sinuosity is over 1.5. However, downstream of their confluence

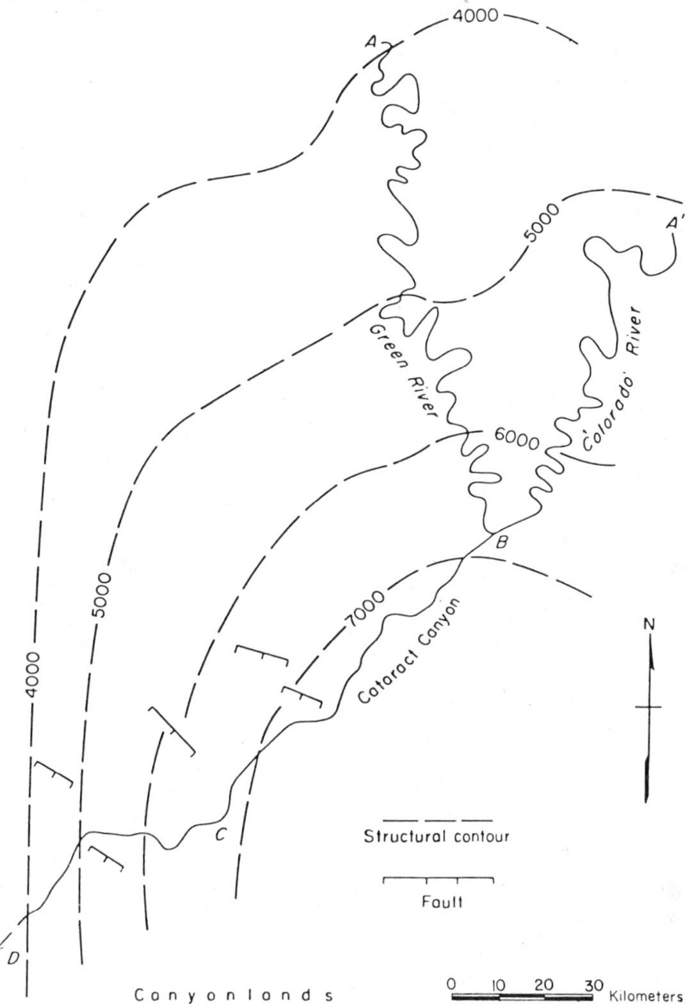

FIGURE 7.19. The course of the Green and Colorado rivers in the vicinity of their confluence. Structural contours (in feet) represent the top of the Chinle Formation.

the river course is nearly parallel to the strike of the bedrock and few meanders have been incised. Sinuosity is about 1.1 (Fig. 7.19, points B–C). Farther downstream, the river flows down the structural dip off the Monument Upwarp, and again there are few incised meanders, with a sinuosity of 1.08 (points C–D). The changes in sinuosity and degree of development of incised meanders are not related to changes in erodibility of the bedrock, although regional fault and joint patterns show some alignment with, and may have aided the development of, straight channel reaches. To the south the San Juan River also crosses the Monument Upwarp (Gardner, 1975). On top of the upwarp, incised-channel sinuosity is high and many deformed incised meanders are well developed in bedrock. Farther west, as the river flows down the structural dip of the bedrock off the upwarp, there are no well-developed incised meanders and sinuosity is low. The similarity in form of the Green, Colorado, San Juan, and experimental channels suggests that similar factors controlled their evolution, and in fact, Hunt (1969, p. 101) suggested that "differences in gradient and meander pattern probably reflect responses of the streams to continuing deformation."

8 Effect of Active Tectonics

Alluvial rivers, those that flow between banks and on a bed that are composed of sediment being transported by the river, are sensitive to changes of sediment load, water discharge, and variations of valley-floor slope (see Chapter 5). Therefore, in addition to the dramatic effects when stream channels and terraces are offset along faults (Stevens, 1974; Wallace, 1967), other, more subtle pattern changes may be recognizable even when deformation is vertical, slow, and aseismic.

Many of the major rivers of the world follow structural lows and major geofracture systems (Potter, 1978). In fact, Melton (1959) suspects that streams that have adjusted to tectonic activity are very numerous. In the United States, the lower Mississippi River (Schumm et al., 1982) and Rio Grande (Reilinger and Oliver, 1976) are clearly in areas of structural instability, as are other major rivers such as the Amazon, Niger, Tigrus, Euphrates, Rhine, and Indus. The high discharges of such major rivers should permit them to maintain their courses in spite of active (ongoing) tectonics, but because of their low gradients, they may be the most significantly affected by the minor changes in slope produced by active deformation.

In spite of the practical significance of tectonic effects on alluvial rivers and the large neotectonic literature, only a few investigators have considered this problem (Adams, 1980; Burnett and Schumm, 1983; Russ, 1982; Schumm, 1977, 1986; Tator, 1958). This may be because variations of channel morphology and behavior are strongly influenced by downstream variations of discharge and the quantity and type of sediment load, and therefore the effects of active tectonics are difficult to detect. In addition, the attention of photogeologists and others generally has been concentrated on the identification of geologic structures that are assumed to be quiescent (Howard, 1967; Ollier, 1981, p. 180) rather than on the effect of ongoing deformation of alluvial channels. Nevertheless, Tator (1958) and DeBlieux (1951,

1962) indicate that fluvial anomalies such as local development of meanders or a braided pattern, local widening or narrowing of channels, anomalous ponds, marshes or alluvial fills, variations of levee width or discontinuous levees, and any anomalous curve or turn are possible indicators of neotectonics or active tectonics. In addition, active tectonics can produce nickpoints, convexities or concavities of the longitudinal profile, depth change, and, of course, either aggradation or degradation.

When uplift is too rapid to be accommodated by a river, there will be disruption of the drainage pattern (Ollier, 1981; Sparling, 1967; Twidale, 1971, pp. 133–136). For example, the Murray River on the Riverine Plain near Echuca, Australia, is an impressive example of channel modification by tectonic activity (Bowler and Harford, 1966). The Cadell Fault block has converted the Murray River from a single-channel stream to an anastomosing system of channels that flow around the obstruction. The old abandoned segment of the Murray River is preserved on the dip slope of the fault block.

The Oakwood Salt Dome in eastern Texas is believed to be actively rising because channels over the central dome are incised up to 4 m, and three abandoned channel reaches suggest lateral movement of the channels away from the dome (Collins et al., 1981).

If an incising river encounters resistant strata, it will either incise into the resistant rock or be diverted. Thus incised meander patterns can indicate deformation (Fig. 7.19). As meanders shift downvalley, their movement is retarded when resistant alluvium or bedrock is encountered (Chapter 7), and a compressed or deformed meander pattern may result (Fig. 7.18). Hence a fault or upwarp may present a barrier, causing compression and deformation of meanders upstream.

The preceding criteria may be evidence for past deformation, but not necessarily for active uplift. Active tectonics are reflected particularly in those geomorphic features that react to the smallest changes of slope, that is, longitudinal profiles of terraces, floodplains, and rivers, river and valley floor morphology, and channel patterns (Bendefy et al., 1967; Neef, 1966; Radulescu, 1962; Zuchiewicz, 1979). Degradation and aggradation and associated changes in channel morphology can be evidence of active tectonics as the channel incises or aggrades. For example, progressive changes of thalweg or water surface elevation, such as that provided by a long gauging-station record, can provide information on vertical changes of the channel (Burnett and Schumm, 1983).

Stream patterns are sensitive indicators of valley slope change. For example, Adams (1980) has demonstrated a relation between measured tilt rates and variations in sinuosity along the Mississippi River between St. Louis and Cairo and along the lower Mississippi River. Twidale (1966) reported that both the Flinders and Leichardt rivers have changed to a braided pattern as a result of the steepening of their gradient by the Selwyn Upwarp in northern Queensland. Unfortunately, it is not always possible to be specific about pattern and channel change, because different types of alluvial channels may respond differently to active tectonics (Fig. 5.2).

The experiments of Shepherd and Gardner (Chapter 7) dealt primarily with the effects of baselevel lowering on bedrock channels. These effects could also be the result of vertical uplift. The discussions of Chapter 5 have shown how valley slope affects several aspects of alluvial channel morphology through its influence on stream power and the channel's ability to transport sediment (Fig. 5.32). Since tectonic movement influences valley slope, channel response is, in principle, predictable from relationships such as that between sinuosity and slope (Fig. 5.33*D*).

Although it is possible to discern the effects of tectonic movement on alluvial channel form, there are many uncertainties introduced by other factors such as type and quantity of sediment load, flow regime pattern thresholds (Figs. 5.33*D*, 5.36), and, particularly, human influence. By holding constant or eliminating such factors, it should be possible to examine the impact of uplift or subsidence in the laboratory, and Ouchi (1983, 1985) and Jin and Schumm (1986) have pursued this possibility.

EXPERIMENTAL STUDIES

The experiments were performed in a wood-framed flume 7.62 m long and 2.44 m wide with a flexible center floor section (Fig. 8.1). The flume was filled with a sand having a median diameter of 0.27 mm, and uplift or subsidence of the sand surface was achieved by raising or lowering a steel beam under the midpoint of the flexible central section of floor. The beam rested on concrete blocks, and uplift and subsidence were accomplished either by placing 1.27-mm-thick metal shims on the blocks or removing them. Water was fed from a hose or recirculated by a small pump, and sediment was fed at the head of the flume with a vibrating sand feeder. Cross sections of the channel were measured using a point gauge mounted on a steel beam spanning the flume.

The channel was divided into four reaches as follows:

Reach A, from 2.0 to 3.5 m, where no significant uplift or subsidence occurred.

Reach B, from 3.5 to 4.65 m, the upstream half of the uplifted or subsided zone.

Reach C, from 4.65 to 5.75 m, the downstream half of the uplifted or subsided zone.

Reach D, from 5.75 to 7.0 m, where no significant uplift or subsidence occurred.

Experiments were conducted by Ouchi (1983) on braided, meandering, and confined (riprapped) channels. Jin (Jin and Schumm, 1986) carried out an experiment on meandering channels with banks that were layered and stabilized with kaolinite.

Braided Channel

Uplift

Ouchi (1983) molded an initial channel 8.9 cm wide and 3.8 cm deep in the alluvial surface, which had a slope of 0.02. The alluvium was composed of a 9:1 mix of

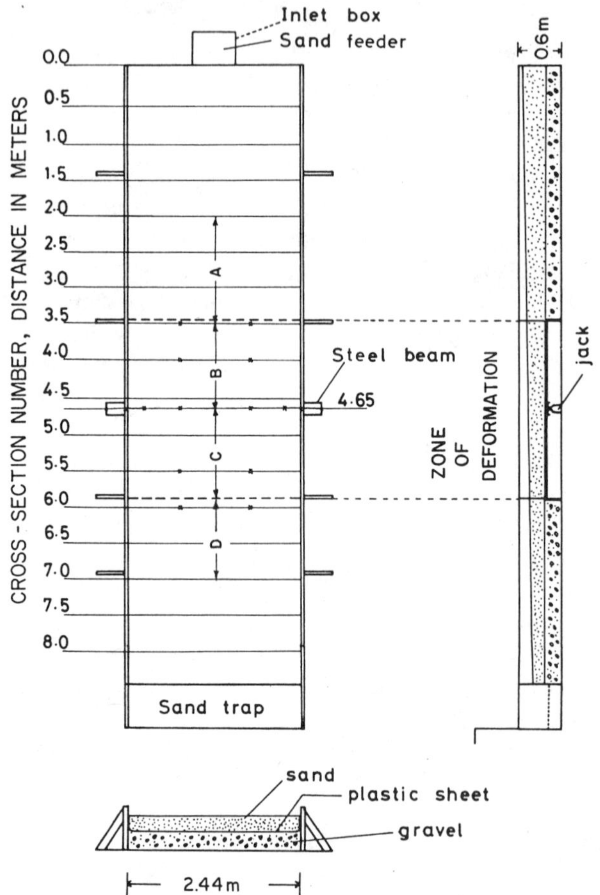

FIGURE 8.1. Flume used during active tectonics experiments. Reaches described in text are shown by the letters A, B, C, D (from Ouchi, 1983).

sand and kaolinite, which provided some bank stability, and sand was fed from the sediment feeder when required by the experimental design.

A braided pattern developed after running for 20 hr at a discharge of 0.1 L/sec, and uplift was started at 26 hr. There was a slight profile convexity before uplift, which was related to scour at the upstream and downstream ends of the flume and not to the uplift. No significant local convexity of the profile appeared in the zone of uplift (Fig. 8.2) because the stream responded to uplift by degradation and thalweg deepening (Fig. 8.3). Degradation started after about 34 hr near cross section 5.0 in reach C after the first uplift, where the slope was steepened the most. By 36 hr the incision had affected reach B. Sediment produced by the degradation caused a slight aggradation downstream in lower reach C and in D (34–40 hr) and the reduction of gradient upstream produced deposition and a convexity in zone A

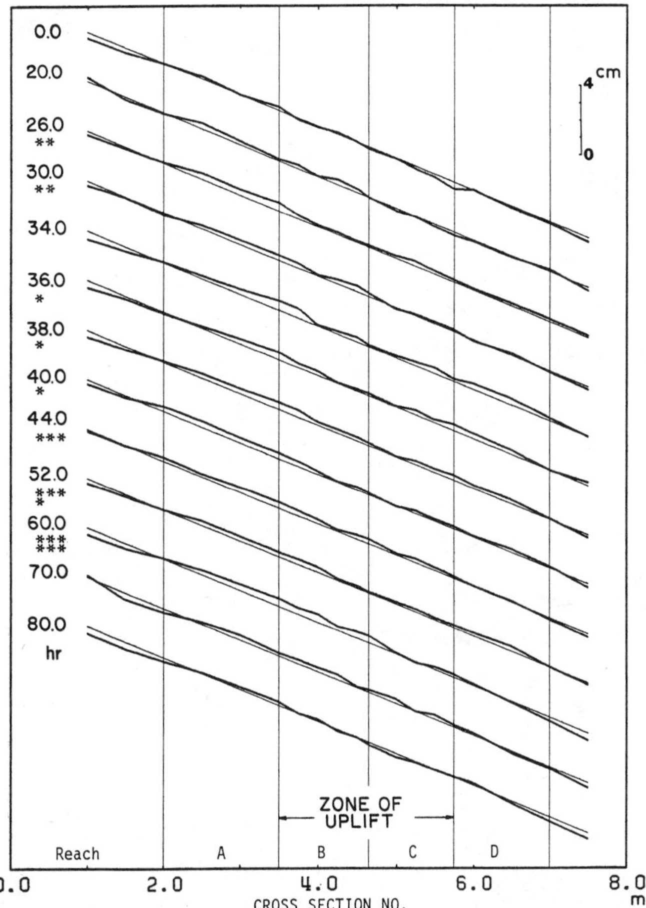

FIGURE 8.2. Bed profiles of the braided channel during uplift. Asterisks indicate number of shims (1.27 mm thick) used to produce uplift after 26 hr. Elevation is a mean bed elevation at each cross section (from Ouchi, 1983).

and upper zone B, although a slight convexity existed in these reaches at the start of the experiments (Fig. 8.2).

After uplift started, channel pattern changed continuously with frequent thalweg shifting, but the effect of uplift was not clear until 28 hr. After 30 hr, terraces were formed in the uplifted zone (Fig. 8.4). Terraces were distinguished from bars by their fixed position and increasing height as uplift continued. These terraces were gradually eroded by thalweg shift, and they were destroyed by 52 hr (Fig. 8.4). After the terraces disappeared, the reaches of slight aggradation upstream and downstream from the uplifted area became reaches of degradation and a relatively well-defined thalweg developed. After 52 hr, when four shims were added, a similar process, beginning with channel incision (60 hr) and terrace formation (70

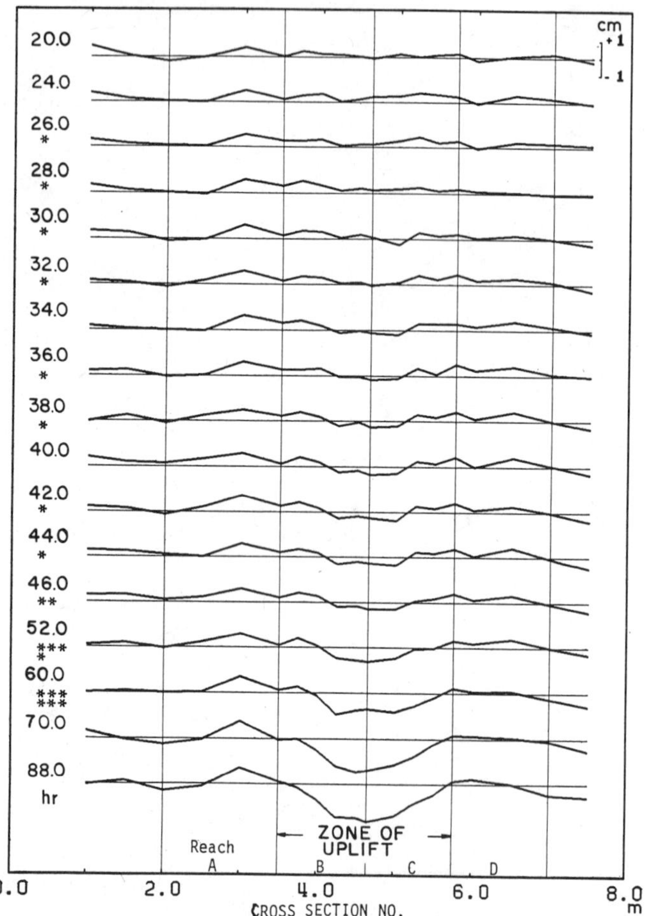

FIGURE 8.3. Change of mean channel depth at each cross section during uplift of the braided channel from 20 to 88 hr. Horizontal lines indicate the depth at time 0, the start of experiment. Distance below these lines indicates amount of degradation. Asterisks indicate number of shims used to produce uplift (from Ouchi, 1983).

hr) was resumed. Twenty hours after the uplift was completed, a weak braided pattern developed throughout the flume (88 hr; Fig. 8.4) and the longitudinal profile was relatively straight (Fig. 8.2).

In summary, the braided stream responded to uplift with degradation that was rapid enough to offset the uplift, so that in the uplifted zone, channel depth increased. Degradation started in reach C where the slope was steepened the most, and it migrated upstream into reach B. As degradation continued, terraces were formed in the central-to-downstream reach of the uplifted area. The thalweg was fixed and downcutting was accelerated in this reach. At the same time, a multiple-thalweg channel with submerged bars, which indicates aggradation, formed in reach

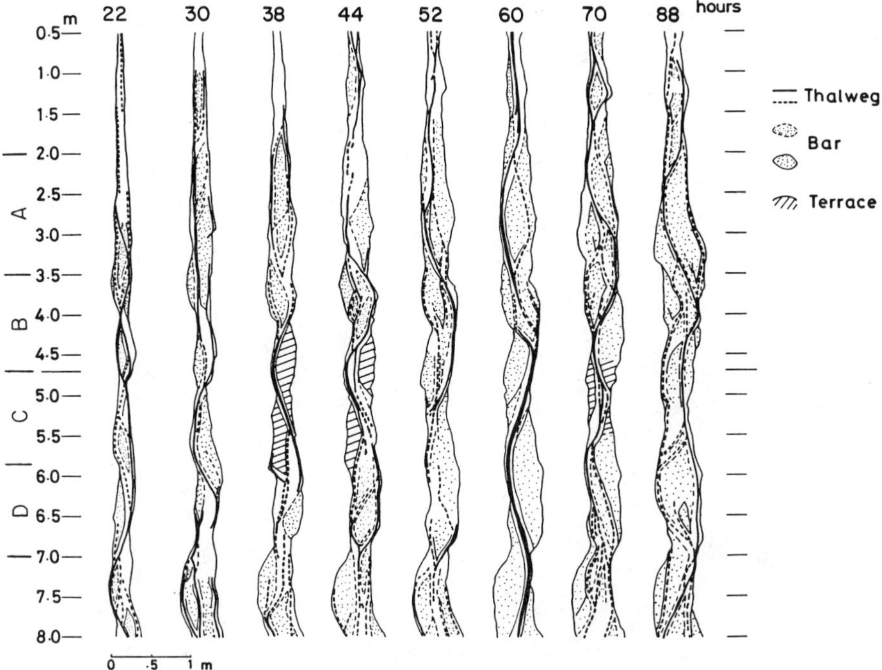

FIGURE 8.4. Pattern change during uplift of braided channel. Axis of uplift is at 4.7 m. Solid line indicates main thalweg and dashed lines represent secondary thalwegs (from Ouchi, 1983).

B. The terraces were eroded and degradation migrated into this reach. A strongly braided pattern, also as a result of aggradation, developed in reach D due to excess sediment supply from the uplifted area. When the terraces were destroyed, reach D developed a single thalweg with alternate bars as a result of degradation caused by the decline of sediment supply. After the uplift ended, the braided pattern slowly redeveloped throughout the entire channel.

The degradation occuring in the uplifted zone did not exactly correspond to the uplift. There were fluctuations and pauses in degradation while uplift continued. These fluctuations and pauses seemed closely related to channel pattern changes. There may have been a certain critical or threshold amount of uplift above which the channel started to respond, but 1.27 mm of uplift over a distance of about 1.2 m (slope change of 1.06×10^{-3}) was apparently large enough to induce a response in this experiment.

Subsidence

The channel existing at the end of the uplift experiments was used by Ouchi (1983) as the initial channel (t = 0 hr) for a subsidence experiment (Fig. 8.5). Subsidence was started at 2 hr by extracting 5 shims (6.35 mm). Four episodes (5 shims each)

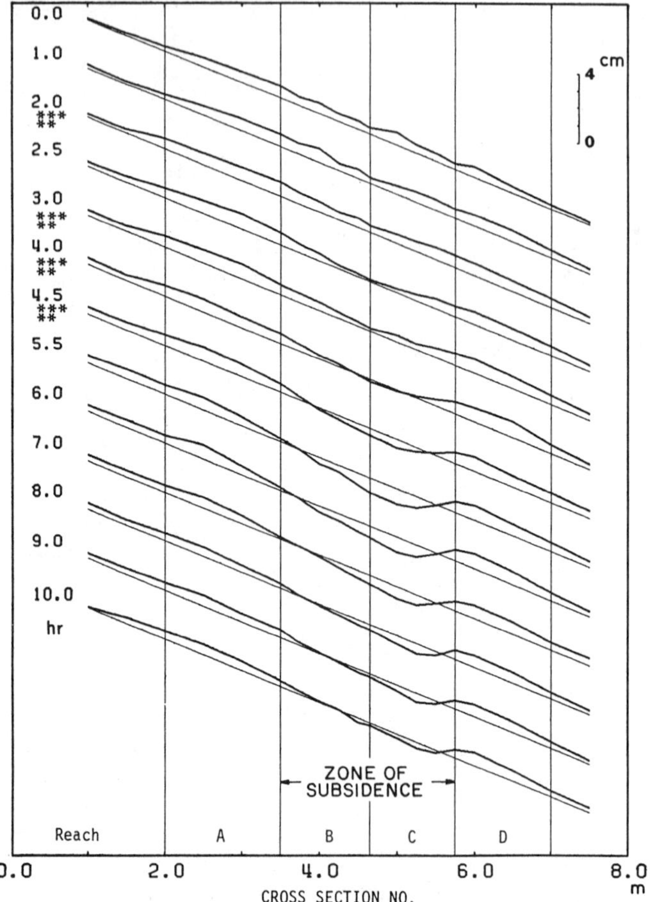

FIGURE 8.5. Bed profiles of braided channel during subsidence. Elevation is a mean bed elevation for each cross section. Asterisks indicate number of shims removed to produce subsidence (from Ouchi, 1983).

of subsidence were performed every hour for 5 hr until a total of 20 shims was removed (Fig. 8.5).

Each period of subsidence was followed by aggradation, but it was not enough to compensate for the subsidence, and there was local concavity in the channel profiles (Fig. 8.5). At the downstream end of reach C, a convexity formed as a result of the reduced slope, and it remained to the end of the experiment. In contrast, degradation reduced the slope convexity at the upstream end of reach B.

Aggradation in response to subsidence decreased mean channel depth from cross section 4.0 to 5.25 after about 3.5 hr (Fig. 8.6). Aggradation started at the central part of the subsidence zone (cross section 4.65) and migrated upstream. The aggradation, however, did not continue upstream beyond reach B.

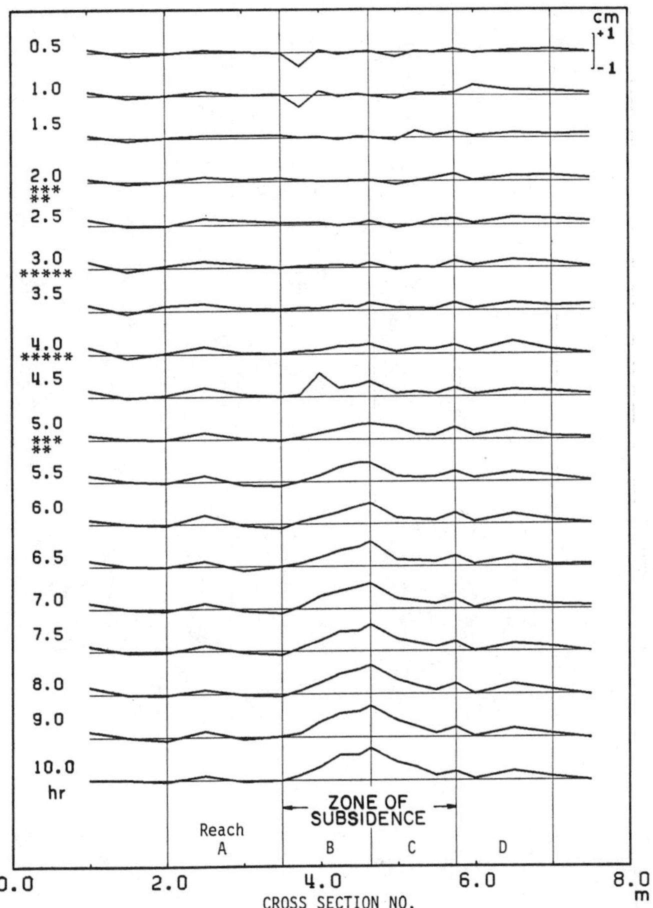

FIGURE 8.6. Change of mean channel depth at each cross-section during subsidence of braided channel. Asterisks indicate number of shims removed to produce subsidence. Distance above line indicates amount of aggradation (from Ouchi, 1983).

Downstream aggradation occurred as transverse bars migrated slowly downstream in reach C, where slope flattening caused flooding of the entire width of channel (Fig. 8.7). In this "flooded" reach, all bed features were underwater and the reach had no distinguishable thalweg (6 hr; cross section 5.0–5.5; Fig. 8.7). The transverse bars did not reach the downstream end of reach C, and the low point in reach C was not filled even after 10 hr (Figs. 8.5, 8.7).

In summary, degradation started in the uppermost reach of the subsided area, where slope was increased by subsidence, and it migrated upstream into reach A, where no subsidence had occurred. This degradation, which occurred mainly as bar destruction, provided sediment to the subsided area. As a result of this increase in sediment discharge and slope steepening by the subsidence, a strongly braided

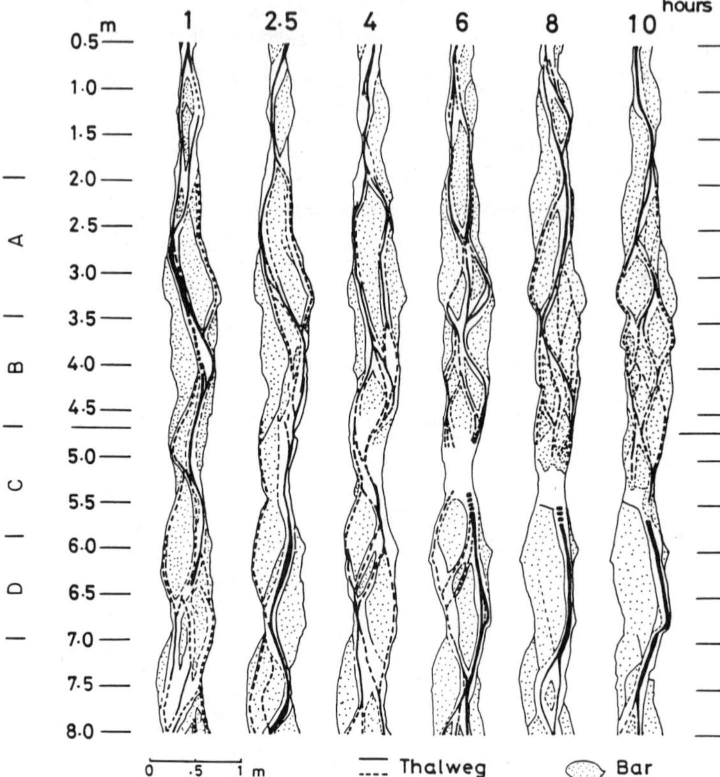

FIGURE 8.7. Pattern change of braided channel during subsidence. Axis of subsidence is at 4.7 m. Solid line indicates main thalweg and dashed lines represent secondary thalwegs (from Ouchi, 1983).

pattern developed in reach B. Deposition occurred quickly in the subsided zone from cross section 4.0 to 4.65. Downstream from this reach transverse bars migrated into the flooded reach, where the flow was slowed by the slope decrease. The transverse bars, however, did not reach the downstream end of the subsided area and, because of a lack of deposition at the lower end of the subsided zone, the slope irregularity at the reach C–reach D boundary remained at the end of the experiment, 5 hr after the subsidence had ended (Fig. 8.5). Downstream from the subsided area, thalweg degradation with alternate bar development occurred due to the deficiency in sediment supply from upstream (Fig. 8.7; 6, 8, 10 hr).

Confined Straight Channel

Uplift

A trapezoidal channel 30 cm and 20 cm wide at the top and bottom, respectively, and 4 cm deep was molded in the sand surface at a slope of 0.015. Both banks were riprapped with gravel in order to prevent widening. Discharge was 0.2 L/sec.

An alternate bar pattern formed (Fig. 8.8) and this was the initial condition for this experiment ($t = 0$ hr). At 2-hr intervals from 4 to 12 hr, uplifts of 3.8 mm were imposed on the channel. Mean bed degradation was about 75% of rapid uplift, so that a convexity in the zone of uplift developed progressively (Fig. 8.9). Terraces formed after about 6 hr (Fig. 8.8). This concentrated the flow in the thalweg and permitted thalweg scour generally to keep pace with uplift (Fig. 8.9).

Slight degradation in the upstream section of and above reach A was in response to a deficiency in sediment supply from the sand feeder, but later degradation in the lower part of reach A was in response to uplift. The alternate bars migrated slowly downstream, causing changes in thalweg and mean bed elevations along the channel (Fig. 8.9). In reach B, the initial response to uplift was lateral erosion of bars, and then accelerating thalweg incision, as flow was concentrated in the thalweg. In reach C, the thalweg incised. Sediment production in the center of the

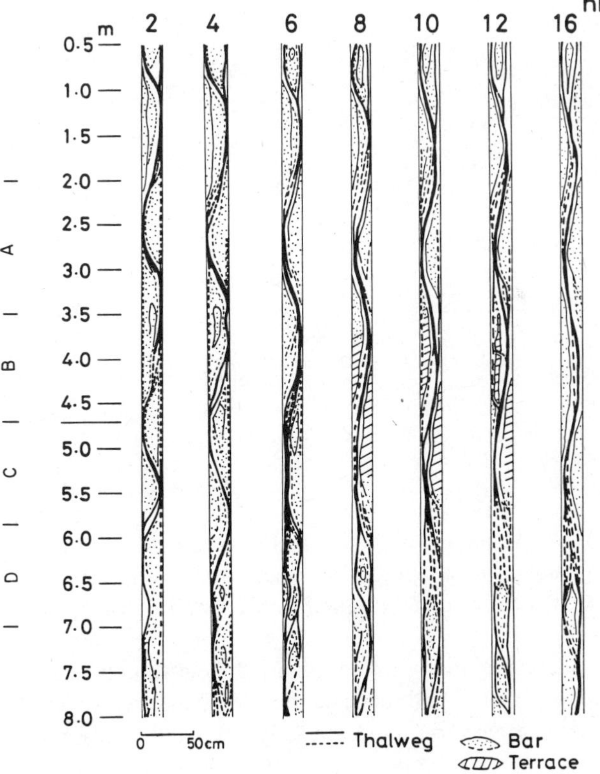

FIGURE 8.8. Pattern change of confined straight channel during uplift. Axis of uplift is at 4.7 m. Solid line indicates main thalweg, dashed lines represent secondary thalweg (from Ouchi, 1983).

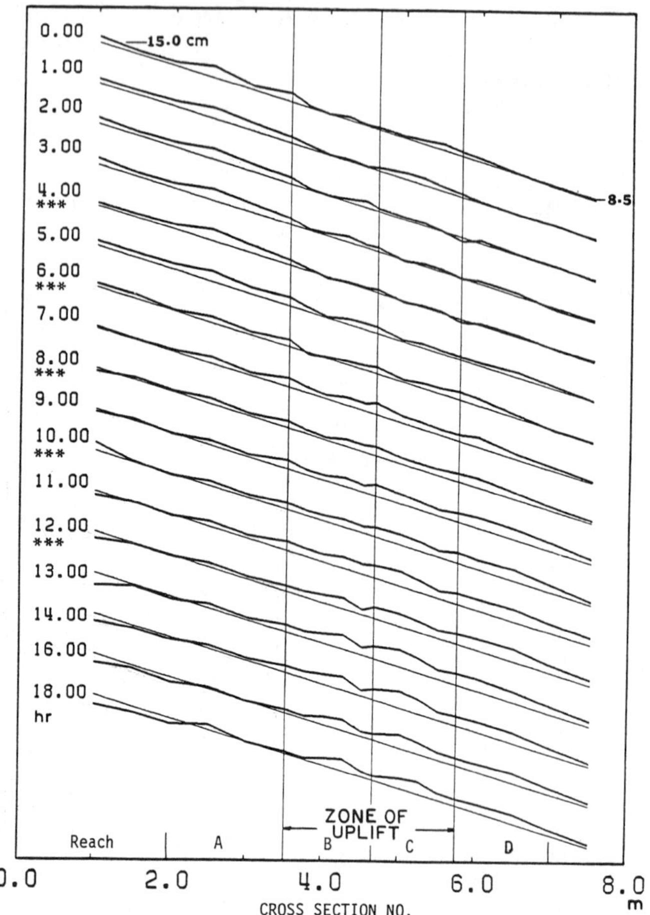

FIGURE 8.9. Bed profiles of confined straight channel during uplift. Elevation is mean bed elevation at each cross section. Asterisks indicate number of shims used to produce uplift (from Ouchi, 1983).

uplift apparently weakened the degradational trend downstream from about 5.0 m, and reach D aggraded during uplift as sediment from the uplifted zone was deposited in the form of central bars (Fig. 8.8). After uplift ended, degradation commenced.

In summary, degradation started in almost the entire uplift zone after the first uplift. The confined stream with concentrated discharge responded by major degradation, and sediment was transported readily beyond the zone of uplift. Terraces were formed in the uplifted zone (Fig. 8.9). Aggradation occurred downstream from the zone of uplift due to the large sediment supply, and central and transverse bars developed in the channel. The confinement of the stream also tended to maintain the alternate bar pattern.

Subsidence

The uplifted channel was used by Ouchi (1983) as an initial channel for a subsidence experiment, with subsidence at the rate of 3.8 mm every 2 hr from 2 to 10 hr. Rapid aggradation in the subsiding zone and slight degradation up- and downstream (Fig. 8.10) were unable to compensate for subsidence, and a concave mean bed profile developed (Fig. 8.10).

After 8 hr, the channel in reach C was drowned, the alternate bar pattern disappeared, and a transverse bar formed as sediment advanced into the pool (Fig. 8.11), finally reaching 5.5 m at 16 hr. In reach D, where no subsidence occurred, there was degradation due to a deficiency of sediment supply from upstream.

Although the ability to carry sediment increased as the discharge was confined

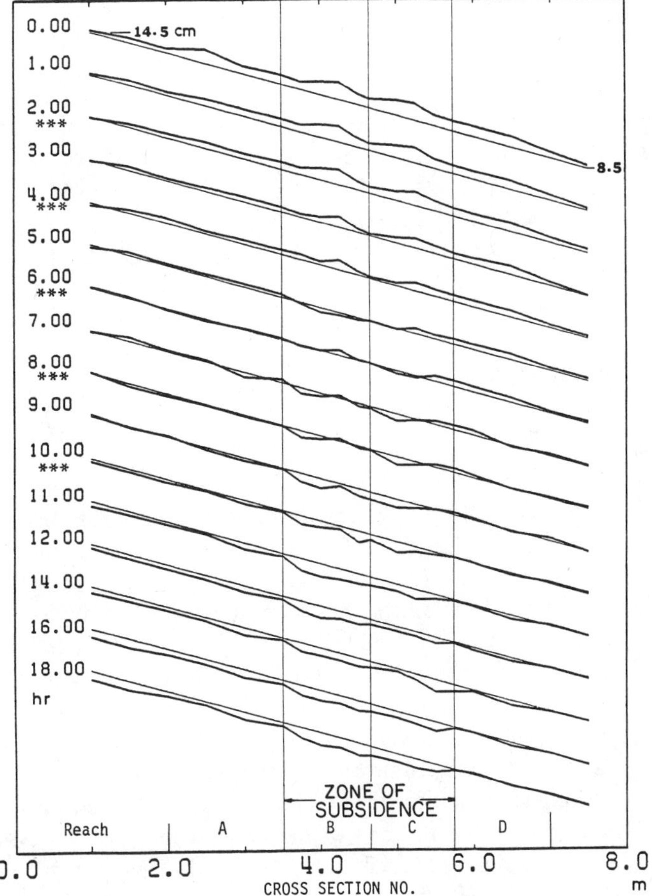

FIGURE 8.10. Bed profiles of confined straight channel during subsidence. Elevation is mean bed elevation at each cross section. Asterisks indicate number of shims removed to cause subsidence (from Ouchi, 1983).

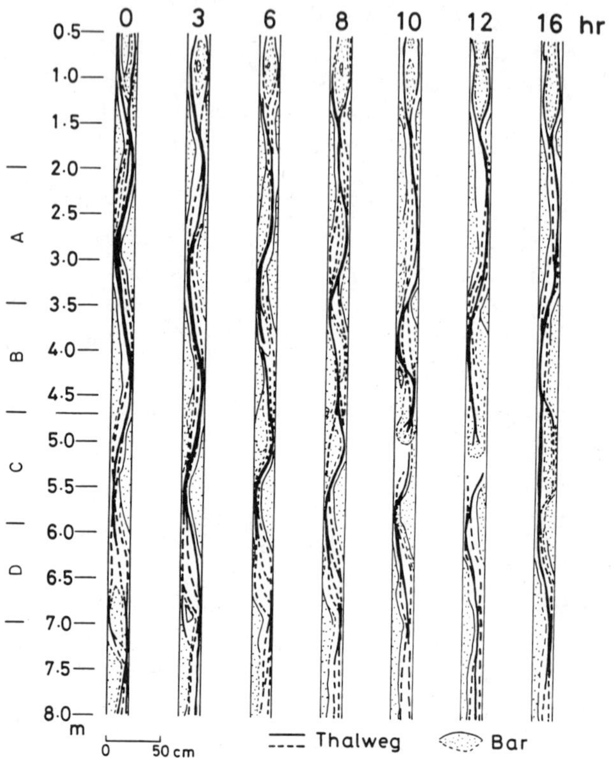

FIGURE 8.11. Pattern change of confined straight channel during subsidence. Axis of subsidence is at 4.7 m (from Ouchi, 1983).

in this narrow-channel, subsidence, which was slower than during the braided-channel experiment, was still too fast for the stream to adjust completely. A slope convexity persisted at the downstream end of reach C (Fig. 8.10), but it was much smaller than that of the braided-channel experiment (Fig. 8.5). Relatively rapid aggradation in the subsided zone and slight degradation both upstream and downstream from the subsided zone (Fig. 8.10) reduced the concavity in reaches B and C.

In summary, in spite of its greater ability to carry sediment and slower rate of subsidence (3.8 mm/2 hr), the confined straight channel did not adjust completely to subsidence. Aggradation at the lower end of the subsided zone was not complete and a concavity remained at the end of the experiment (Fig. 8.10). There was degradation in and upstream of reach A as a result of slope steepening.

As the channel aggraded, the alternate bar pattern was destroyed (8 hr, Fig. 8.11). A transverse bar formed in reach C and migrated downstream into the flooded area. Downstream from the zone of subsidence there was degradation as a result of the decrease of sediment from upstream.

Meandering Channel

Uplift

A trapezoidal straight channel 4 cm wide and with a 30° entrance bend was molded into a surface with a slope of 0.008 and composed of a 9:1 sand–kaolinite mixture. After 150 hr at 0.1 L/sec discharge, a meandering thalweg had developed. No sand was fed into the channel at the head of the flume, but during uplift kaolinite was added to the water. Uplift was 2.54 mm at 158 hr and 162 hr, then 1.27 mm at four intervals (Fig. 8.12).

A convexity in the thalweg profile developed in response to uplift (Fig. 8.12) and changes in channel pattern occurred, particularly in reach C, where slope

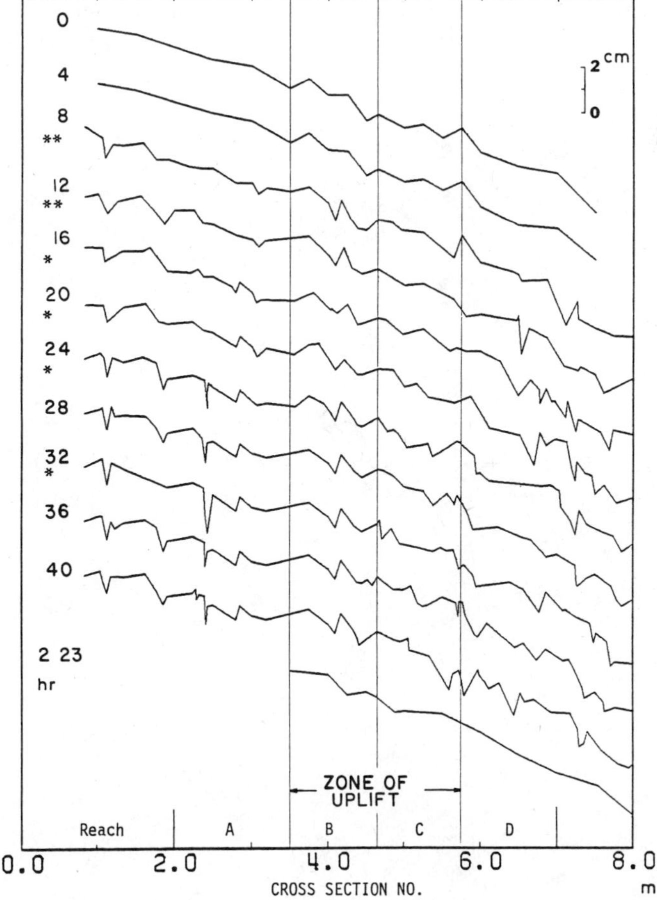

FIGURE 8.12. Projected channel profiles (channel elevation plotted along straight-line valley distance) of meandering channel during uplift. Asterisks indicate number of shims used to produce uplift (from Ouchi, 1983).

increased (Fig. 8.13). There were no significant changes in reach A, and in reach B sinuosity was unchanged, although slope decreased from 0.01 to 0.005 as a result of uplift. Reach B was flooded by 28 hr, and deposition of suspended sediment rendered the thalweg indistinct. In reach C, sinuosity was increased slightly by bank erosion, but the resistant banks prevented sufficient lateral migration for a sinuosity increase to offset the slope increase caused by uplift. In reach D, the channel tended to braid as thalweg shift occurred in response to sediment supply from the uplifted area.

The main response of the meandering channel to uplift was the increase in sinuosity in reach C where slope was steepened (Figs. 8.12, 8.13). When slope was increased, flow eroded the outer bank on the lower half of a bend. Sediment produced there was deposited on the edge of the next point bar. The growth of point bars induced further bank erosion, and this process resulted in an increase in sinuosity. This response was observed after the first uplift, although the rate of uplift (1.27–2.54 mm/4 hr) was slow. Slight thalweg aggradation was observed in

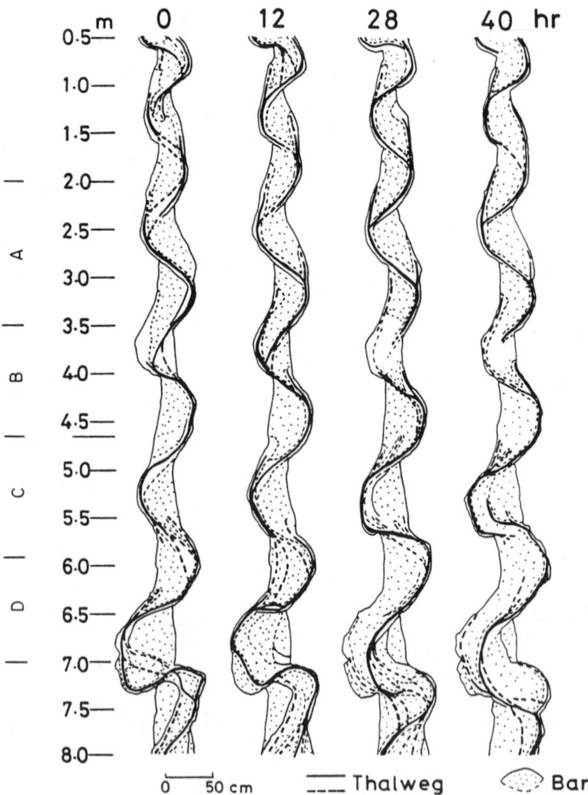

FIGURE 8.13. Pattern change of meandering channel during uplift. Axis of uplift is at 4.7 m (from Ouchi, 1983).

reach C, and some of the sediment produced by bank erosion was deposited in the thalweg of reach D, while the main deposition occurred on point bars. Because the bank or bar erosion was the source of bed load, no significant deposition occurred upstream from the uplift. Clay deposition occurred in reach B where slope was flattened, but this was not enough to increase bed elevation significantly. Water flow was distributed over the point bars and the thalweg became indistinct.

The channel had not fully adjusted its thalweg slope by an increase of sinuosity, even at the end of the experiment. Thalweg slope increased with valley slope steepening, although the rate of increase was lowered by the sinuosity increase in reach C. The relatively stable banks probably prevented, or at least delayed, the complete adjustment of slope.

Subsidence

The meandering pattern was used as the initial stage ($t = 0$ hr) of the subsidence experiment. The subsidence was made by extracting one shim (1.27 mm) every 4 hr from 8 to 36 hr (total eight shims) (Fig. 8.14).

The convexity of projected thalweg profiles, which remained from the uplift experiment (Fig. 8.12), eventually disappeared with subsidence (Fig. 8.14). A slight concavity appeared in the subsided zone at about 24 hr, and it increased with continued subsidence (Fig. 8.14). The main response to the subsidence was a sinuosity increase upstream in reach B, where slope was steepened (Fig. 8.15). In reach C the channel was flooded (Fig. 8.15; 5.5 m).

In summary, the main response to the subsidence was an increase of sinuosity in reach B. The sinuosity increase was similar to that in reach C during the uplift experiment (Fig. 8.13). Part of reach C was flooded, and the point bar in this reach was completely submerged.

Floodplain Simulation

Jin (Jin and Schumm, 1986) used the same flume to continue experiments on the effect of uplift on meandering patterns. Unlike other experiments, a layered floodplain was constructed which was composed of 40% silt and clay (kaolinite) and 60% fine sand. The silt–clay mixture was placed over fine sand and then a straight channel that extended into the underlying sand was cut (Fig. 8.16). This provided cohesive and stable banks, a source of sediment, and an opportunity for the banks to be undercut and erode. A discharge of 0.2 L/sec was introduced into the channel at a gradient of 0.0078. After 400 hr, a regular meandering channel developed, and the center of the flume was uplifted 31 mm at 424 hr, 58 mm at 448 hr, and 79 mm at 550 hr, for a total of 168 mm.

The channel was mapped at different stages of adjustment to uplift (Fig. 8.17), and although the results are not greatly different from those of Ouchi, the layered floodplain sediments provided a situation more analogous to that in the field. The resulting channel changes are striking, with flooding and chute cutoffs in reaches A and B, incision in reach C, and aggradation in reach D. The net effect was a

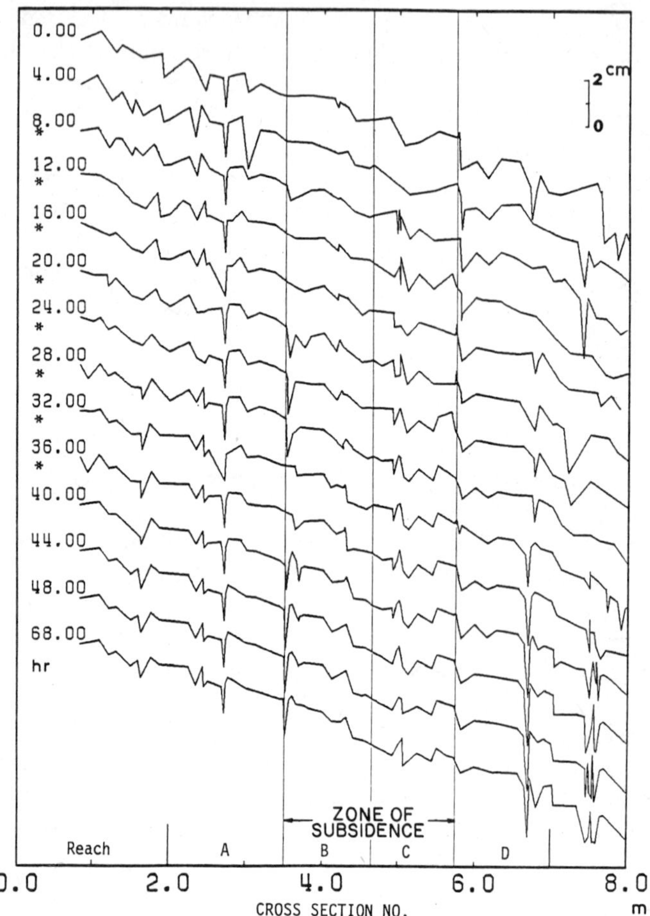

FIGURE 8.14. Projected channel profiles of meandering channel during subsidence. Asterisks indicate number of shims removed to produce subsidence (from Ouchi, 1983).

remarkable alteration of the regular meandering pattern between 400 and 500 hr of run time.

In reach A, the valley floor was not deformed by uplift, but backwater effects from reach B reduced sinuosity and fine sediment was deposited on point bars and in pools. In reach B, the valley slope was reduced and the reduced velocity of flow caused sediment deposition, overbank flooding, and cutoffs. As incision progressed upstream through the axis of uplift, the floodplain was converted to a terrace in the downstream part of reach B (453 hr; Fig. 8.17).

Reach C was affected by uplift in three ways: the gradient was increased; a nickpoint moved upstream toward the axis of uplift and into reach B; and the incision increased the exposure of sand beneath the floodplain sediment, which

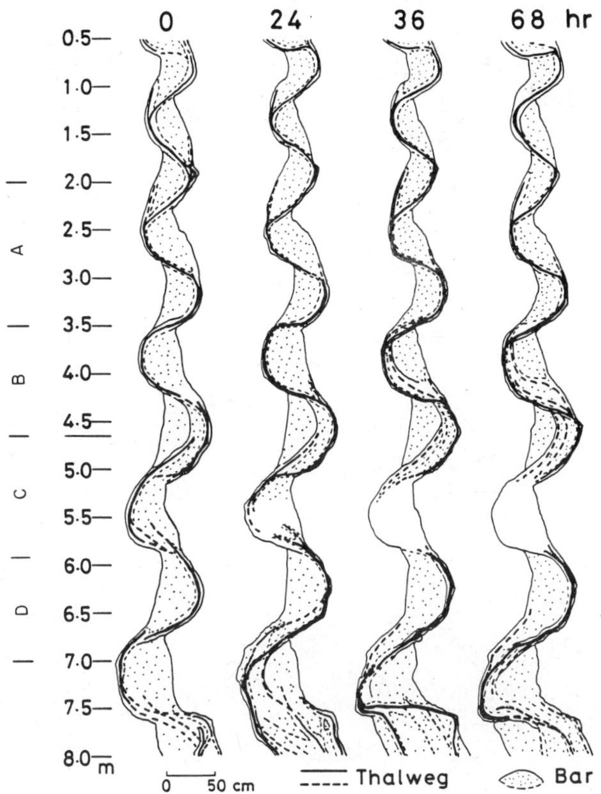

FIGURE 8.15. Pattern change of meandering channel during subsidence. Axis of subsidence is at 4.7 m (from Ouchi, 1983).

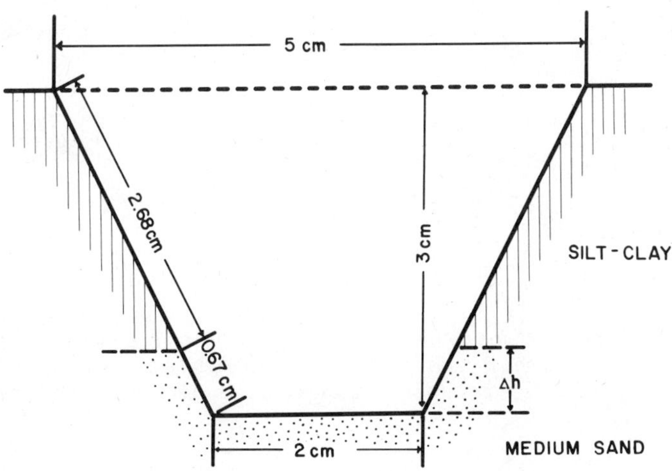

FIGURE 8.16. Cross section of Jin's channel showing exposure of silt–clay mix and sand in channel perimeter.

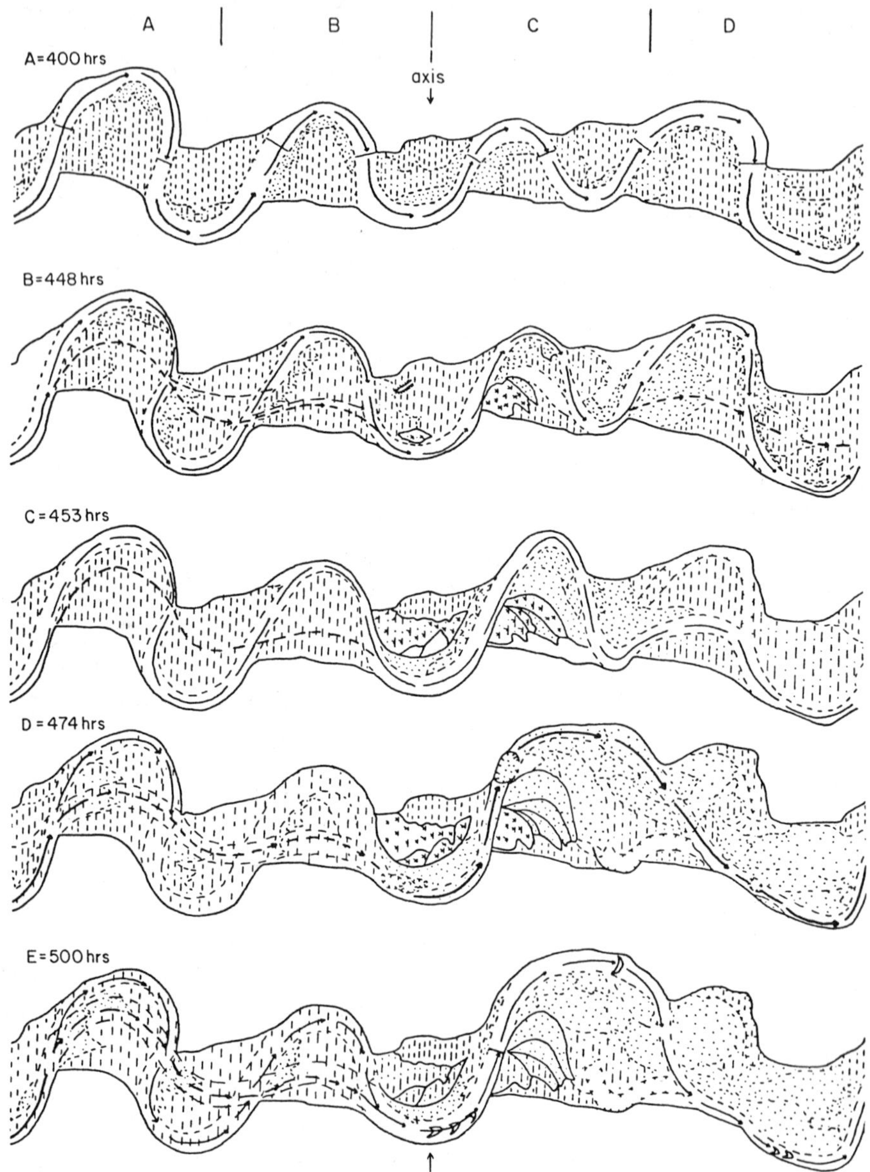

FIGURE 8.17. Channel patterns as a result of uplift (from Jin and Schumm, 1986). Figures show 4.5-m section of channel. Reaches B and C are directly affected by uplift at axis. Patterns represent the following: dots represent sand, dashes represent silt and clay, checks represent a terrace remnant.

caused bank instability and erosion and terrace formation. Each uplift produced a new terrace, until there were three. The increased bank instability and sediment production led to cutoffs and greatly altered meander amplitude, wavelength, and sinuosity (Fig. 8.18). In reach D, the increased sediment delivery from upstream caused deposition and alternation of the channel pattern, but, unlike the results of Ouchi's experiment, sediment delivery was not sufficient to form a braided pattern.

As Fig. 8.18 reveals, there were initial variations among the reaches. For ex-

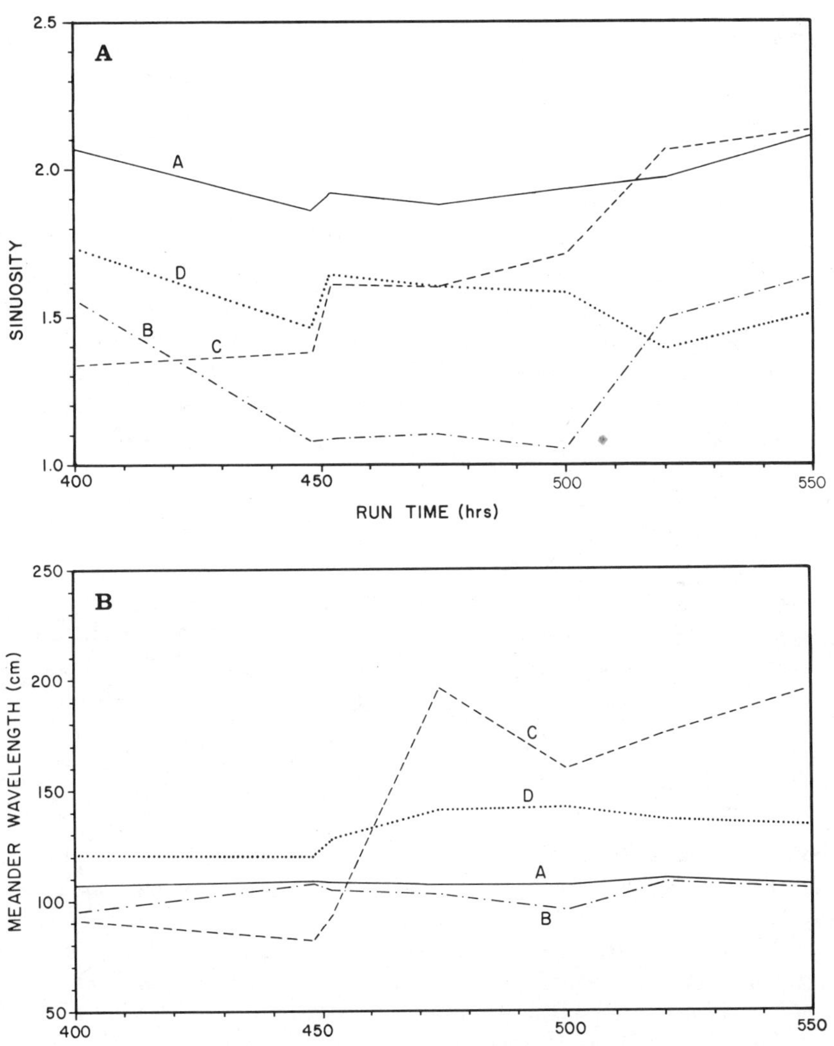

FIGURE 8.18. Change of channel pattern during uplift in the four channel reaches (from Jin and Schumm 1986). (*A*) sinuosity, (*B*) meander wavelength, (*C*) meander amplitude.

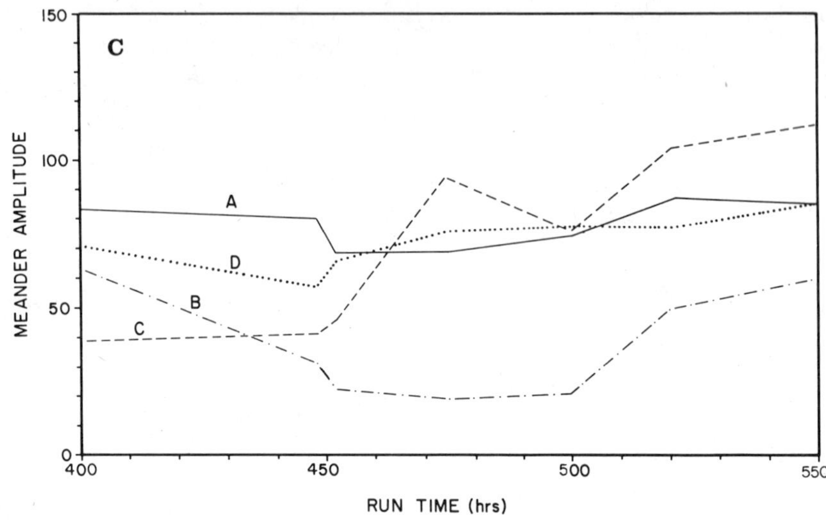

FIGURE 8.18. *(Continued)*

ample, at the beginning of deformation (400 hr) sinuosity was high in reach A and considerably lower in reach C. Channel adjustment to uplift was significant except in reach A, although there were slight decreases of sinuosity and meander amplitude as a result of deposition in reach A and a tendency for avulsion by chute cutoff (Fig. 8.17). The decreased slope of reach B caused a marked decrease of sinuosity and meander amplitude, but meander wavelength was unchanged (Figs. 8.17, 8.18). The increased slope in reach C caused a marked increase of sinuosity, meander wavelength, and amplitude. In spite of the dramatic channel changes in reach D, (Fig. 8.17) there were only slight changes of sinuosity, wavelength, and amplitude (Fig. 8.18).

DISCUSSION

The responses of the experimental channels are summarized by Ouchi (1983) in Table 8.1. Braided and confined straight channels responded to uplift in the same way, with incision and terrace formation in the uplifted area. This was preceded by a short period of lateral instability and erosion. There was aggradation downstream of the uplift and a tendency for aggradation upstream. However, when uplift ended, the channels returned to their original pattern.

The meandering channel responded to uplift primarily by an increase in thalweg sinuosity where slope was steepened, and flooding, clay deposition, and obscuring of the formerly distinct thalweg occurred where slope was flattened.

TABLE 8.1 Effect of Uplift and Subsidence on Channel Morphology

ZONE		A	B	Axis C	D
BRAIDED CHANNEL	UPLIFT		Aggradation Thalweg Shift Submerged Bars	Degradation Terrace Formation Single Bars	Aggradation Braided
BRAIDED CHANNEL	SUB-SIDENCE		Degradation Single Thalweg	Aggradation Braided Flooding	Degradation Single Thalweg
MEANDERING CHANNEL	UPLIFT		Aggradation Flooding Multiple Channels	Degradation Sinuosity Increase Bank Erosion	Aggradation
MEANDERING CHANNEL	SUB-SIDENCE	Degradation	Sinuosity Increase Bank Erosion	Aggradation Flooding, Cutoffs Multiple Channels	Local Scour

Both braided and straight channels responded to subsidence with rapid aggradation as a result of degradation of the steepened slope upstream, and a braided pattern resulted. Since sediment did not move through the subsided, ponded area, there was a tendency for degradation of the sediment-starved reach D. The meandering channel responded to subsidence in the same way as to uplift, with an increase in sinuosity in the steepened reaches and flooding and deposition where the slope was reduced.

The responses of the channels were conditioned both by rates of uplift and subsidence and by other factors such as sediment load and type of load. The response to uplift in the confined-straight channel was most rapid because of the greater energy of the confined flow, but because of its rapid uplift rate (3.8 mm/2hr) it was unable to adjust fully to the uplift, as did the braided channel (with an uplift rate of only 1.27 mm/2hr). On the other hand, extremely rapid subsidence (6.35 mm/hr) prevented the braided channel from adjusting fully. The rate of subsidence in the straight channel was the same as for uplift, but its effect was more marked than that of uplift, which indicates that aggradation occurred more slowly than did incision. Aggradation could only proceed at a rate controlled by sediment availability, whereas degradation was conditioned by stream power (that is, by increasing slope). In addition, in order to restore the valley slope, aggradation was required over the entire valley floor, but adjustment by incision was in only a narrow part of the valley floor. Therefore, more sediment was needed for slope adjustment by aggradation than was required to be removed by incision. Also, the adjustment of the meandering stream to slope change by changing sinuosity was considerably slower than by the adjustment of the braided stream by degradation or aggradation.

It appeared that the meandering stream adjusted to the uplift and subsidence by lateral changes, modifying channel slope by changing sinuosity, whereas the braided and straight channels responded primarily by vertical changes. In the latter case there was, however, an initial period in Ouchi's experiments when lateral changes accommodated uplift of up to 6 mm, before vertical adjustment occurred.

The experimental observations may be combined with existing knowledge of the relationship between channel morphology and controlling factors (Chapter 5) to propose models for the response of braided and meandering channels (Fig. 5.2) to tectonic movement. Pattern threshold conditions are important; for example, a meandering channel that is very close to the meandering–braided slope threshold will change its channel pattern with only a small slope increase due to uplift, whereas a meandering channel that is well inside the "meandering channel domain" (Fig. 5.33*D*) may never do so.

The observations made by Ouchi (1983) are complementary to those by Gardner (1973, 1975), and they may help to explain under what circumstances meanders may be incised from an alluvial cover into a rising bedrock mass or will be swept out and replaced by an incised straight channel.

The numerous factors affecting alluvial rivers, such as human activities, influence of tributaries, effects of rare hydrologic events, and the differences among alluvial rivers (Fig. 5.2), may make it difficult to identify the effects of tectonic movement. For example, slow movements of the earth's surface may be accommodated by small, essentially continuous river adjustments (Fig. 8.19*a*), or the

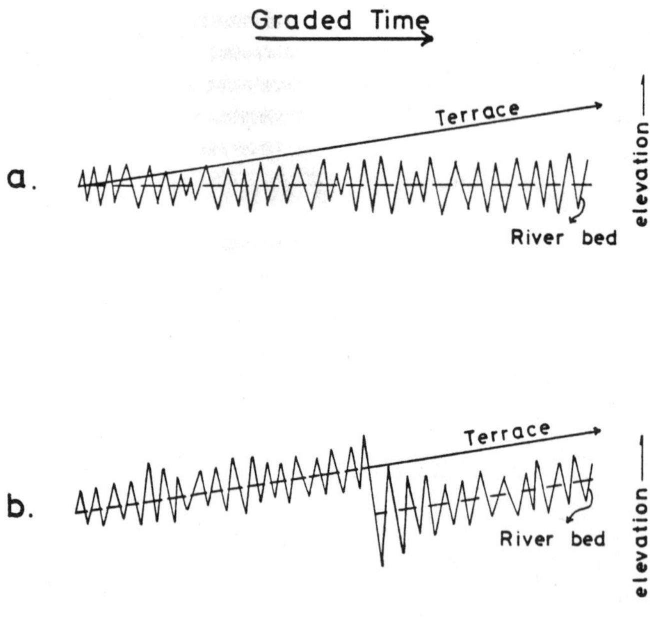

FIGURE 8.19. Diagrams showing two types of river response to uplift: (*a*) continuous adjustment, (*b*) episodic adjustment.

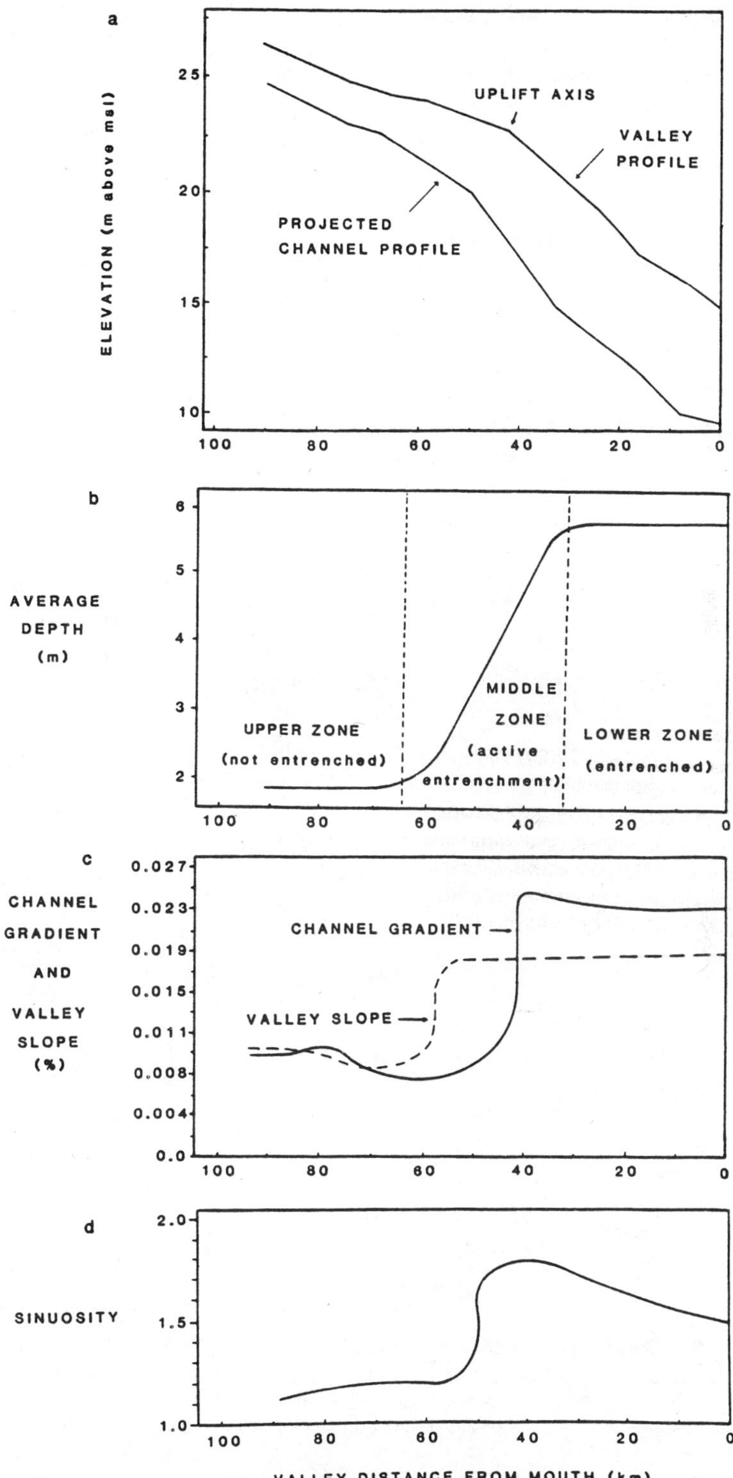

FIGURE 8.20. Change of channel morphology associated with uplift, Big Colewa Creek, Louisiana (from Burnett and Schumm, 1983).

adjustment may not occur until the deformation becomes large enough for the river to exceed a threshold (Fig. 8.19*b*).

In addition to these primary influences on channel and valley gradient and configuration, there will be secondary effects as the rivers respond to the changed gradient (aggradation or degradation) and tertiary effects as decreased or increased sediment loads influence reaches downstream of the deformed reach and decreased or increased stream power influences transporting capacity upstream of the deformed reach (Table 8.1).

There are several excellent examples of river response to active tectonics, such as the Lake County Uplift area near New Madrid, Missouri (Russ, 1982). In this area the longitudinal profiles of old Mississippi River channels, natural levees, and the modern floodplain are warped. Above the uplift the Mississippi River is relatively straight, whereas downstream it is highly sinuous. Investigations in the area of the Monroe Uplift in northern Louisiana and the Wiggins Uplift of central Mississippi reveal channel behavior (Fig. 8.20) that conforms to that expected from the experimental studies (Burnett, 1982; Burnett and Schumm, 1983). Ouchi (1985) also found examples of channel adjustment on the coastal plain of Texas, the Rio Grande valley of New Mexico, and the central valley of California (Ouchi, 1985; Schumm, 1986).

Another aspect of active tectonics is the uncovering of more resistant materials in the channel. More resistant rock will confine the channel and retard meander shift and bank erosion. The result should be deformed or compressed meanders upstream and a change of meander character at the contact (Figs. 5.42, 7.18). Recent studies of Mississippi River morphology, where the river crosses the Monroe and Wiggins structures, indicates the effect of relatively resistant Tertiary clays in the channel. These outcrops have caused meander compression (Greenville Bends) and reaches of relative channel stability where the clay prevents channel shift (Schumm et al., 1984).

Although terraces were formed during the experiments, they were soon de-

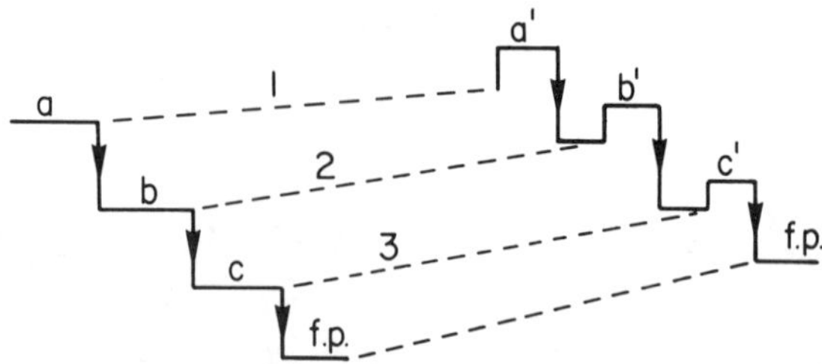

FIGURE 8.21. Sketch showing channel behavior upstream and downstream of an axis of uplift. Incision downstream following uplift forms terrace a, while aggradation upstream forms surface a′, which becomes terrace a′ after incision. The sequence is repeated during uplifts 2 and 3 (after Coleman, 1958) (f.p. indicates floodplain).

stroyed. What might have been expected if they were preserved is described by Coleman (1958) for terrace flights that formed in two segments of the Salzach River valley in Austria during Holocene time as a result of uplift of the valley in the area of the Salzachöten gorge. She showed that uplift produced deposition and formation of an alluvial deposit upstream while incision occurred downstream. Then, when the axis of uplift is incised during a period of tectonic stability, the upstream deposit is also incised, to form a terrace that is really equivalent to the downstream terrace but formed at a different time (Fig. 8.21).

III DEPOSITIONAL LANDFORMS AND SEDIMENTOLOGY

The outline of this book was designed to lead the reader through the fluvial system from the drainage basin (Part I), through the channels (Part II), to the piedmont or coast, where depositional processes predominate. The emphasis in Part III will be on morphology, dynamics, and sedimentology of alluvial fans.

The studies of alluvial fans (Chapter 9) and fan deltas (Chapter 10) were done by Weaver (1984) and Jackson (1981), respectively, while Macke (1977) evaluated their sedimentologic and stratigraphic characteristics (Chapter 11), which provides an important link between the internal composition of depositional landforms and the geomorphic processes observed at the surface.

As with other landforms, alluvial fans often reveal a great range of morphologic characteristics within one area, providing an example of landform diversity and complexity that often eludes traditional explanations. The experimental studies are therefore a valuable adjunct to fieldwork and permit documentation of evolutionary changes that cannot be observed on the prototype because of the infrequent occurrence of geomorphically significant events. They also allow isolation of the effects of the various influences on fan behavior in a controlled fashion.

9 Alluvial Fans

An alluvial fan is an accumulation of sediment that has been deposited where a debris-laden stream emerges from the confined valley of an upland area onto the piedmont, where it is free to spread laterally and deposit its load. The ideal form of an alluvial fan is semicircular in plan (Murata, 1966; Rachocki, 1981). Because of their excellent exposure and ease of investigation, alluvial fans in arid and semiarid areas have received the greatest attention in recent scientific literature (for example, see Bull, 1964a, b; Denny, 1965; Hooke, 1967, 1968a; Lustig, 1965). However, fans are also common features of more humid regions such as Japan (Murata, 1966; Saito, 1981, 1982), Canada (Ryder, 1971a, b; Winder, 1965); New Zealand (Carryer, 1966), arctic Scandinavia (Hoppe and Ekmar, 1964), the Canadian arctic (Leggett et al., 1966), Alaska (Anderson and Hussey, 1962), the Himalayan mountains (Drew, 1873), Australia (Wasson, 1974, 1977), Spain (Harvey, 1984a, b), and India (Gole and Chitale, 1966; Williams, 1977). While it is now generally accepted that fluvial processes and debris flows are important, humid region fan-building processes (McGowan, 1974; Kochel and Johnson, 1984), fans in glacial and arid regions may also be composed of virtually all mudflow and debris flow material (Beaty, 1970, 1974; Bluck, 1964; Ryder, 1971a), various proportions of mudflow and streamflow deposits (Bull, 1964a; Ryder, 1971a; Wasson, 1977), or essentially all fluvially deposited sediment (Blissenbach, 1954; Funk, 1976; Melton, 1965). Even neighboring fans, such as those along the base of the Panamint Mountains in Death Valley, California, and southeast Spain (Harvey, 1984b), show wide ranges in internal composition that suggest a complicated pattern of mixed-mode deposition. Similarly, alluvial fans in the same geomorphic region can be actively growing (Beaty, 1970), undergoing dissection (Hunt and Mabey, 1966), or be in a steady-state equilibrium (Denny, 1967). Beaty (1974) argues strongly for the importance of chance catastrophic discharges in producing mudflow alluvial fans in the White Mountains of California and Nevada, whereas observations by Hooke and Rohrer (1979, p. 152) indicate that alluvial fans are adjusted to the more frequent moderate-magnitude episodes.

Geomorphic Thresholds

Hypotheses that attempt to explain the dynamic and seemingly variable nature of depositional systems have been primarily concerned with alterations imposed on the landforms from outside the system, including such forces as climatic fluctuations, tectonism, baselevel changes, or intensified land use activities. However, these hypotheses fail to explain adequately why in some regions landforms that have been subject to identical environmental forces during their evolution are not in the same stage of geomorphic development. The obvious differences in the morphologic details of landforms within the same limited area suggest that at least the most recent phase of evolutionary development has not been a continuous process that is interrupted only by the operation of external controls. To account for the observed complexity, it was assumed that the evolution of many depositional landforms is controlled not only by the action of external driving forces but also by the existence of intrinsic thresholds of geomorphic stability (Schumm, 1973, 1977). When a geomorphic threshold is exceeded, sudden and dramatic landform response may occur even during a period in which external factors remain constant, resulting in marked geomorphic dissimilarity and stratigraphic complexity.

Fanhead Trenches

Because alluvial fans are depositional landforms, no single feature has received more attention than the incised channel that traverses the proximal region or apex of many alluvial fans. These fanhead trenches, which contain all the upstream flow and convey sediment downfan, were described as early as 1873 by F. Drew in the upper Indus basin. He observed abundant unincised alluvial fans as well as trenches resulting from toe trimming by adjacent rivers and from baselevel lowering caused by valley downcutting beyond the fan margins.

Since Drew's (1873) investigation, numerous descriptions and studies have dealt with the characteristics, implications, and causes of fanhead trenches (Table 9.1). Wasson (1977) has outlined two types of causes: (1) those which represent fundamental changes in regime that are usually derived from varying external conditions and (2) those which result in apex incision without the operation of external catalysts.

Hooke (1967, p. 458) recognized two basic types of fanhead trenches that result from these two groups of causes. The first group is composed of those that are produced under prevailing discharge and sediment caliber conditions and are not influenced by other factors (e.g., Table 9.1, causes 11, 12, and 13). Such trenches are considered to be short-term features that are subject to frequent overbank flooding. In the second case, trenches that are attributed to major external influences such as tectonism and climate change are so deep as to prevent overbank flooding.

During geologically significant periods of time, continued erosion, sediment

Table 9.1 Identified Causes of Fanhead Entrenchment

Cause of Apex Incision	Selected Reference(s)
1. Climate change towards more arid conditions	Lustig (1965)
2. Deglaciation	Funk (1976), Ryder (1971a), Wasson (1977)
3. Regime change from predominance of mudflows towards more frequent streamflow conditions	Bluck (1964)
4. Regime change resulting in an increased frequency of mudflows	Lustig (1965)
5. Stream capture in the drainage basin resulting in increased discharge to fan	Dzurisin (1975), Eckis (1928)[a]
6. Tectonism (uplift and/or tilting)	Bull (1964b), Hooke (1968b)
7. Decreased load	Eckis (1928)[a], Ryder (1971a)
8. Increase in the frequency of high-magnitude rainfall events with corresponding decrease in frequency of low-magnitude rainfalls	Bull (1964)
9. Extreme event of intense rainfall, runoff	Beaty (1970, 1974), Bull (1964), Denny (1967)
10. Destruction of drainage basin vegetation, resulting in increased surface runoff	Bull (1964), Eckis (1928)[a]
11. Alternation of debris flows and water flows	Hooke, (1967)
12. Erosion of fan surface, headward gully erosion, followed by capture	Denny (1967), Rich (1935)
13. Lateral channel migration to steeper areas on the fan surface	Hooke (1967), Rich (1935)
14. Baselevel lowering (adjacent valley incision)	Drew (1873), Ryder (1971b)
15. Toe trimming (valley stream encroachment)	Drew (1873), Wasson (1977)
16. Basin downwearing over geologic time	Eckis (1928)

[a] Mentioned without elaboration.

production, and sediment transport should diminish as the morphology of a drainage basin is altered. Eckis (1928), in his study of the Cucamonga fan at the base of the San Gabriel Mountains in southern California, recognized that the general diminution of channel gradients and sediment yield through geologic time will affect the evolution of alluvial fans. He suggested that as sediment yields decrease, alluvial fans should become deeply entrenched.

Fanhead incision causes shifting of sediment from the apex to downfan areas. It results in the planimetric growth of young fan segments and in the replacement of material lost by surface erosion on older fans. Because it is recognized that apex deposition cannot continue in an uninterrupted fashion for indefinite periods of time (Lustig, 1974, p. 617), natural fanhead incision must be integral to any reasonable evolutionary model of alluvial fan growth.

EXPERIMENTAL STUDIES

An alluvial fan is an ideal landform to study experimentally (Hooke, 1967; Hooke and Rohrer, 1979; Rachocki, 1981). It is relatively easy to model the conditions responsible for fan formation and growth, which are a sediment source area, a well-defined depositional area, and kinetic energy in the form of artificial precipitation to erode and transport sediment.

Such gross external variables as climate change, tectonism, and changes in baselevel and land use practices can also be roughly modeled, but, perhaps even more important, these variables can be eliminated. Strict control over source material variability and erodibility, energy inputs (precipitation and relief), and external controls such as tectonics, baselevel, and land use allows the experimenter to evaluate the effect of any one of these variables without the complications inherent in the natural system. This is in marked contrast to the prototype where a multitude of variables of unknown importance and magnitude combine to produce the fan.

Weaver's (1984) first experiment doucmented the evolution of a single alluvial fan formed exclusively by streamflow as sediment was discharged from an eroding and evolving drainage basin that was subjected to a constant and steady rate of artificial precipitation. This was the fluvial fan. Then fans were studied that formed in response to discrete rainfall events or storms. Depending on initial conditions and the character of sediment delivery, the fans were termed episodic fluvial, mixed-mode, or mudflow fans.

The rainfall-erosion facility (REF) was modified for the alluvial fan experiments (Fig. 2.10) by constructing a wall with a central outlet across the center. This provided a 56.1-m^2 sediment source area (drainage basin) and a 69.7-m^2 depositional area.

The sprinkler system was the same as that used for the drainage network experiments (Chapter 2). The mean precipitation rate was 5.54 cm/hr, which compared closely with the equilibrium discharge of 51 mm measured at the watershed outlet. The areal distribution of precipitation over the experimental area was not uniform (Fig. 9.1), but the replication of the pattern was excellent. The depositional area was protected from drop impact by a plastic sheet suspended above it.

The same experimental material that was used for the drainage network studies (Chapter 2) was used during the alluvial fan experiments (Table 2.1). During the experiments, sediment samples were taken at regular and frequent intervals at the outlet of the experimental drainage basin.

A point gauge mounted on the mobile carriage, which spanned the width of the REF, was used to measure changes in source area morphology. However, this system could not be used to measure alluvial fan morphology, and as a result a network or grid of 0.25-cm-diameter, 0.30-m-long pins was inserted into the depositional area. The growth of the alluvial fans was monitored by recording the gradual burial of these pins.

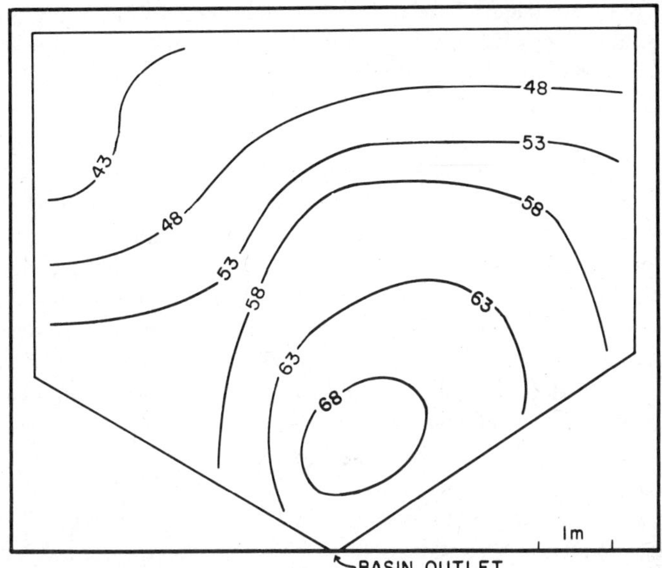

FIGURE 9.1. Precipitation (mm/hr) measured over experimental drainage basin, averaged for five 1-hr precipitation episodes (from Weaver, 1984).

FLUVIAL FAN EXPERIMENT

For the first fluvial fan experiment, a silt–sand mixture was placed in the upper, or western, half of the REF to a depth of nearly 2 m. The sediment was graded and slightly compacted to form a symmetrical, slightly concave drainage basin.

During the experiment, precipitation was applied to the entire source area at full intensity (55.4 mm/hr). At about 2-hr intervals the sprinkler system was shut down, the drainage network and alluvial fan were measured by use of the point gauge and the erosion and deposition pins, and photographs were taken. The duration of each run or precipitation event depended on the rate of change of the landforms of interest.

Because of the unconstrained nature of the fluvial fan, it eventually became necessary to restrict its lateral expansion during the first experiment. This was accomplished by placing a permeable gravel barrier roughly 5.5 m beyond the drainage basin outlet and around the perimeter of the developing fan. Water and suspended sediment were free to pass through the barrier, but silts and sands remained behind on the fan surface. Such a boundary may be analogous to the physical constraints imposed by coalescing alluvial fans in nature. The artificial boundary conditions did not appear to affect the outcome of the experiment.

Baselevel was lowered three times, at 8, 11, and 13 hr, until basin relief was

1.36 m. A drainage pattern incised in the source area while the fan grew in the depositional area, until the fan surface and the channel at the outlet became accordant at the end of run 8A. Data collection began with run 9A (Table 9.2).

In addition to morphological changes occurring within the system, total sediment load was measured at 10-min intervals by capturing discharge at the outlet of the developing drainage network with a specially constructed sediment and water sampler. During the experiment, a total of 450 samples were collected (Fig. 3.9).

An alluvial fan develops in direct response to the discharge of erosional products from an evolving drainage basin upstream. Those processes that play a key role in the production and subsequent yield of sediment from the source area are therefore of primary importance in fan growth. Three fundamental processes were observed to control sediment yield as follows: (1) drainage network expansion, (2) in-channel scour and channel-slope interactions, and (3) slope processes. The relative contribution of each of these processes varied with time during the evolution of the drainage network; the first two processes were dominant in the early stages.

During the latter part of the experiment, the down-wearing of slopes and divides, together with occasional channel widening by lateral scour, accounted for the greatest sediment production. Sediment derived from raindrop-impact erosion remained relatively constant, with perhaps a gradual reduction stemming from the continually diminishing transport efficiency of low-gradient hillslopes.

Growth of the alluvial fan and the consequent rise of baselevel accounted for the accumulation of thick (8 cm) valley fills in the lower 2.5–4.5 m of the three main stream channels. Fanhead trenching caused frequent reworking of this deposit.

Growth Patterns

The experimental fluvial fan grew in area and volume with time (Fig. 9.2). The rate of areal growth declined rapidly from 3.3 m^2/hr initially to 0.46 m^2/hr after 20 hr as sediment yield declined (Fig. 3.9) and as increasing quantities of sediment contributed to vertical growth of the fan.

The growth of the alluvial fan was recorded by the repeated measurement of the 225 pins on the depositional surface. A number of reoccurring patterns of fan deposition are visible on the maps that were prepared from the pin data. They include (1) widespread, relatively uniform deposition over the entire fan surface (Fig. 9.3*A*); (2) single concentrations of deposition located at some point on the perimeter of the fan (Fig. 9.3*B*, *C*); (3) single concentrations of deposition centered near the midfan or apex areas (Fig. 9.3*D*); (4) two centers of deposition, with a relatively high angle of separation relative to the apex (Fix. 9.3*E*); and (5) two centers of deposition with a very low separation angle (Fig. 9.3*F*).

Deposition at the fan perimeter was associated with periods of channelization of the upper fan and midfan. Frequently, a single trench several meters long developed that conveyed sediment and water to the toe of the fan while gradually shifting laterally and reworking the fanhead. Sediment was thus moved downfan and the trench slowly backfilled from toe to apex. This general sequence of events—

Table 9.2 Run Durations and Drainage Basin Relief during the Fluvial Fan Experiment[a]

Run No.	Duration (min)	Duration (m³)	Cumulative Duration (min)	Cumulative Duration (m³)	Drainage Basin Relief (m)	Run No.	Duration (min)	Duration (m³)	Cumlative Duration (min)	Cumlative Duration (m³)	Drainage Basin Relief (m)
1	60	2.9	60	2.9	1.03[b]	10B	120	5.7	2343	111.5	1.36
2	60	2.9	120	5.7	1.03	10C	110	5.2	2453	116.8	1.36
3	60	2.9	180	8.6	1.03	10D	61	2.9	2514	119.7	1.36
4A	60	2.9	240	11.4	1.03	10E	120	5.7	2634	125.4	1.36
4B	60	2.9	300	14.3	1.03	10F	121	5.8	2755	131.2	1.36
5A	60	2.9	360	17.1	1.03	10G	95	4.5	2850	135.7	1.36
5B	60	2.9	420	20.0	1.03	10H	61	2.9	2911	138.6	1.36
6A	60	2.9	480	22.8	1.13[b]	10I	121	5.8	3032	144.4	1.36
6B	60	2.9	540	25.7	1.13	10J	121	5.8	3153	149.3	1.36
7A	120	5.7	660	31.4	1.25[b]	10K	54	2.6	3207	152.7	1.36
7B	120	5.7	780	37.1	1.25	10L	122	5.8	3329	158.5	1.36
8A	70	3.3	850	40.5	1.25	10M	122	5.7	3451	164.3	1.36
9A[c]	60	2.9	910	43.3	1.36[b]	10N[d]	119	5.7	3570	170.0	1.36
9B	60	2.9	970	46.2	1.36	11A	121	5.8	3691	175.7	1.36
9C	60	2.9	1030	49.0	1.36	11B	121	5.8	3812	181.5	1.36
9D	51	2.4	1081	51.5	1.36	11C	121	5.8	3933	187.2	1.36
9E	83	3.9	1164	55.4	1.36	11D	121	5.8	4054	193.0	1.36
9F	92	4.4	1256	59.8	1.36	11E	121	5.8	4175	198.8	1.36
9G	100	4.8	1356	64.6	1.36	11F	121	5.8	4296	204.5	1.36
9H	89	4.3	1445	68.8	1.36	11G	121	5.8	4417	210.3	1.36
9I	120	5.7	1565	74.5	1.36	11H	121	5.3	4528	215.6	1.36
9J	120	5.7	1685	80.2	1.36	11J	130	6.2	4658	221.7	1.36
9K	86	4.1	1771	84.3	1.36	11K	121	5.8	4779	227.5	1.36
9L	90	4.3	1861	88.6	1.36	11L	121	5.8	4900	233.3	1.36
9M	91	4.3	1952	92.9	1.36	11M	121	5.8	5021	239.0	1.36
9N	90	4.3	2040	97.2	1.36	11N	121	5.8	5142	244.8	1.36
9O	60	2.9	2102	100.1	1.36	11O	122	5.8	5264	250.6	1.36
10A	121	5.8	2223	105.8	1.36	11P	120	5.7	5384	256.3	1.36

[a] Expressed as time (min) and as water discharged from the source area (m^3).
[b] Baselevel lowering.
[c] Fan apex becomes accordant with main channel of drainage basin.
[d] Initiation of phase two: addition of plastic squares on source-area interfluves.

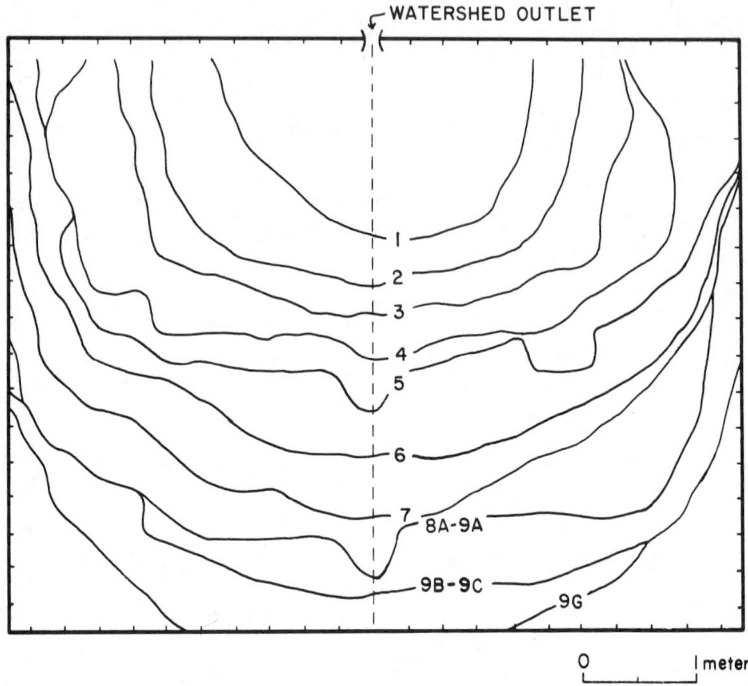

FIGURE 9.2. Growth of the fluvial fan through run 9G when a permeable gravel barrier was placed around the fan perimeter. Tick marks indicate the location of rows and columns of measurement pins (from Weaver, 1984).

apex trenching, lateral channel migration, and channel backfilling—was repeated numerous times during the evolution of the experimental fan (Fig. 9.4). Following channel backfilling there was widespread, dispersed deposition over the midfan and apex, which caused vertical fan building as opposed to lateral expansion (Fig. 9.4*D*). At this time water and sediment from the source area spread over the fan apex in a number of distributary channels, which continued to bifurcate in the downstream direction. During these periods of deposition on the midfan and apex, the fanhead was rapidly built up relative to the lower fan. Slopes near the apex increased in gradient with time and fanhead trenching occurred frequently. During the evolution of the fan, such periods of rapid accumulation at and near the apex occurred following each backfilling event.

The growth and evolution of the experimental alluvial fan was dominated by repeated fanhead trenching and the areal extent of flowing water and active channels, whether aggrading or degrading, was in large part dependent on flow conditions at the fanhead (Fig. 9.5). When numerous streams were spread over the fan surface (Fig. 9.6), there was rapid deposition near the apex. In contrast, the

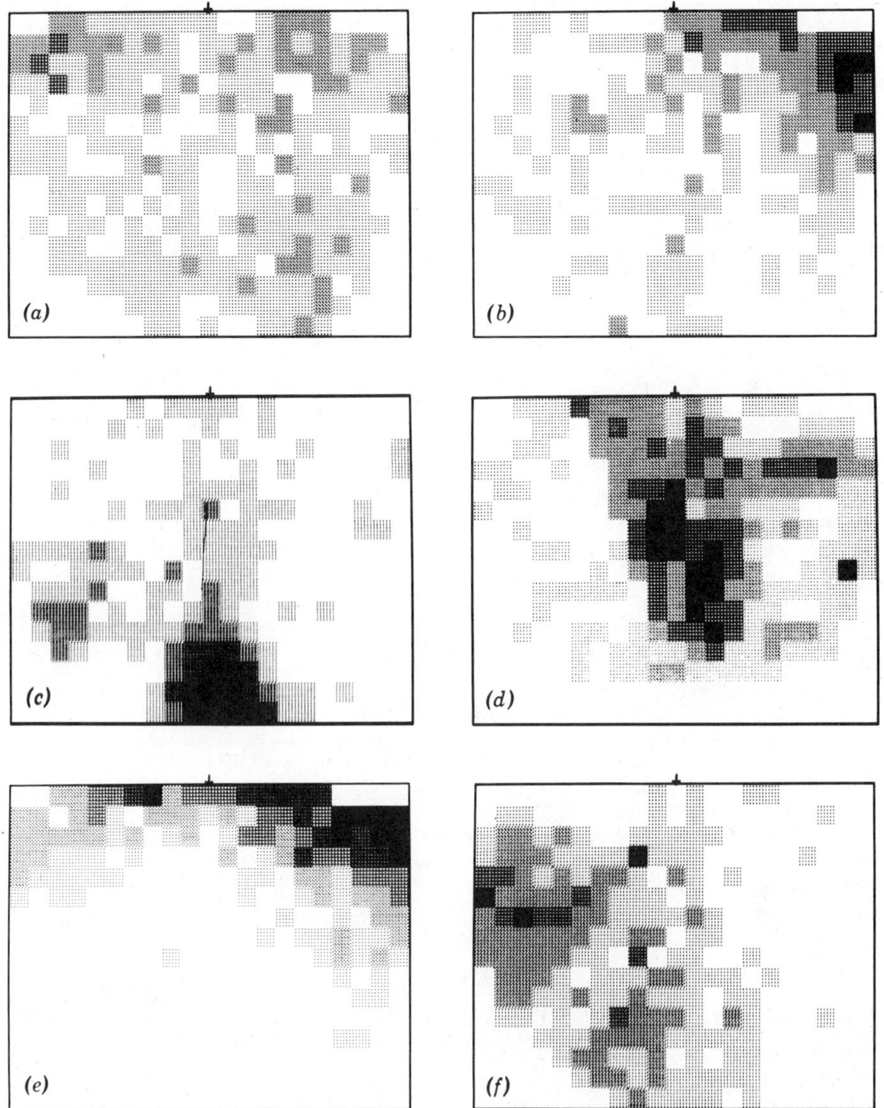

FIGURE 9.3. Computer-generated maps of relative depositional rates showing the six different depositional patterns observed during fan growth. The darkest pattern indicates 4.6–6.1 cm of deposition, the white areas received less than 3 mm of deposition during a run. The eight other patterns represent 6-mm increments of deposition: (*A*) Widespread deposition over the entire fan (run 10A); (*B*) single concentration of deposition near toe at the right fan margin (run 10M); (*C*) single concentration of sediment near toe on fan axis (run 11M); (*D*) single concentration of deposition in central midfan and apex regions (run 9*C*); (*E*) two centers of deposition displaying a wide angle of separation (run 10B); (*F*) two dispersed centers of deposition displaying a low angle of separation (run 11C). Blank areas in the lower left and lower right corners of the figure were beyond the fan margin (from Weaver, 1984).

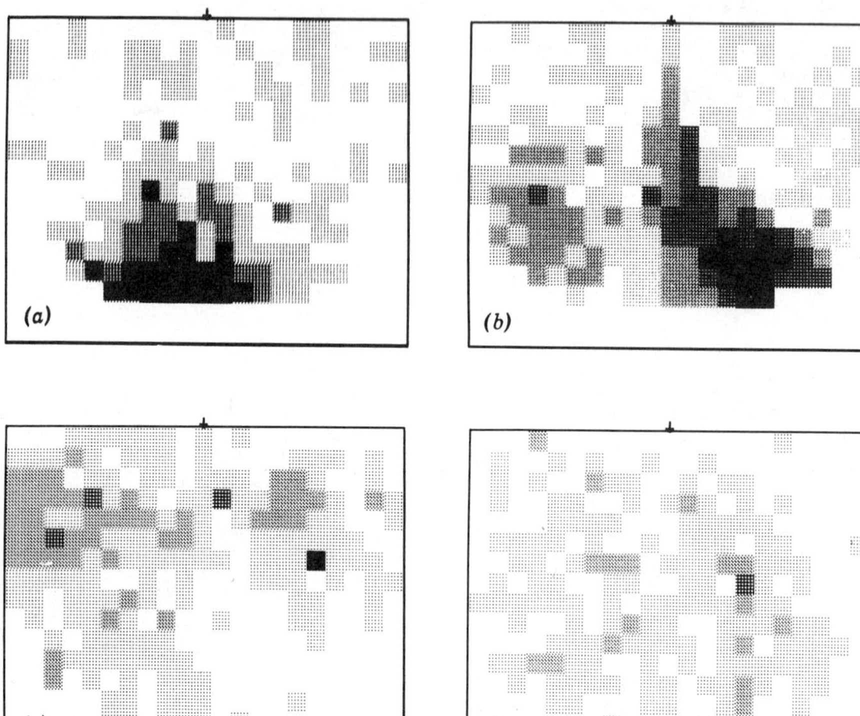

FIGURE 9.4. A typical cycle of fanhead trenching, lateral migration, and backfilling, runs 9H–10A (13 hr). (A) Deposition at single location near toe during period of fanhead trenching (run 9H); (*B*) lateral shift of depositional centers and beginning of trench backfilling (run 9I); (*C*) widely separated centers of deposition as channel filled (run 9N); (*D*) widespread uniform deposition (run 10A) (from Weaver, 1984).

formation of a fanhead trench (Fig. 9.7) resulted in periods of accelerated lateral fan expansion and aggradation at the toe. Even with constant discharge, geomorphic changes on the fan surface were extreme and often dramatically episodic (Fig. 9.5), with only 5 min separating conditions of widespread deposition from fanhead trenching.

The experimental alluvial fan was symmetrical about the source area outlet (Fig. 9.8), and innumerable small channels occupied its surface. The slope of the fan surface following run 9A was approximately 0.074 as measured down the axis of the fan and 0.086 at right angles to the axis. By the conclusion of the experiment (run 11P), the general topographic characteristics of the alluvial fan had changed little (Fig. 9.8*B*). The apex had broadened slightly, but the overall pattern of contours and degree of surface irregularities remained essentially unchanged.

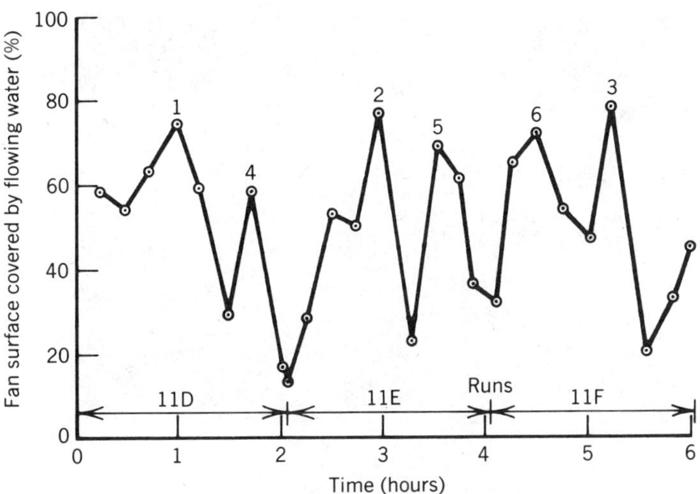

FIGURE 9.5. Percentage of fan surface covered by flowing water during runs 11D–11F. This figure shows three periods of apex incision (1, 2, 3) after all channels were backfilled and water was spread over the fan (Fig. 4.6), and three periods of incision as a result of channel avulsion (4, 5, 6) to a steeper part of the fan surface (from Weaver, 1984).

Growth Rates

There are three distinct periods of accumulation for the fan (Fig. 9.9): (1) an initially high but rapidly declining growth rate during the first 28 m^3 of water discharge (about 10 hr); (2) a relatively constant average growth rate to approximately 142 m^3 of water discharge (50 hr); and (3) slowly declining volumetric growth rate after 142 m^3 of water discharge. The curve is, of course, comparable to the sediment yield curve (Fig. 3.9).

During period 2, which is defined by a relatively constant mean growth rate, at least four cycles of increasing rates, followed by decreasing rates, are evident (Fig. 9.9), and the amplitude of these fluctuations decreases with time. These apparently regular, long-term perturbations are somewhat masked on the corresponding sediment yield curve (Fig. 3.9) as a result of the large natural fluctuations of sediment discharge measured over short time periods. Data-smoothing techniques (moving averages) remove much of the inherent variability in the sediment discharge data and expose comparable trends on a slightly delayed time frame (Fig. 9.3). Finally, the decreasing trend following 142 m^3 of water discharge (50 hr) was directly attributable to the artificial reduction of sediment yield by the placing of plastic sheets in the sediment source area (see Chapter 3).

In order to characterize the downfan patterns and long-term trends of deposition on the fan, the surface was divided into four concentric regions of approximately

FIGURE 9.6. Photographs of fluvial fan. (*A*) Dispersed flow at the apex, and shallow flow in a large number of small distributary stream channels; (*B*) close-up view of dispersed flow at the apex (run 11A). Flow direction is from top to bottom of picture (from Weaver, 1984).

equal area (Fig. 9.10), as follows: (1) the apex or proximal region, (2) the upper midfan, (3) the lower midfan, and (4) the toe or distal region.

The fan apex had the most variable rate of deposition (Fig. 9.11). For example, during five runs the apex was trenched and previously deposited sediment was shifted downfan. In contrast, the lower midfan showed comparatively little varia-

FIGURE 9.7. Photographs of fluvial fan. (*A*) Fanhead trench extending from apex to midfan region. Below the intersection point, streamflow is spread evenly over the fan surface. (*B*) Close-up of fanhead trench (run 11A) (from Weaver, 1984).

tion (63% less) in the average rate of deposition, with virtually all the extreme values occurring during the earliest periods of fan evolution, when watershed sediment yield was high and the fan was actively prograding.

The variability of sedimentation rates within the four regions reflects the major processes at work in each area (Fig. 9.11). The extreme variability of depositional

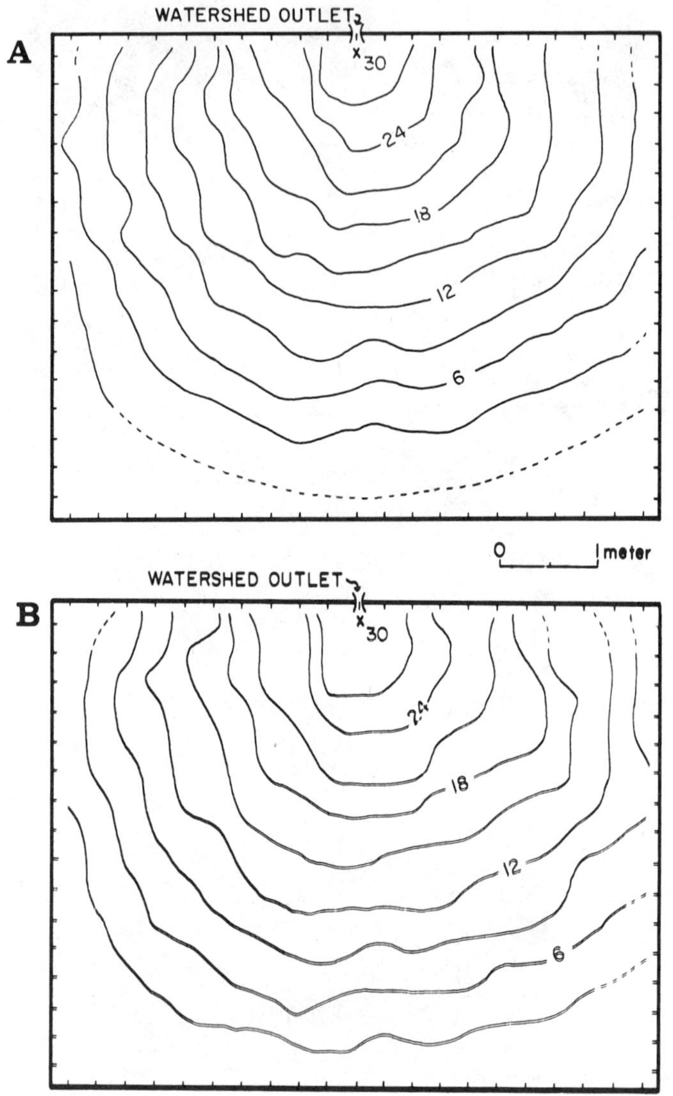

FIGURE 9.8. Topography of the fluvial fan following run 9A (*A*) and run 11P (*B*). Contour interval is 3.0 cm (from Weaver, 1984).

rates at the apex was the result of cyclic repetition of fanhead trenching and channel backfilling. In contrast, the midfan served primarily as a zone of transport and was not subject to extreme erosional or depositional events. Finally, the toe of the fan was the primary receptor of sediment, and it was not in phase with the apex. Erosion of the apex and upper midfan caused significant deposition at the lower midfan and toe. Thus a wide range in depositional and erosional rates was char-

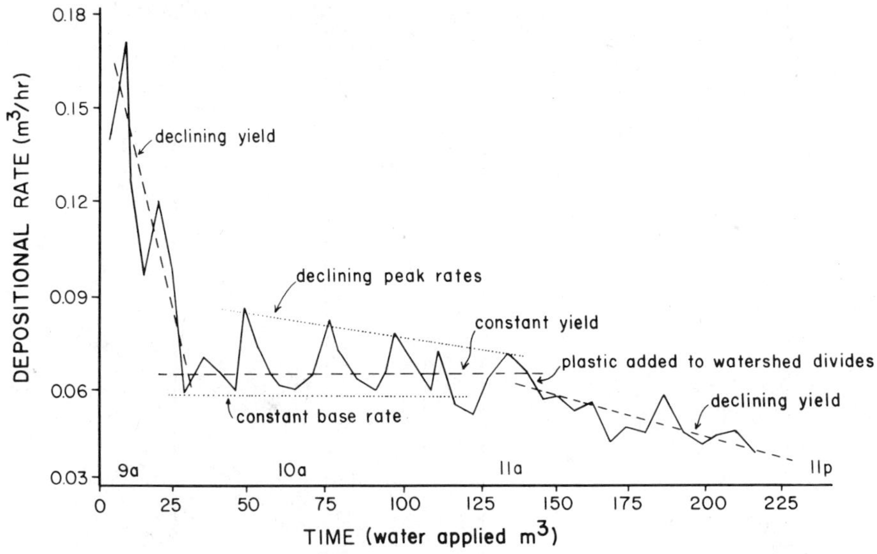

FIGURE 9.9. Volumetric accumulation rate of the fluvial fan (from Weaver, 1984).

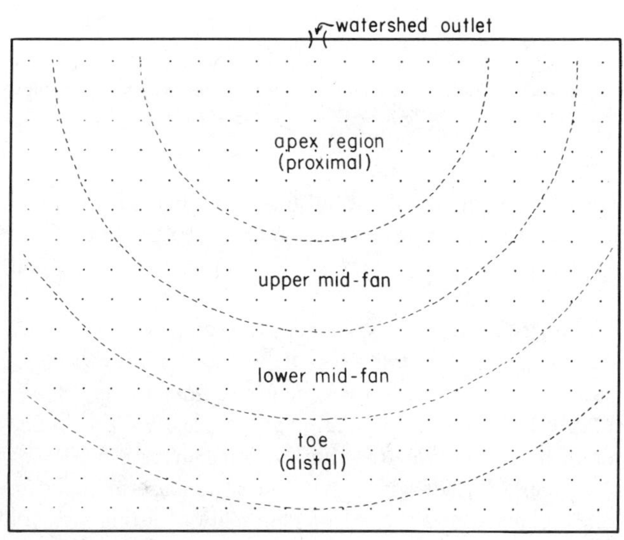

FIGURE 9.10. Concentric division of the fluvial fan into equal areas. Measurement pins marked by dots (from Weaver, 1984).

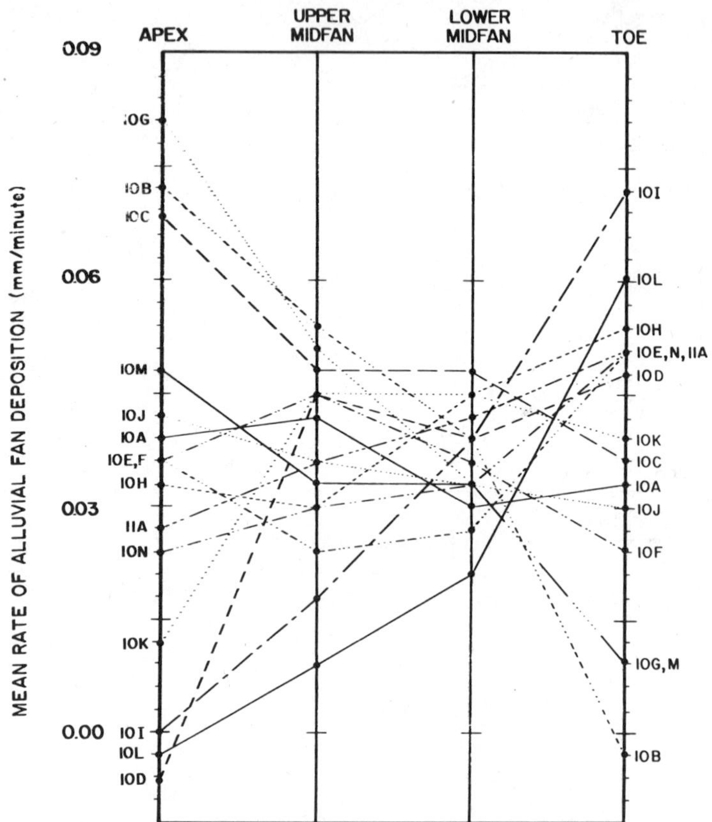

FIGURE 9.11. Depositional rates of the fluvial fan surface, runs 10A–11A. Note an especially wide range of rates in the apex and toe regions and low variability of sedimentation rates in the midfan areas (from Weaver, 1984).

acteristic of the lower reaches of the alluvial fan, but these rates were directly dependent on processes upfan. Therefore, events at the apex of the alluvial fan largely determined the dominant processes and rates of sedimentation and erosion downfan.

Figure 9.12 illustrates a single sequence of events that characterized the evolution of the fluvial fan. During run 10A, deposition in all four regions of the fan was nearly equal and evenly spread over the entire fan surface (Figs. 9.4*D*, 9.12). During run 10B there was rapid aggradation at the apex while virtually no sediment reached the toe of the fan (curve 10B, Fig. 9.12). During run 10C there was rapid deposition at the apex together with a substantial increase in deposition at the toe. Finally, the apex underwent a period of degradation during run 10D, while the midfan and toe of the fan received the eroded, reworked deposits removed from the apex (curve 10D, Fig. 9.12).

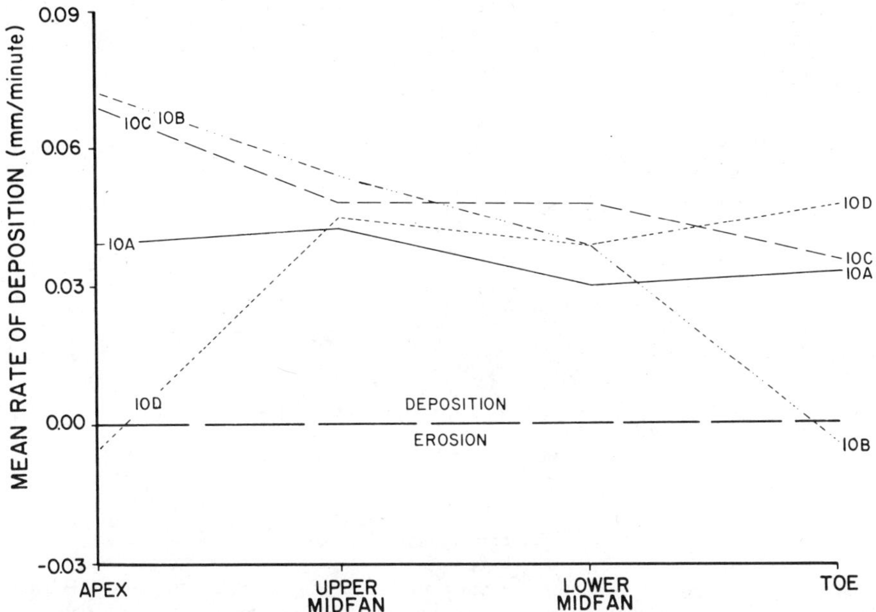

FIGURE 9.12. Typical cyclic pattern of deposition on the fluvial fan surface (see text for discussion) (from Weaver, 1984).

This sequence of events was repeated numerous times during the evolution of the alluvial fan. Each major downcutting event (e.g., run 10D) was preceded by major aggradation at the apex (e.g., runs 10B, 10C). During periods of accumulation at the fanhead the average depositional rate at the apex was always greater than downfan; hence slopes were continually increasing at the apex. Conversely, incision of the apex was always coincident with relatively greater rates of deposition in the midfan and distal segments of the fan. This effectively reduced slope and stream channel gradients at the apex.

The long-term alternation of depositional patterns from fan apex to toe and back to apex is depicted in Fig. 9.13. There are four maxima in the rates of deposition for both the apex and distal fan segments after about time 100. In each case, periods of fanhead building were also periods of reduced deposition downfan.

Large fluctuations in both the location and rate of deposition over the fan surface persisted during the period of relatively constant sediment yield from approximately run 9E through run 11A (Fig. 9.9). Therefore, it does not appear that the events occurring on the fan were directly controlled by sediment delivery (Fig. 9.9). Similarly, minor fluctuations of sediment yield did not correlate well with periods of fanhead aggradation or incision. Thus, while downfan patterns of alluvial fan sedimentation were controlled by events at the apex, events at the apex were not uniquely controlled by events in the drainage basin.

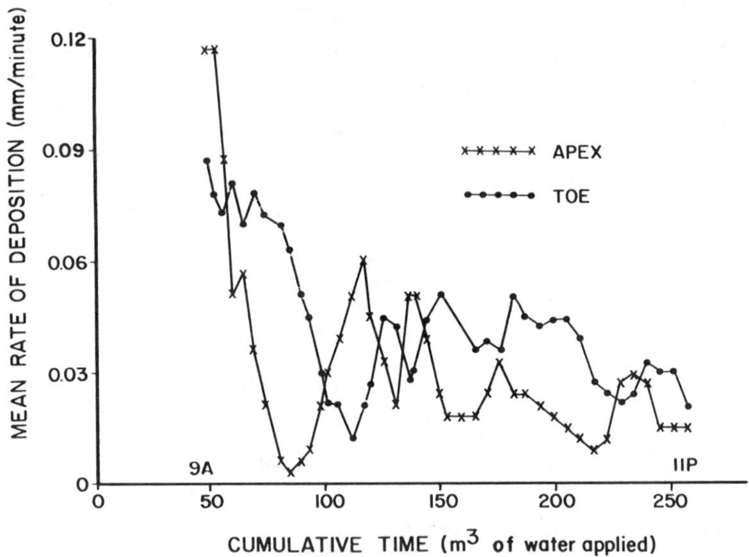

FIGURE 9.13. Depositional rates at the apex and toe of the fluvial fan, runs 9A–11P. Note the generally inverse relationship and rapidly fluctuating values (from Weaver, 1984).

Lateral Growth. The changing lateral patterns of deposition on the fan were examined with reference to flow lines that represent typical paths of flowing water over the fan surface (Fig. 9.14). Depending on the angle at which water and sediment are discharged from the source area to the fan surface, paths ranging from flow line 1 to flow line 20 were followed. Similarly, depending on morphologic conditions at the apex, the momentum of the flow frequently resulted in water and sediment being carried some distance onto the fan apex before it succumbed to direct gravitational flow down the steepest, most direct course (normal to topographic contours).

When data collection began (run 9A), deposition was centered over the right half of the fan (flow lines 11–20, Fig. 9.15). A major division of the centers of deposition occurred after eight runs (run 9H) and by run 10A (16 hr) the two depositional centers were widely separated, with virtually no aggradation in the center of the fan. During the next seven runs, the location of the major depositional sites remained fixed, but significant deposition eventually ceased on the left. For a short time after 34 hr there were three depositional centers.

The overall pattern portrays a readily identifiable trend in the long-term evolution of the alluvial fan (Fig. 9.15). Broad shifts of the major sites of deposition resulted from the migration of channels in a regular side-to-side pattern. Minor sites of active deposition did not persist and they were quickly abandoned (e.g., 37 hr).

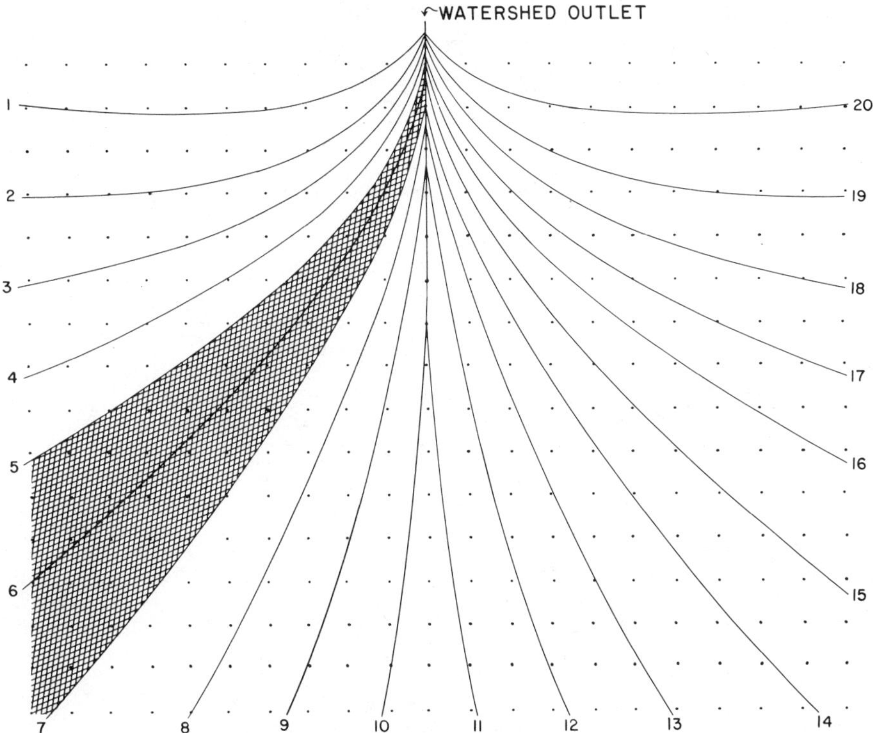

FIGURE 9.14. Flow lines divide the fan surface into 20 overlapping, equal-area segments. The segments flanking flow line 6 are shaded as an example. Each flow line represents the semiarcuate course typically taken by a concentrated flow of water and sediment discharge (from Weaver, 1984).

Probability of Erosion and Deposition

The growth of the fluvial fan consisted of a number of reoccurring short- and long-term patterns of deposition as well as certain random or less well-defined components of development. In order to define the probabilities of erosion and deposition on the fan surface, frequency matrices were constructed from data collected at each measurement pin. Changes were classified as: (1) E, erosion (greater than or equal to 6 mm/run); (2) NC, no significant change (+3 to −3 mm/run); (3) MD, moderate deposition (6–9 mm/run); and (4) HD, heavy deposition (greater or equal to 12 mm/run). This matrix of changes records the probability of various amounts of erosion or deposition between runs 9A and 11F (Table 9.3), depending on the nature of the event that preceded it. For example, there was 27% chance (n = 414) that heavy deposition at any location would be followed by moderate depo-

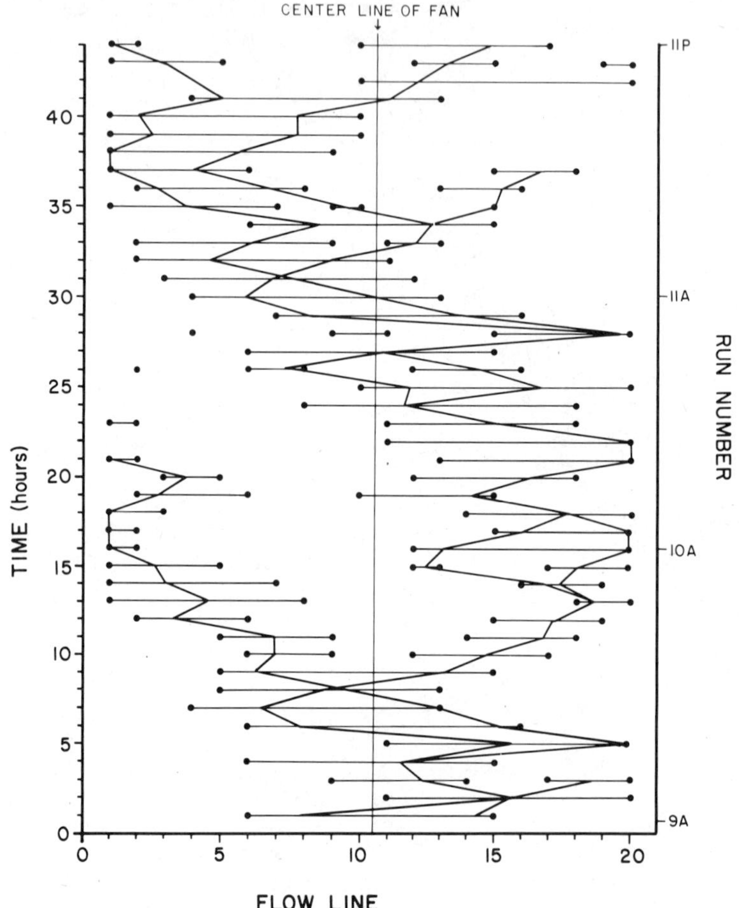

FIGURE 9.15. Radial patterns of deposition on the fluvial fan. Horizontal ranges depict the 10 most rapidly aggrading segments (Fig. 9.14) during each run. Major depositional locations are connected and traced through time (from Weaver, 1984).

sition at the same point. This matrix can be considered to be a matrix of probabilities (Till, 1974, p. 11; Hooke and Rohrer, 1979) from which the likelihood of one type of event following another can be predicted.

For the fan as a whole, there was never more than a 2.0% chance of erosion occurring at a given location, regardless of the nature of the preceding event. However, once erosion was recorded at a spot, there was an excellent probability (76%) that it would be followed by moderate (41%) to heavy (35%) deposition during the following period.

Areas showing heavy deposition during one run had a 57% chance of recording

TABLE 9.3. Probability Matrices for the entire fluvial fan surface, Runs 9A–11P[a]

	Following State				
Former State	Erosion	No Change	Moderate Deposition	Heavy Deposition	Total
Erosion	0.01	0.24	0.41	0.35	1.0 (n = 187)
No change	0.01	0.67	0.21	0.11	1.0 (n = 7745)
Moderate deposition	0.02	0.57	0.28	0.13	1.0 (n = 2932)
Heavy deposition	0.02	0.57	0.27	0.14	1.0 (n = 1554)
Total	0.02	0.63	0.24	0.12	1.0 (n = 12418)

[a]For definitions of *erosion, no change, moderate deposition*, and *heavy deposition*, see text.

no significant change during the following period. Of the 12,418 measurements used to prepare Table 9.3, 63% recorded no substantial change while 36% recorded deposition and only 2% recorded significant erosion. Regions of the fan surface with no change during any given run had a 67% chance of remaining relatively unchanged in the next. It becomes apparent that during relatively long periods of fan evolution, most areas on a fan showed little or no change, while regions that did show measurable deposition were likely to stabilize during the next run.

Relatively large, rapid fluctuations of deposition were more common at the apex and toe of the fan, whereas the mid-portions displayed comparatively less variation in depositional rates, as noted above (Fig. 9.11).

Moderate or heavy depositional episodes at the fan apex had a significantly greater chance of being followed by erosion (trenching) than did similar depositional events anywhere downfan (Table 9.4*A*). For example, the probability of significant surficial erosion immediately following a period of moderate deposition at any given location at the apex was roughly 6% (Table 9.4*A*), on the upper midfan it was only 3% (Table 9.4*B*), while there was little likelihood of any widespread erosion occurring as far downfan as the toe (Table 9.5*B*).

Successive periods of heavy deposition had a 13% chance of occurring both at the apex and near the fan margins, whereas periods of moderate or no deposition were twice as likely to be followed by heavy deposition at the toe than at the apex (Tables 9.4*A*, 9.5*B*). Deposition at the apex was also more likely to be followed by no change (or even slight erosion) than near the toe of the fan.

Regardless of which segment is described, the frequency matrices indicate that the most common or probable occurrence on the surface of the fan was for no significant change. Least likely was major erosion.

An appreciation for the potential effect of changes in sediment yield on erosional and depositional events on the fan surface is indicated by a matrix (Table 9.6) that compares the probability of successive events at the fanhead within the initial period

Table 9.4. Probability Matrices for the Apex (A) and the Upper Midfan (B)

Former State	Following State: Erosion	No Change	Moderate Deposition	Heavy Deposition	Total
			A. Apex		
Erosion	0.01	0.21	0.45	0.33	1.0 (*n* = 87)
No change	0.03	0.68	0.21	0.08	1.0 (*n* = 1374)
Moderate deposition	0.06	0.63	0.24	0.07	1.0 (*n* = 483)
Heavy deposition	0.04	0.61	0.21	0.13	1.0 (*n* = 206)
Total	0.04	0.64	0.23	0.09	1.0 (*n* = 2150)
			B. Upper Midfan		
Erosion	0.00	0.26	0.43	0.31	1.0 (*n* = 54)
No change	0.02	0.64	0.25	0.09	1.0 (*n* = 1638)
Moderate deposition	0.03	0.56	0.32	0.10	1.0 (*n* = 734)
Heavy deposition	0.03	0.59	0.24	0.14	1.0 (*n* = 283)
Total	0.02	0.60	0.27	0.10	1.0 (*n* = 2709)

Table 9.5. Probability Matrices for the Lower Midfan (A) and Toe (B) Regions of the Experimental Alluvial Fan

Former State	Following State: Erosion	No Change	Moderate Deposition	Heavy Deposition	Total
			A. Lower Midfan		
Erosion	0.00	0.27	0.27	0.47	1.0 (*n* = 30)
No change	0.01	0.69	0.19	0.11	1.0 (*n* = 2212)
Moderate deposition	0.01	0.56	0.28	0.14	1.0 (*n* = 775)
Heavy deposition	0.01	0.55	0.27	0.17	1.0 (*n* = 466)
Total	0.01	0.64	0.22	0.13	1.0 (*n* = 3483)
			B. Toe		
Erosion	0.00	0.31	0.38	0.31	1.0 (*n* = 16)
No change	0.00	0.66	0.20	0.14	1.0 (*n* = 2521)
Moderate deposition	0.00	0.55	0.28	0.16	1.0 (*n* = 940)
Heavy deposition	0.01	0.57	0.29	0.13	1.0 (*n* = 599)
Total	0.00	0.62	0.23	0.14	1.0 (*n* = 4076)

Table 9.6. Probability Matrices for the Alluvial Fan Apex for Runs 9A–9F and 9G–11A

Former State	Following State				
	Erosion	No Change	Moderate Deposition	Heavy Deposition	Total
		A. Apex, Runs 9A–9F			
Erosion	0.00	0.00	0.50	0.50	1.0 (n = 2)
No change	0.01	0.46	0.27	0.27	1.0 (n = 134)
Moderate deposition	0.00	0.53	0.32	0.16	1.0 (n = 57)
Heavy deposition	0.00	0.53	0.21	0.26	1.0 (n = 57)
Total	0.00	0.48	0.27	0.24	1.0 (n = 250)
		B. Apex, Runs 9G–11A			
Erosion	0.00	0.21	0.37	0.39	1.0 (n = 60)
No change	0.04	0.65	0.22	0.08	1.0 (n = 746)
Moderate deposition	0.08	0.63	0.22	0.06	1.0 (n = 275)
Heavy deposition	0.07	0.66	0.18	0.09	1.0 (n = 119)
Total	0.05	0.63	0.23	0.10	1.0 (n = 1200)

of high, rapidly diminishing sediment yields (Fig. 9.9, runs 9A–9F) and within the following period of lower, relatively constant average sediment yield (Fig. 9.9, runs 9G–11A). Although the predominant pattern was still one of "no change," heavy deposition was more common and erosion less common during runs 9A–9F than during the lower depositional rates of runs 9G–11A.

The construction of probability matrices to describe alluvial processes provides a basic framework for the identification of hazardous areas on the evolving alluvial fan. Significantly, patterns and rates of erosion and deposition on the fan surface are neither strictly random nor strictly controlled by previous conditions. Rather, events appear to exhibit a large component of memory spanning more than a single measurement period.

This is confirmed by statistical evaluation of the probability matrix of Table 9.3 for the Markov property, according to the method discussed by Till (1974), which indicates that the process of successive events can be assumed to be Markovian in nature. Thus long-term patterns in planimetric and volumetric growth of the fan, as well as the shorter-term events recorded by changes at each measurement pin following individual runs, suggest that Markov processes control alluvial fan evolution (Hooke and Rohrer, 1979).

Fanhead Trenching

Throughout the development of the fluvial fan, fanhead trenching was a common, if not regular, occurrence. It was generally possible to predict major trenching events just prior to their occurrence. However, there were times when downcutting was initiated considerably earlier or later than would have been expected. Several factors appeared to play a part in determining the precise timing, initial location, and subsequent behavior of fanhead trenches.

Trench Location

Local topography and slope were instrumental in determining the specific location of trenching at the fan apex. Once developed, the future behavior of the incised channel was controlled by the momentum of sediment and water within the trench and flow deflections at or near the drainage basin outlet.

Fanhead trenches typically were curved, as the tendency for water to flow directly downslope overcame the tendency to continue in the direction in which it left the source area (Fig. 9.16). The dominance of local downfan topographic controls on streamflow directions of several trenches is evident. Figures 9.17 and 9.18 show the planimetric characteristics of seven trenches formed during the experiment. In each example, an originally straight or axial orientation was increasingly dominated by topographic influences in the downfan direction, resulting in generally increasing curvature of the trench. Trenches that developed along the

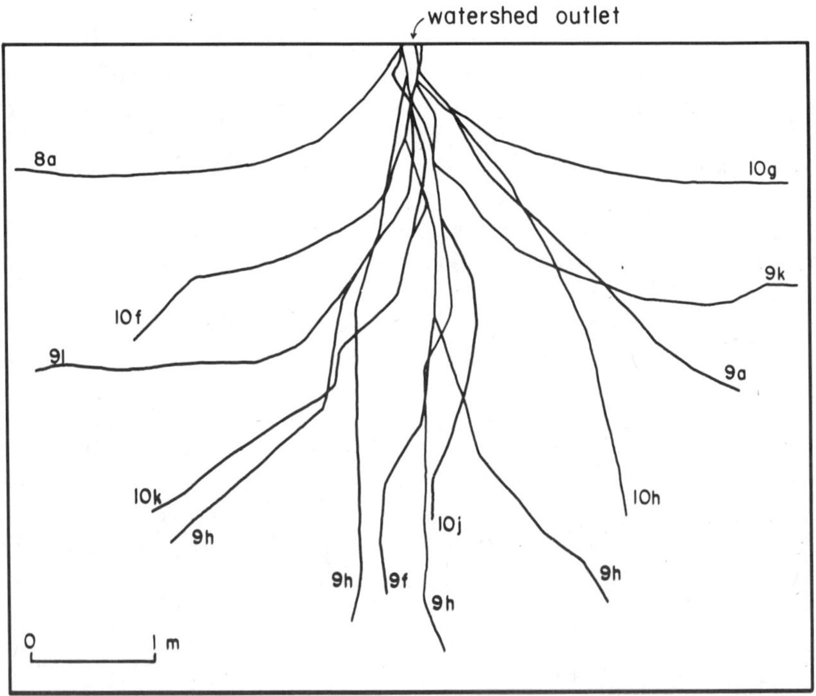

FIGURE 9.16. Maps showing position of 14 fanhead trenches. Numbers indicate runs (from Weaver, 1984).

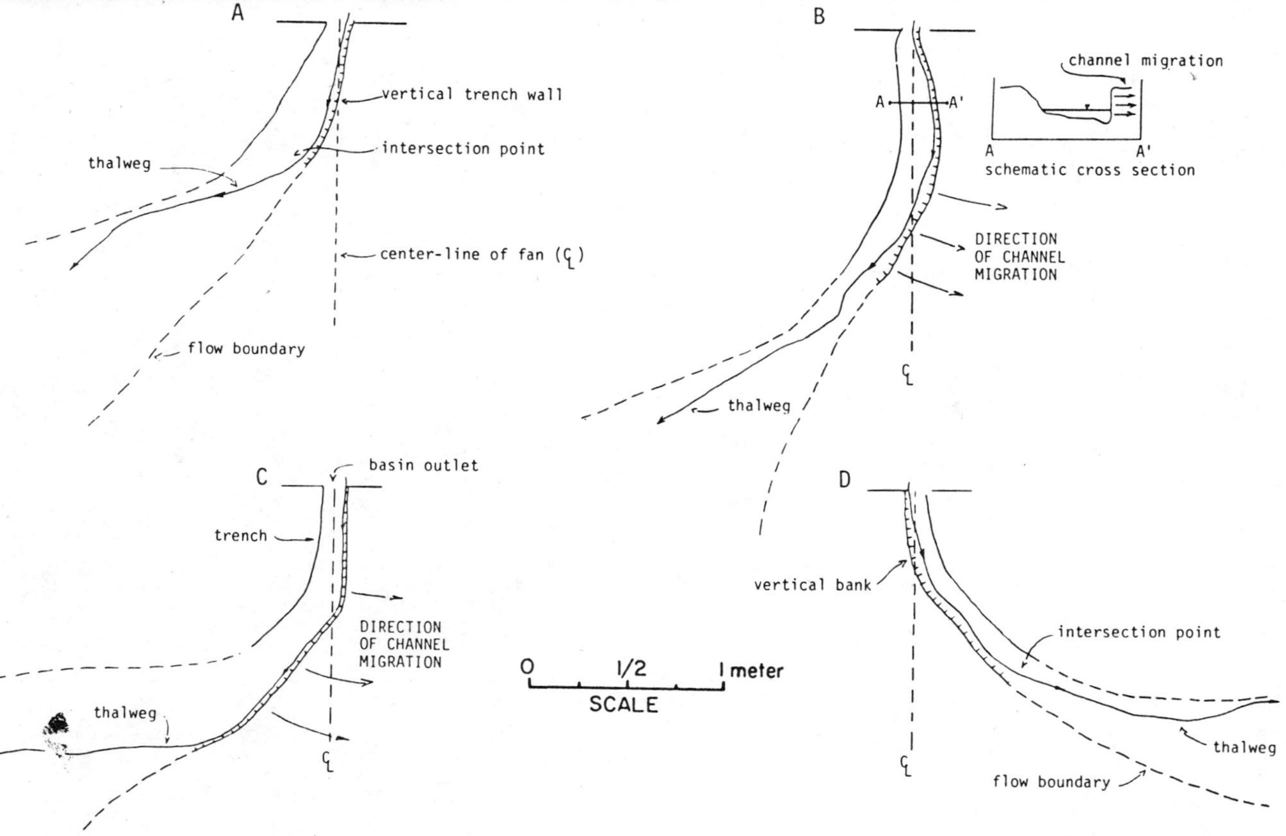

FIGURE 9.17. Maps of fanhead trenches. (*A*) Trench developed during run 10F. Note thalweg meander caused by gravitational influences (slope) at the intersection point. (*B*) Trench formed during run 10K. Orientation of the channel is close to the axial line of the fan. (*C*) Trench actively migrating toward the centerline of the fan formed during run 9L. (*D*) Trench formed during run 9K (from Weaver, 1984).

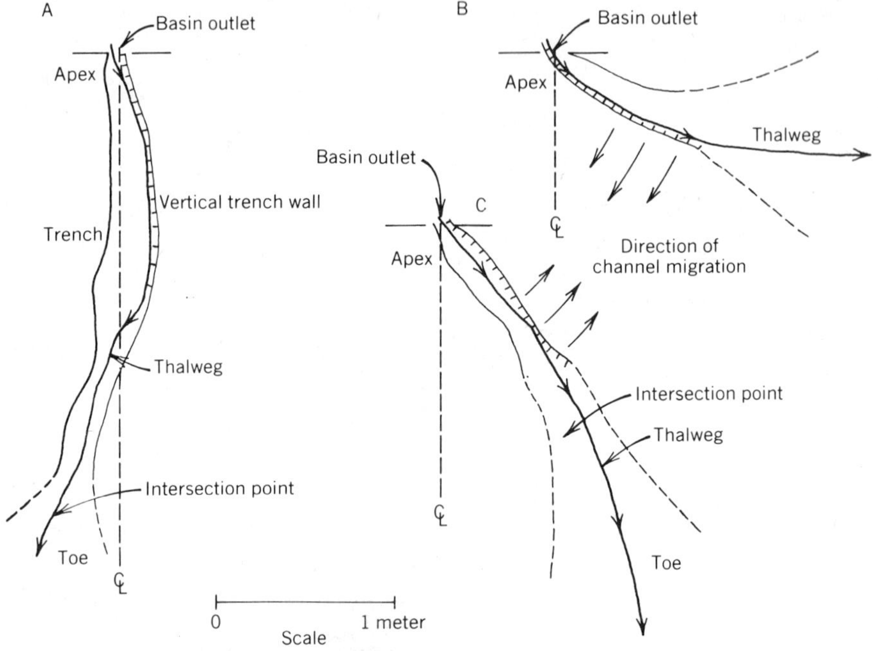

FIGURE 9.18. (*A*) Trench formed during run 9F. Note the axial orientation of the channel and the extreme length of the incision. (*B*) Trench formed during run 10G. One of the principal causes of the large angle between the axial line and the thalweg was the deflection of streamflow immediately above the basin outlet. Note also that the direction of active channel migration is toward the center, or axial, line of the fan. (*C*) Trench formed during run 10H. Atypically, this channel developed and then migrated away from the axial line of the fan because flow deflection at the basin outlet forced the thalweg away from the fan apex (from Weaver, 1984).

axis of the fan did not show marked curvature in their distal sections because gravity was acting with flow momentum (Fig. 9.18*A*).

Flow momentum commonly forced the channel thalweg against the streambank (trench wall) closest to the fan axis. Selective erosion against this side resulted in lateral migration toward the more stable fan axis position. For example, most of the trenches (Figs. 9.17, 9.18) displayed a long, actively eroding trench wall along the fan axis side of the trench, with the thalweg directly impinging on this bank. Where the influence of topographic or gravitational forces exceeded flow momentum, the thaleweg shifted toward the center of the trench.

In summary, the most stable orientation of a fanhead trench is along the axis of the fan. During the experiments, trenches migrated toward this axial position by lateral shift. However, flow deflections in the main channel above the drainage basin outlet sometimes changed the direction of flow on the fan and resulted in other orientations (Figs. 9.18*B*, *C*).

Timing

The relative duration of each phase of the fanhead trenching cycle is shown by the histograms of Fig. 9.19. The average time required to backfill a trench until discharge was again spread over the fan was 31 min for 58 events (Fig. 9.19*A*), but the average length of time required to progress from spread to incised flow conditions at the apex was approximately 11 min (Fig. 9.19*B*). For 34 events, 68% of the trenching episodes required less than 10 min.

Changes at the fan apex are shown by the photographs of Fig. 9.20. During 208 min, flow at the fan apex evolved from being confined in a deeply incised fanhead trench (Fig. 9.20*A*), to incipient overflow as the channel was backfilled (Fig. 9.20*B*), to spread flow over the entire fanhead (Fig. 9.20*C*). In comparison, during run 11A the conversion from spread flow to apex incision was accomplished in only 16 min. Within an additional 3 min, trenching of the fan apex had progressed into the valley deposits of the main trunk stream of the source area, thereby creating well-defined terraces along the channel and flushing previously stored sediment to the lower reaches of the fan (Fig. 9.20*A*).

Fan Slope

As trenches formed, sediment was shifted towards the toe of the fan and slope gradients at the apex were substantially reduced. In contrast, channel backfilling and sheetflow at the apex caused deposition and the local slopes increased rapidly, which eventually led to renewed trenching.

During run 10C, 61 slope angles were measured on the surface of the fan; they were evenly distributed among the apex, upper midfan, lower midfan, and distal regions. The average gradient was 2.6°. The mean gradient of 166 slope measurements taken along the bottom of fan trenches was a comparable 2.5°. The similarity of the gradients indicates that the trenched channels are not likely to experience significant additional downcutting.

During five trenching episodes, mean fan slopes were 2.1° (Fig. 9.21), but trenching reduced the slope to roughly 1.8° (13% reduction, on average). Most important, the mean of the steepest flowline segment (Fig. 9.14) in each of these five situations, where trenching immediately followed, was 2.8°. Thus there was always a single steep flow path at the fanhead just prior to incision. On the fan, slopes of 2.5°–2.6° appeared to be relatively stable, but less steep areas typically experienced rapid aggradation. Fan surface slopes in excess of 2.6°, and especially those reaches steeper than 2.8°, were characterized by incision. Therefore, in the vicinity of the fan apex, the average surface slope did not appear to be the key factor in controlling eventual apex incision. Rather, individual excessively steep reaches of fan surface, which developed by rapid aggradation, were the sites where downcutting was initiated.

When steep reaches were located downfan, local headcuts developed and a midfan trench formed as the channel incised and captured what flow was available. Since the quantity of accessible discharge was relatively low, these midfan erosional

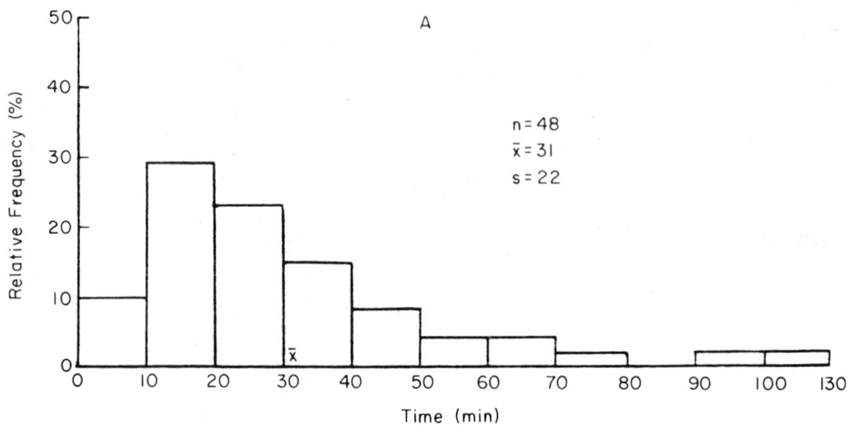

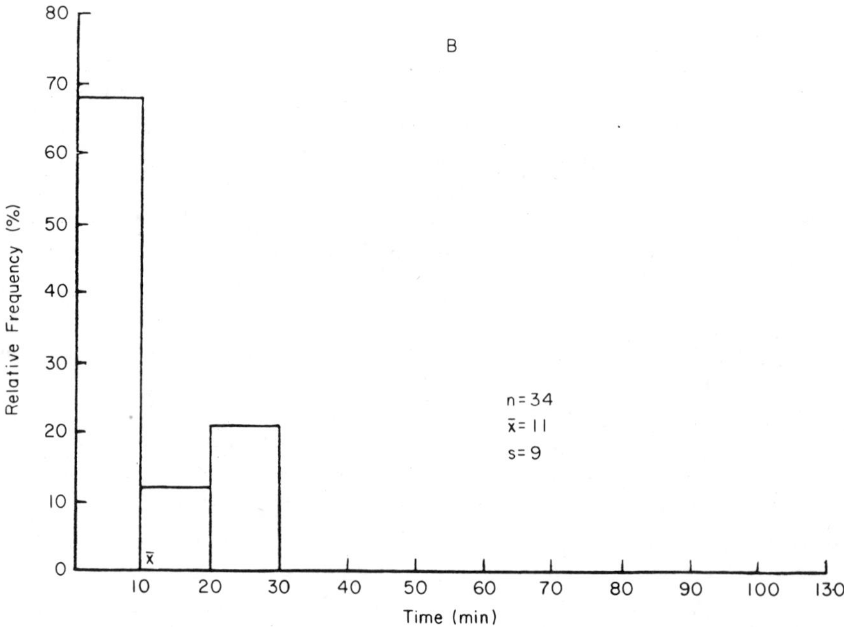

FIGURE 9.19. Frequency distribution of time required to evolve (*A*) from trenched to spread-flow conditions at fan apex and (*B*) from spread flow to confined (trenched) flow conditions at fan apex (from Weaver, 1984).

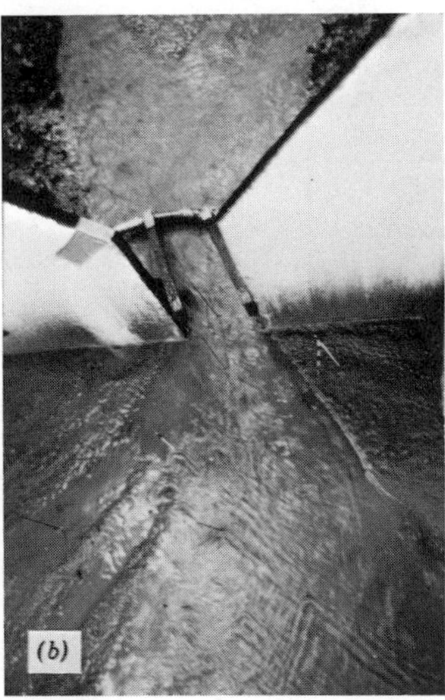

FIGURE 9.20. Photographs of fluvial fan showing apex changes during 208 min of run 10L. (*A*) Well-developed fanhead trench at t = 25 min (note terraces in source area). (*B*) Sixty-five minutes later (t = 90 min) trench has backfilled nearly to the watershed outlet and flow is spilling over the channel banks. (*C*) After 121 min, flow was widely spread over the fan surface, and at 208 min streamflow was totally spread upon emerging from watershed outlet. Sediment was being deposited at the apex and slopes were locally increasing (from Weaver, 1984).

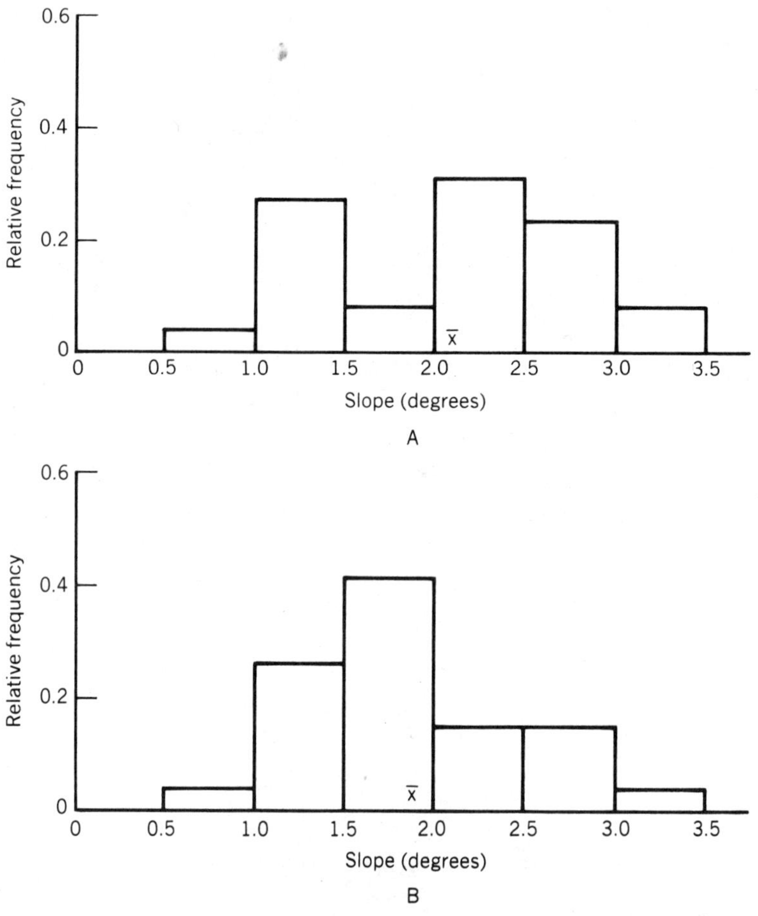

FIGURE 9.21. Relative frequency of (*A*) fan-surface slopes prior to trenching and (*B*) channel-bottom slopes following trenching. (*A*) includes 26 measurements adjacent to five trenches ($\bar{x}$ = 2.07°; *s* = 0.64°). (*B*) includes 27 channel bed measurements in five trenches ($\bar{x}$ = 1.81°; *s* = 0.57°) (from Weaver, 1984).

episodes were short-lived and the channels rarely extended themselves more than 1 m upfan.

In contrast, when comparable erosion was initiated in the apex area, large quantities of discharge were captured and funneled into the already oversteepened reach. Consequently, erosion was rapid and within a short time the channel had advanced across the apex and captured all available streamflow. Thus, although overall fan-head slope steepness did not directly trigger channel incision, the buildup of the entire apex area during periods of spread flow was a requisite condition that allowed the formation of localized, excessively steep gradients.

Sediment Load

Changes of sediment yield were not the most important factor influencing trenching of the fluvial fan. During the initial period of high but declining sediment yield, 10 trenches formed between times 40 and 70 (Fig. 9.22), and trenching frequency was relatively high with one trench per 3 m^3 of runoff. During the period between times 70 and 120 there were only eight recorded trenching events, representing one trench per 5 m^3 of runoff. This decrease in the frequency of trenching corresponded closely with a major sediment yield high, which peaked at about time 110 and then rapidly decreased. It is likely that this increase of sediment discharge caused aggradation at the apex, but the high sediment loads did not permit incision of the fanhead. However, it was not long after sediment yield again decreased that the frequency of trenching episodes once again rose sharply.

For the period from roughly times 120 to 195 (Fig. 9.22), 27 trenches were recorded for a frequency of about 1 trench per 3 m^3. It appears that although the frequency of trenching events may be influenced by changes in sediment yield (actually sediment–water ratio), periods of constant water and sediment discharge were also characterized by relatively high frequencies of fanhead trenching.

Once slopes were reduced at the fan apex to lesser, more stable gradients by downcutting and lateral shifting, fanhead trenches were slowly obliterated by deposition which eventually filled the channel with sediment and returned the locus of active sedimentation to the fan apex (Fig. 9.20).

Fan-Lobe Development

Although fanhead trenching was the most obvious erosional mechanism operating on the fluvial fan, a secondary erosional process involved erosion of depositional lobes by distributary channels that cross these steeper areas of the fan surface.

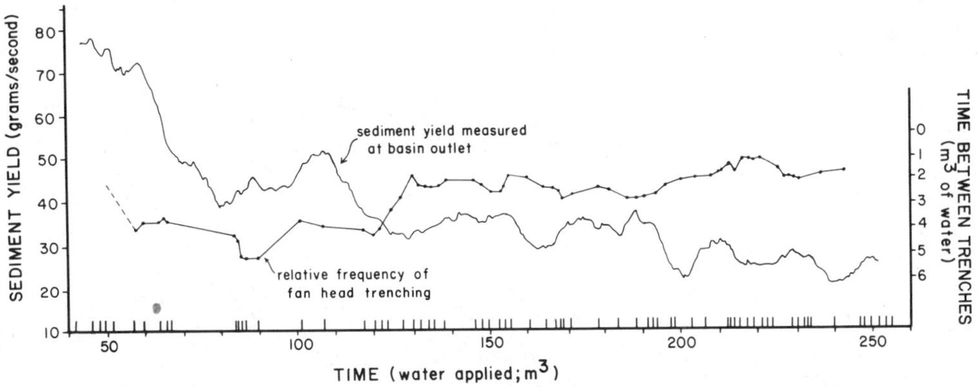

FIGURE 9.22. Sediment yield and the relative frequency of fanhead trenching. Runs characterized by trenching are shown by ticks along horizontal axis; each may represent more than one trench (from Weaver, 1984).

Increased flow velocities on the steeper lobe fronts result in incision, headward erosion, and channel widening, with the eroded sediment being carried a short distance downfan and deposited as a lobe in an area of lesser gradient. These small fan-shaped deposits were common on the surface of the larger alluvial fan (Fig. 9.23).

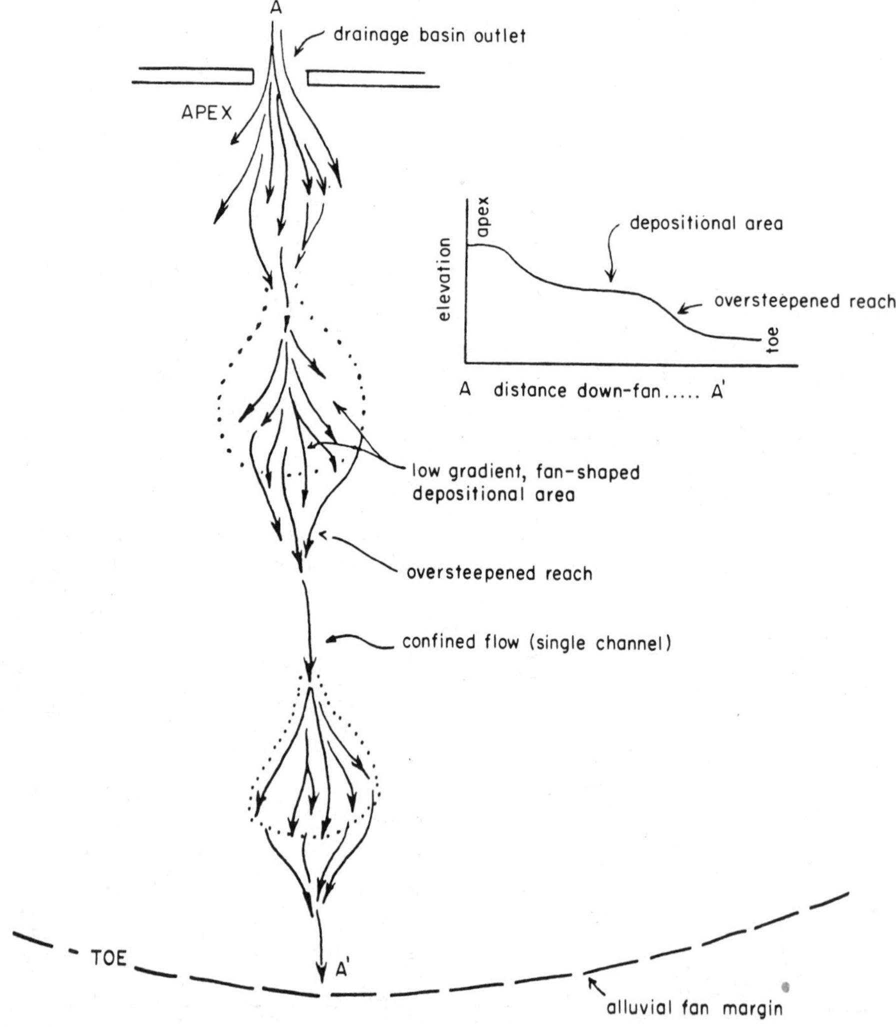

FIGURE 9.23. Idealized development of small fan lobes on the surface of the fluvial fan. Low gradient reaches are separated by steep slopes where flow concentrates and scours through previously deposited material (see inset cross section A–A′). This mechanism effectively reworks sediment deposited at the apex and gradually shifts it toward the fan margin between periods of fanhead trenching (from Weaver, 1984).

The downfan margin of each of these secondary fans is steep. Small nickpoints develop on this steep surface and headward erosion forms a channel that captures the remaining flow and scours portions of the newly deposited fan. In this manner, sediment is moved downfan, briefly stored, and again remobilized and transported farther towards the toe.

Most distributary channels on an alluvial fan have a very limited life span because rapid accumulations of sediment and subsequent bar formation choke streams and cause channels to widen and shallow and migrate laterally. Typically, these channels are filled and abandoned well before they can become dominant erosional features and capture significant quantities of flow.

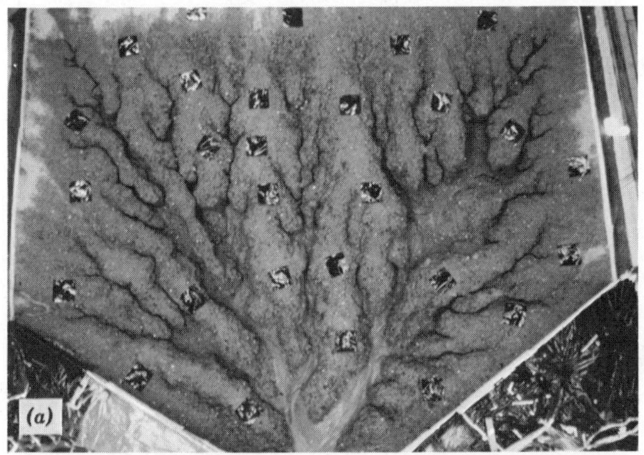

FIGURE 9.24. Photographs of REF showing progressive coverage of surface by impermeable plastic squares beginning with run 10N. Coverage was added in increments of 5% of the basin area to artificially reduce hillslope sediment production. (*A*) run 10N; 5% coverage; (*B*) run 11P; 40% coverage (from Weaver, 1984).

Reduction of Sediment Yield

After documenting the growth of the experimental alluvial fan during 56.6 hr of precipitation (162 m^3 of water discharge), another phase of the experiment was initiated to test the effect of reduced sediment yield on fan morphology. In order to reduce the sediment yield reaching the fan, 0.1-m^2 squares of impermeable, 0.5-mm-thick plastic were placed randomly on the sediment source area in increments of 5% of the total drainage area. Beginning with run 10N, roughly 4 hr of precipitation separated each new addition of squares (Fig. 9.24*A*).

Although sediment yield to the alluvial fan was reduced, it was discovered that while sediment production from raindrop impact was eliminated in those areas covered by plastic, sediment that was no longer supplied by slope processes was derived from channel widening and removal of stored alluvium in the drainage network. Nevertheless, reductions of sediment yield had little noticeable effect on the alluvial fan until roughly 40% of the watershed had been covered at about 190 m^3 time (Figs. 3.9, 9.9, 9.24*B*), when the decrease of sediment yield (Figs. 3.9, 9.9) caused the incision of a major trench into the fan (Fig. 9.25). Unlike the fanhead trenches discussed previously, this channel began in the midfan as a small channel that eroded upfan. It captured water as it did so, enlarging and deepening to form the type of channel envisioned by Eckis (1928) when he postulated that a

FIGURE 9.25. Fanhead trench formed at the end of run 11N. Trench developed in response to diminishing sediment yields from upstream drainage network (flow is from right to left) (from Weaver, 1984).

major trench would form as sediment yields decreased during the erosional evolution of the source area.

One might expect landforms to respond gradually if 10–30% of basin sediment production is progressively eliminated, but the long lag before there was a major response of the alluvial fan suggests that some fans may be relatively stable and will not respond readily to external influences.

ALLUVIAL FANS FORMED BY EPISODIC EVENTS

In nature, alluvial fans develop under a wide range of environmental and geomorphic conditions and in response to highly varied intensities, frequencies, and durations of fan-building events. Fans in many arid areas are constructed by infrequent, large-magnitude mudflows triggered by episodes of short-duration, high-intensity rainfall. In other regions, seasonal streamflow derived from glacial or annual snowpack melting develops alluvial fans that are constructed primarily of fluvial deposits. Many alluvial fans appear to be composed of a complex combination of fluvial and mudflow sediments.

Fluvial and mudflow deposition produces alluvial fans that are grossly comparable in overall form yet different in certain physical characteristics, including surface morphology, slope, relative size, and sedimentology and stratigraphy (Beaty, 1970; Bull, 1968, 1972; Hooke, 1967, 1968b, Lustig, 1974).

After the processes responsible for alluvial fan formation under fluvial conditions were investigated, three additional experimental fans were developed in the REF under conditions of increasing sediment concentration, culminating in the construction of a fan formed predominately by mudflows.

The three alluvial fans, designated (A) episodic-fluvial, (B) mixed-mode, and (C) mudflow fans, were developed from sediment and water discharged from basins of increasing drainage density and channel and valley side slope gradients. All were formed by episodic hydrologic events, but the ratio of mudflow to streamflow deposition increased from fan A to fan C. The source material that was used in the fluvial fan experiment (roughly 25% silt and clay, 75% sand) was also used to produce the three types of fans. In addition, coarse sand was placed on the floor of the REF where these fans would be deposited. It was assumed that this would promote dewatering of the flows and restrict fan growth.

Procedure

Prior to the episodic-fluvial-fan experiment, the source area was divided into eight roughly equal-area parcels and a crude Y-shaped channel system was established on the surface (Fig. 9.26). In order to generate more frequent high-density, viscous flows, hillslope gradients and drainage densities in the source area were increased.

In the initial phases of the several experiments and prior to each event, the surface of the hillslopes was slightly loosened by raking. This simulated the development of weathered debris and colluvial accumulations along the hillsides. It

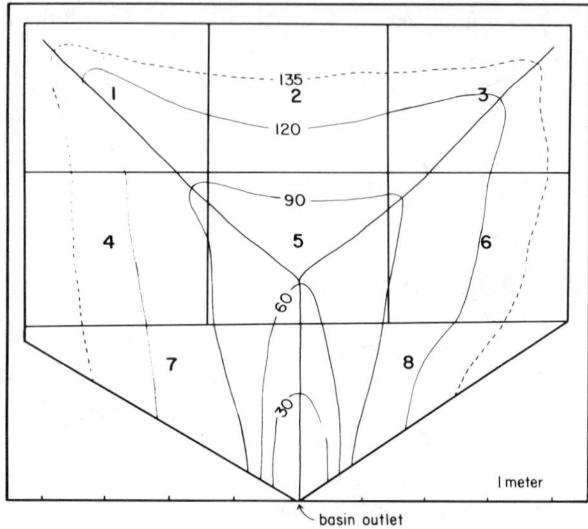

FIGURE 9.26. Map of the source area surface prior to the beginning of the episodic-fluvial-fan experiment. Elevations are in centimeters above basin outlet. Large numerals are centered in the equal-area segments (from Weaver, 1984).

generally assured maximum sediment yield and reduced the variability of sediment discharge between runoff events from any one area within the basin.

The episodic-fluvial fan was the result of 549 precipitation events that produced either (1) no runoff, (2) runoff that left the area receiving direct rainfall but did not reach the fan, or (3) runoff that moved through the channel network and reached the fan surface. Ten-second bursts appeared to produce the most satisfactory discharges, and these were randomly applied to the eight basin parcels by a burst of artificial rainfall applied by a hand-held nozzle.

The mixed-mode fan developed from a total of 426 precipitation-runoff events. To increase the frequency of mudflows, the duration of high-intensity precipitation events as well as channel and slope gradients, local relief, and drainage density were all increased.

The basin was divided into 10 roughly equal 5.8-m^2 parcels (Fig. 9.27) and included a dendritic valley system. Random numbers were again used to dictate the order in which specific areas received precipitation bursts. It was found that the duration of the rainfall events, in combination with the distance of an area from the basin outlet, controlled the relative frequency, magnitude, and character of depositional episodes on the fan surface. Following initial experimentation, rainfall bursts of 12, 15, and 18 sec were used to generate runoff from areas 9 and 10, 5–8, and 1–4, respectively. No less than 20 min was allowed to lapse before any individual area received two successive precipitation events, and there was at least 5 min between each precipitation event. It was felt this represented sufficient time

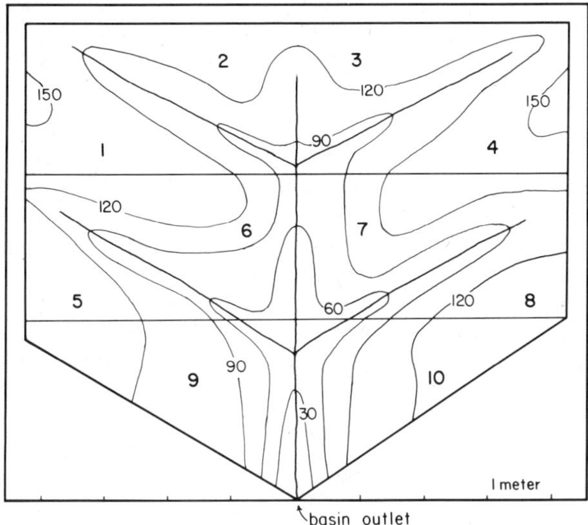

FIGURE 9.27. Map of the source area prior to the start of the mixed-mode-fan experiment. Elevations are in centimeters above the basin outlet. Large numerals are centered in each of the 10 equal areas. Note the increased drainage density and steepened channel sideslopes as compared to the episodic-fluvial-fan configuration (Fig. 9.26) (from Weaver, 1984).

for complete infiltration of surface water remaining on hillslopes or in mudflow deposits.

The mudflow fan was produced by 216 precipitation-runoff events. Each event was generated by a 10-sec burst of artificial rainfall of uniform intensity from a hand-held nozzle. There was a minimum of 10 min between rainfall events and not less than 20 min between precipitation events on the same parcel. To increase sediment yield and the frequency of mudflow events, drainage basin relief was increased while the eight equal sized areas receiving precipitation during each event was reduced to 2.1 m^2 (Fig. 9.28). Precipitation intensities were thereby increased by a factor of three, which produced 4–6-l runoff events. Following initial experimentation during the first 24 events (run 1), each of the eight areas received 10-sec bursts of rainfall applied in the same fashion as earlier experiments. During the first four runs, precipitation was applied uniformly over the etched surface of each area. However, to increase the frequency and sediment concentration of mudflows, rainfall was then directed to the upper slopes of each area. Runoff generated on these swept loose sediment from lower areas as it traveled toward the channels, thereby increasing the concentration of the flow. During 192 precipitation events, 86% resulted in sediment delivered to the alluvial fan. Once the rainfall-application techniques had been perfected (run 4), 97% of the events resulted in measurable sediment yield and each of the eight areas produced viscous mudflows.

For the first four runs, mudflows composed 46% of the total number of events

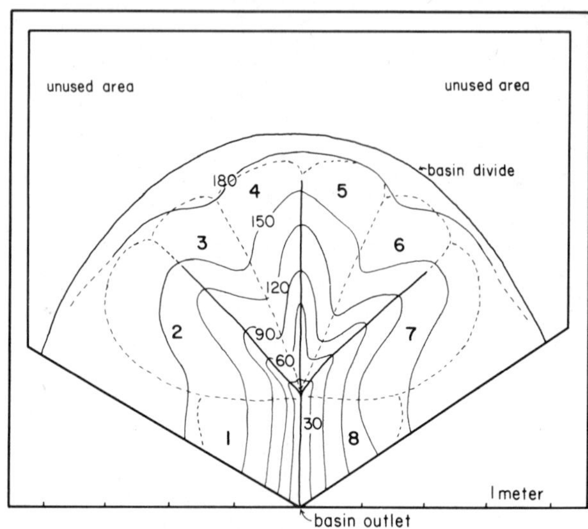

FIGURE 9.28. Map of drainage basin constructed for the mudflow fan experiment. Elevations are in centimeters above the basin outlet. Large numerals and dashed lines depict the eight sediment source areas (from Weaver, 1984).

exiting the basin. With the slight change in rainfall application procedures, about 80% of the events of runs 5–15 were mudflows. As in the other two experiments, the alluvial deposit accumulated on a highly permeable, coarse sand surface.

Following each series of runoff events, the fan surface was allowed to dry prior to mapping. In addition to the measurement grid, cross-fan profiles and radial traverses extending from the toe of the fan up and into the main channel of the source area were recorded after each run.

Episodic-Fluvial Fan

Each of the eight equal-area regions (Fig. 9.26) within the drainage basin responded differently to the same episodic application of precipitation (Table 9.7). For example, the closer an area was to the basin outlet, the more rapidly peak runoff was achieved.

Varying topographic conditions from area to area also affected sediment production and yield. For example, areas 7 and 8 (Fig. 9.26) displayed the greatest relative reflief and hillslope gradients. These areas also produced relatively large mudflows (Table 9.8). However, area 5, although displaying less relief, contained more channel length; and sediment from area 5 was efficiently conveyed to the main channel and to the fan (Table 9.8).

Flows to the fluvial fan were typically 4% sediment by weight. In contrast, the sediment concentrations of flows to the episodic-fluvial fan were 9–19% by weight. These flows were highly charged streamflows rather than mudflows.

Table 9.7 Episodic-Fluvial Fan: Hydrologic Characteristics of Runoff from the Eight Areas

Area	Number of Applications	First Runoff (sec)	Range (sec)	Time to Peak (sec)	Range (sec)	Mean Sediment Discharge (g)
1	18	28.2	32–33	38.8	25–48	1913
2	9	38.0	30–46	48.2	37–62	966
3	6	32.2	29–39	43.0	36–50	1588
4	10	29.6	26–33	42.3	35–46	1601
5	8	15.3	12–17	22.1	21–24	3063
6	15	27.8	24–32	40.5	30–43	1840
7	18	9.1	3–12	25.4	19–30	1892
8	12	10.2	5–14	20.7	15–26	2028

[a] Fig. 9.26.

Fan Growth and Dynamics

Although sediment concentrations were high, fluvial processes dominated the episodic-fluvial fan. Because of strong similarities with the fluvial fan, only a few aspects of the growth of this fan will be discussed. As streamflow left the confined main channel of the source area, it spread over the apex of the alluvial fan and deposited the coarsest fraction of its sediment load. Finer sediments were conveyed to the toe of the fan, where they were deposited as water rapidly infiltrated into the permeable sand layer. The continued lateral expansion of the alluvial fan produced a deposit characterized by an impervious, clay-rich basal layer capped by a relatively coarser deposit. These gross stratigraphic relationships are comparable to field situations where playa and fine distal sediments are progressively buried by coarser debris during fan growth.

Table 9.8 Episodic-Fluvial Fan: Characteristics of Sediment Yield from the Eight Areas[a]

Area	n	Sediment–Water Ratio $\bar{x}$ (by weight)	Total Sample Size (sediment/event)		
			$\bar{x}$ (g)	s^2 (g)	Volume (cm^3)
1	7	0.18	2258	639	2.3
2	7	0.09	1062	504	0.3
3	7	0.19	1897	463	2.0
4	7	0.19	1838	610	1.4
5	7	0.22	3777	648	4.5
6	7	0.16	2149	658	2.0
7	7	0.18	2236	514	2.3
8	7	0.18	2411	679	2.3
					$\bar{x}$ = 2.0

[a] Fig. 9.26.

Lateral growth of the episodic-fluvial fan was primarily controlled by three factors (1) morphologic condition (degree of channelization) of the fan apex (2) total volume of runoff events (sediment plus water) and (3) sediment discharge. Significant areal growth of the fan occurred during periods of fanhead incision. The largest-magnitude (volume) events produced the most rapid rates of horizontal expansion. When apex entrenchment was minor or absent, significant deposition at the toe of the alluvial fan could only occur in response to the largest 10% of events of any run. As the fan enlarged and the apex-to-toe distances increased, most of the sediment was dropped in upper- and midfan regions. Long-term growth of the fan was comparable to that produced by the constant discharge of sediment and water during the fluvial-fan experiment (Fig. 9.29).

Small fanhead trenches capable of containing minor runoff events were common, and they persisted until obliterated by major discharge events (Fig. 9.30). Moderate runoff episodes frequently incised small channels during their rising stage and

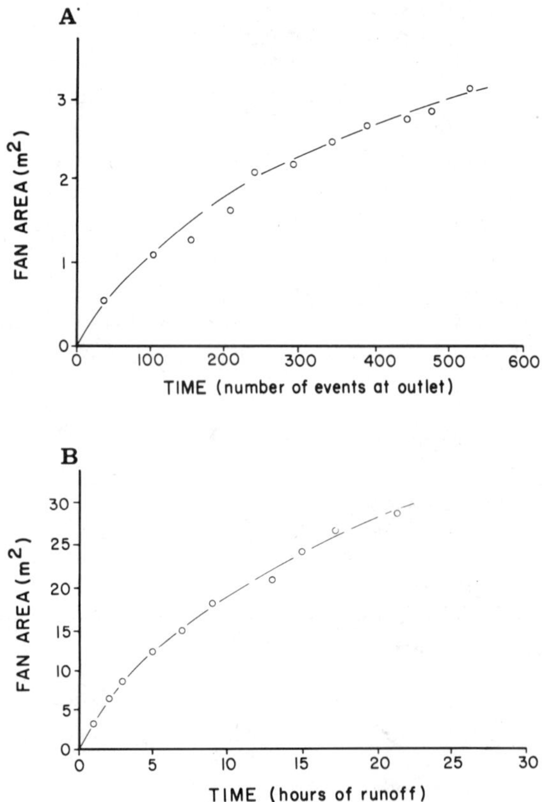

FIGURE 9.29. (*A*) Growth of episodic-fluvial fan through time. Time is indexed by the number of depositional events. (*B*) Growth of the fluvial fan (from Weaver, 1984).

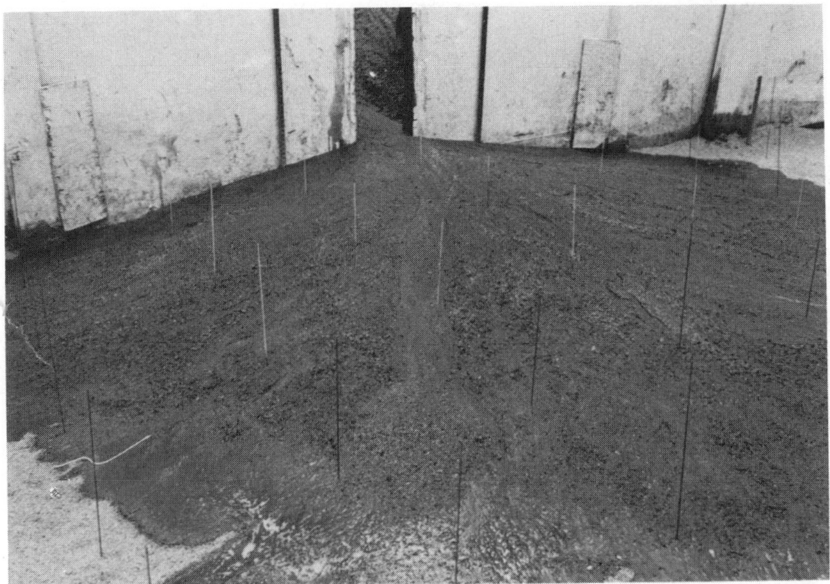

FIGURE 9.30. Small fanhead trench funnels runoff to toe of episodic-fluvial fan (from Weaver, 1984).

completely filled the same channels as flow subsided. Both types of channels caused deposition in the midfan region.

Major, deeply incised trenches at the fanhead typically formed abruptly and in response to one or a series of large runoff events following extended periods of apex deposition (Fig. 9.31). Because they were triggered by major high momentum discharges, they were oriented close to the axis of the fan.

The episodic-fluvial fan went through a series of stages involving aggradation and oversteepening of apex slopes, trenching and flushing of previously stored sediment, and channel backfilling and regrading of the fan surface. Except for several complicating modifications, the fan built by episodic rather than continuous fluvial processes showed the same identifiable stages as the fluvial fan described earlier. The fact that only a small percentage of the major runoff episodes (5%) resulted in significant fanhead entrenchment indicates that these floods were the triggering mechanisms rather than the causes of downcutting.

Large flood events with high momentum and high sediment concentrations deposited the bulk of their loads on the midfan area, even during periods of spread flow at the apex. In contrast, small-and moderate-scale flows deposited virtually all their load immediately upon reaching the unconfined slopes at the fanhead. These flows, together with larger flows that contained high sediment loads (i.e., those generated on steep slopes near the basin outlet), were responsible for rapid buildup of the fanhead. Frequent high-magnitude, high-momentum runoff events

FIGURE 9.31. Major fanhead trench triggered by large runoff events on the episodic-fluvial fan (from Weaver, 1984).

that were low in relative sediment content caused little or no fanhead deposition. Typically they eroded sediment from the fan apex and retarded the apex-building process.

Channels rarely migrated laterally over substantial distances. Instead, the episodic flood flows of high sediment discharge and momentum caused channels to be abandoned by avulsive shifts. The greater the angular divergence of a channel or trench from the fan axis, the more likely it would be abandoned in favor of a straighter path.

Long-term depositional patterns were primarily controlled by the topography of the fan surface, and the result was essentially uniform deposition overall. Although sediment concentrations were generally high, fluvial processes dominated. Therefore, many of the evolutionary characteristics of the episodic-fluvial fan were closely comparable with those of the fluvial fan formed under more continuous, constant-discharge conditions.

Mixed-Mode Fan

The mixed-mode alluvial fan was developed from runoff generated in a basin of greater local relief and higher drainage density (Fig. 9.27) than either the fluvial-fan or the episodic-fluvial-fan sediment source areas. Rainfall events in the source area resulted in alluvial fan deposition 82% of the time. The sediment content of individual flows (33–64%) was approximately three times that of the flows that built the episodic-fluvial fan (9.0–19%). The volume of sediment delivered to the fan surface varied from 2565 to 7096 g per event. Extremely large standard deviations of sediment yield (Table 9.9) indicate the high variability in event size and sediment content. However, these values still average being 3 to 15 times larger than episodic-fluvial-fan depositional episodes.

A total of 311 fan building discharge events were produced from 426 precipitation events. Of these discharges, 20% were classed as mudflows, 44% as heavily charged streamflows, and the remaining 36% as small or minor flows of a mixed variety. Mudflows deposited about 0.003 m^3 per event, while streamflows deposited 0.0004 m^3 per event. The streamflows were heavily charged with sediment and deposited approximately 1.86 times the average volume of the episodic-fluvial-fan flow events. The average depositional rate, as measured by volumetric increase of the fan following each of the 17 runs, was 0.0008 m^3 per runoff event.

Fan Growth and Dynamics

With sedimentation rates four times those of the episodic-fluvial fan and with styles of sediment movement distinctly different from normal fluvial transport processes, the geomorphic characteristics, surface processes, and evolutionary trends of the mixed-mode fan were very different from both fluvial fans. The mixed-mode fan expanded episodically (Fig. 9.32*A*), and periods of low lateral expansion were separated by periods of more rapid growth. Although the average growth rate was three times that of the episodic-fluvial fan (0.02 m^2/event), slow growth periods (runs 5–7 and 12–14) were characterized by lateral expansion rates of only 0.007 m^2 per runoff event.

The volumetric growth curve for the fan (Fig. 9.32*B*) does not show an episodic growth pattern and both periods of rapid lateral growth and the intervening intervals of very slow expansion had accumulation rates of 0.0008 m^3 per event, which was exactly the average rate of deposition. Therefore, variation in rates of lateral growth of the mixed-mode fan cannot be accounted for by changes of sediment yield, but they must be attributed to processes acting on the alluvial fan. Periods of rapid lateral growth were dominated by comparatively efficient mudflow transport to the toe. These mudflow events were directly responsible for roughly 75% of the lateral growth of the fan. Nearly every mudflow event that reached the toe passed through

Table 9.9 Mixed-Mode-Fan Sediment Yield

Drainage Areas	*n*	Sediment–Water Ratio (by weight)	Sediment Yield (g)
1–4	12	0.54 (s^2 = 0.27)	2565 (s^2 = 1618)
5–8	11	0.75 (s^2 = 0.31)	3212 (s^2 = 1881)
9–10	6	1.84 (s^2 = 0.59)	7096 (s^2 = 5826)

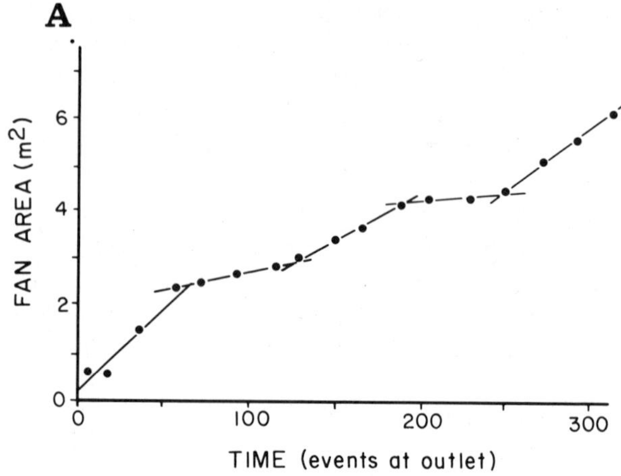

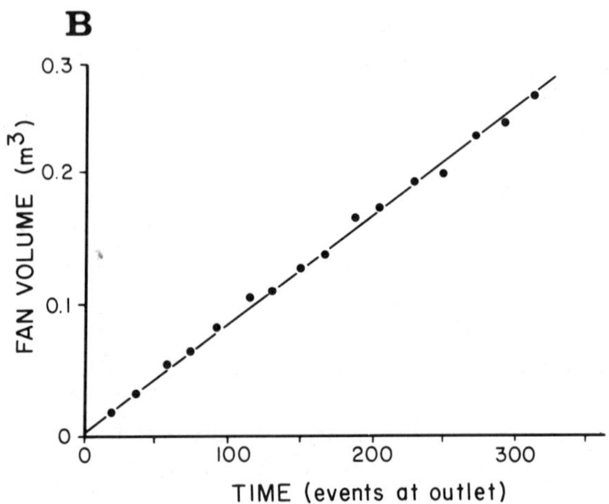

FIGURE 9.32. Growth of mixed-mode fan. (*A*) Areal growth. Note the episodic growth pattern. Time is indexed by the number of runoff events that reached the fan surface. (*B*) Volumetric growth of mixed-mode fan. While only 25% of the discharges were mudflows, each mudflow generated over eight times the volume of sediment as compared to fluvial events. Time is indexed by the number of events reaching the fan (from Weaver, 1984).

a fanhead trench developed by preceding periods of streamflow erosion (Fig. 9.33). Large-magnitude streamflows were over 2.5 times as common as mudflow events and were relatively efficient at scouring the oversteepened fanhead. However, major mudflows transported over eight times the volume of sediment to the fan surface as compared to streamflow episodes, and therefore mudflows dominated the overall aggradational process.

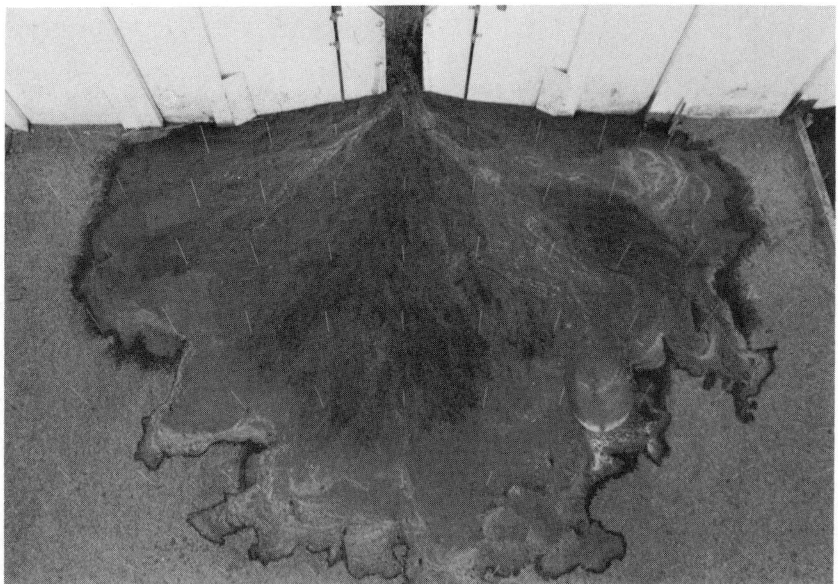

FIGURE 9.33. Streamflow incision at the fan apex funneled mudflows to the midfan and toe of the mixed-mode fan (from Weaver, 1984).

The fan surface was fashioned by the frequent occurrence of fluvial events and by the volumetrically more important but less common mudflow episodes. Both processes not only produced their own distinctive features on the fan surface, but each strongly influenced the operation of the other. For example, mudflows produced a topographically irregular fan surface and relatively steep slopes near the apex. Streamflow activity formed fanhead trenches, reworked and partially sorted the mudflow deposits, and created conditions favorable for the eventual delivery of large quantities of sediment to the lower fan.

Gradients at the midfan and apex were controlled by rapid mudflow deposition, whereas fan gradients near the toe were primarily controlled by fluvial processes. Figure 9.34 illustrates the continuity of slope profiles, which extended from the lower source area channel to the toe of the alluvial fan. Gradients in the apex and midfan areas averaged 0.130, about three times steeper than the fluvial fan. A distinct break in the slope profile near the toe marks the point at which silts and clays were deposited at the leading edge of the prograding deposit (Fig. 9.34). Their flat depositional gradient (0.035) attests to their fluvial origin.

The evolution of fan surface slopes during part of run 10 is graphically displayed in Fig. 9.35. One profile depicts the fan surface following several closely spaced mudflow events. Nine successive streamflow episodes then scoured through these deposits, moving the locus of deposition over 60 cm downfan to the newly estab-

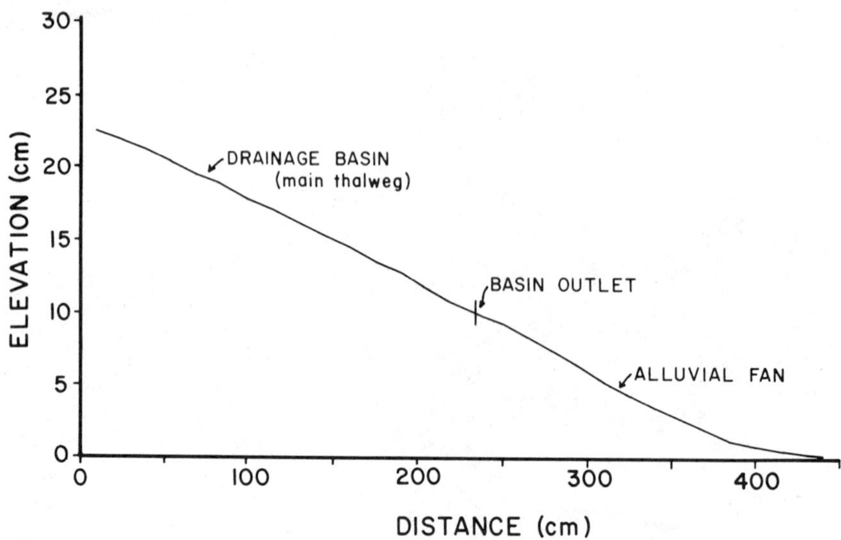

FIGURE 9.34. Profile of the main channel thalweg from the outlet of the drainage basin to the toe of the mixed-mode alluvial fan. Profile measured following run 16 (from Weaver, 1984).

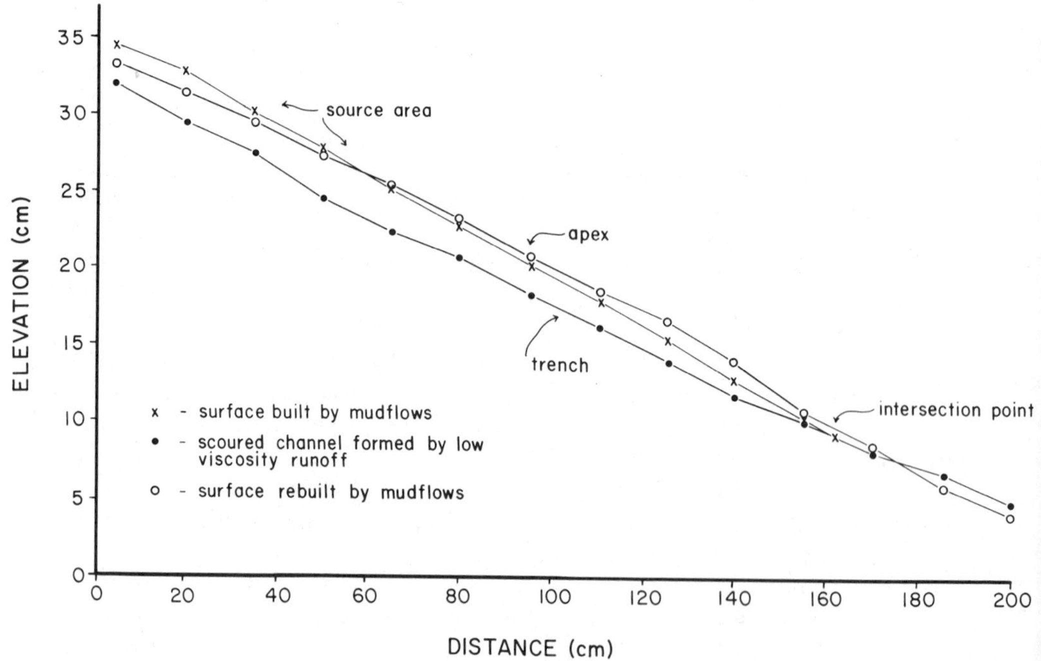

FIGURE 9.35. Evolution of mixed-mode fan surface profile under alternating periods of mudflow and fluvial activity, run 10. Profiles shown extend from the midfan into the drainage basin (from Weaver, 1984).

lished intersection point. The next mudflow rebuilt the surface and totally filled the incision by following the course created by preceding fluvial events.

The direction of fluvial discharges, and hence fanhead trenches, was determined by the fan topography as well as by streamflow momentum. The path of mudflows leaving the source area was, in turn, determined by trench orientation, trench capacity in relation to mudflow volume, mudflow viscosity as well as flow momentum, and fan topography. For example, highly viscous flows filled and essentially ignored the position of small- or moderate-size fluvial channels, whereas comparatively fluid mudflows could be totally contained and directed by a larger fanhead trench.

The mixed-mode fan experienced repeated cycles (Fig. 9.36) of mudflow deposition followed by numerous less viscous fluvial events and, finally, renewed mudflow activity, which shifted sedimentation sites laterally across the fan surface as well as horizontally from apex to toe.

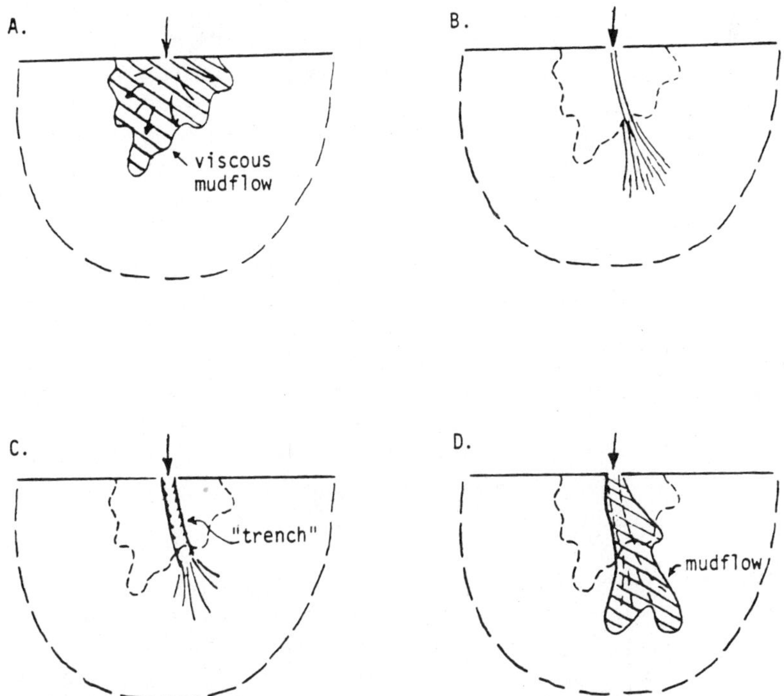

FIGURE 9.36. Schematic growth model for mixed-mode alluvial fans. (*A*) Deposition of a viscous mudflow in the apex region obliterates any preexisting channels. (*B*) Streamflows move over the deposit towards topographically low areas. (*C*) Eventually, similar discharges scour a channel through the mudflow deposits. (*D*) A volumetrically large mudflow generally follows the scour path developed by preceding fluvial runoff and may obliterate and entirely fill the channel (from Weaver, 1984).

Mudflow Fan

The final experimental fan was formed primarily by mudflows from a small, steeper sediment source area (Fig. 9.28). Mudflows, which accounted for 80% of the 128 runoff events reaching the fan following run 4, deposited 97% of the total accumulated sediment. That is, although streamflow events may have been relatively numerous, they were insignificant in terms of the rates of alluvial fan deposition. As in the mixed-mode fan, mudflows contributed roughly 0.003 m^3 per event. This similarity indicates that while the frequency of mudflow events reaching the mudflow fan was much higher, the relative contribution of sediment from a single event of either type was not significantly changed from the previous experiment, and mudflows still produced five to eight times the volume of sediment produced by streamflows per discharge event.

Fan Growth and Dynamics

The mudflow fan grew at a relatively constant rate of 0.04 m^2 per event (Fig. 9.37), or roughly 5.6 times more rapidly than the episodic-fluvial fan. The constant rate of areal expansion contrasts with the declining and fluctuating trends for the other two fans. The unchanging growth rate was primarily the result of (1) constant sediment yield through time and (2) the dominance of mudflow processes over fluvial (streamflow) activity.

The volumetric growth curve for the mudflow fan shows an average rate of growth following run 4 of 0.0025 m^3 per event, 11 times that of the episodic-fluvial fan (Fig. 9.37*B*), with streamflow events contributing an insignificant quantity of sediment to the fan surface.

During the experiment, roughly 52 cm of deposition occurred at the apex, and the main channel of the source area built up at the same rate, as mudflows deposited portions of their loads before reaching the fan surface.

Patterns of vertical and areal fan growth were dominated by mudflow deposition. Although fluvial events accounted for 25% of the total number of events reaching the fan surface, they were neither substantial nor frequent enough to significantly modify previously deposited sediments. Rather, a combination of highly variable mudflow viscosities and the long-term development of topographic anomalies on the fan surface accounted for the observed spatial distribution of sedimentation through time.

The mudflow fan, evolving in a manner similar to the fluvial fans, progressed through a series of regular stages that were repeated during the course of the experiment (Figs. 9.38, 9.39). In general, this pattern consisted of the following sequence of events:

1. As mudflows crossed the fan apex, deposition occurred along their lateral margins and formed natural levees (Fig. 9.38). Future flows were guided by these levees, but large events overtopped the channels and added thick mudflow deposits to the already elevated banks. The efficient channel system permitted mudflows to reach the more distal regions of the fan.

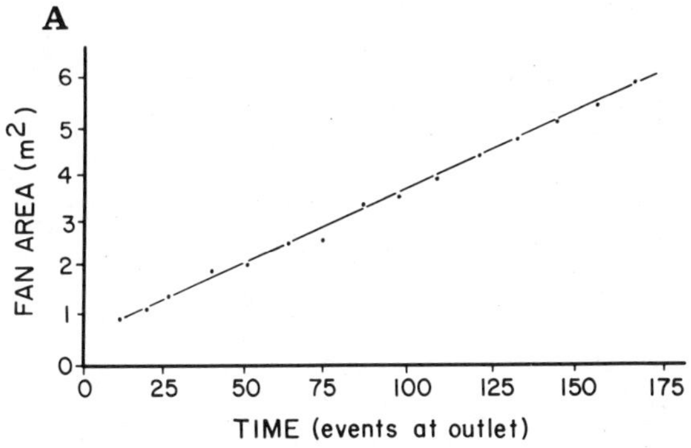

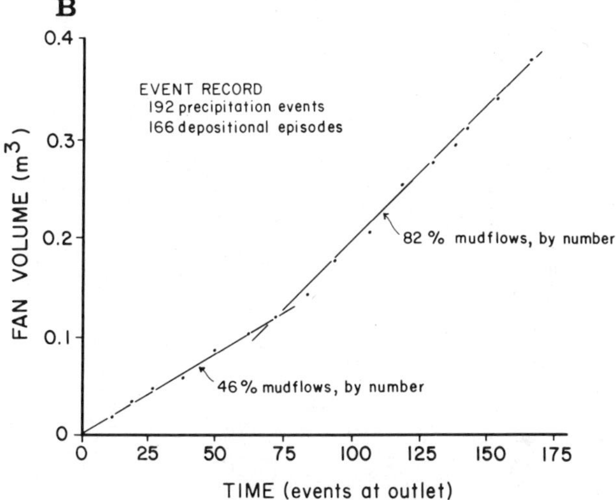

FIGURE 9.37. Growth curves of mudflow fan. (*A*) Areal growth curve. Time is indexed by the number of flows that exited the drainage basin. (*B*) Volumetric growth curve. At roughly t = 75, the procedure was slightly modified to yield more frequent mudflows (from Weaver, 1984).

2. Continued deposition in one area or segment of the fan eventually resulted in its having a higher elevation than adjacent fan surfaces (Fig. 9.38).

3. As deposition continued, mudflow velocities were gradually reduced on these elevated, more gently sloping areas, and large, viscous mudflows deposited their loads at the apex without being funneled farther downfan.

4. The filling and obliteration of the channels near the fanhead permitted subsequent flows to follow paths down the adjacent, steeper slopes to the topographically lower parts of the fan. Low-viscosity flows scoured through the recently deposited mudflow deposits at the apex to form a shallow trench, which channelized the flow, and the cycle started again (Fig. 9.39*B*).

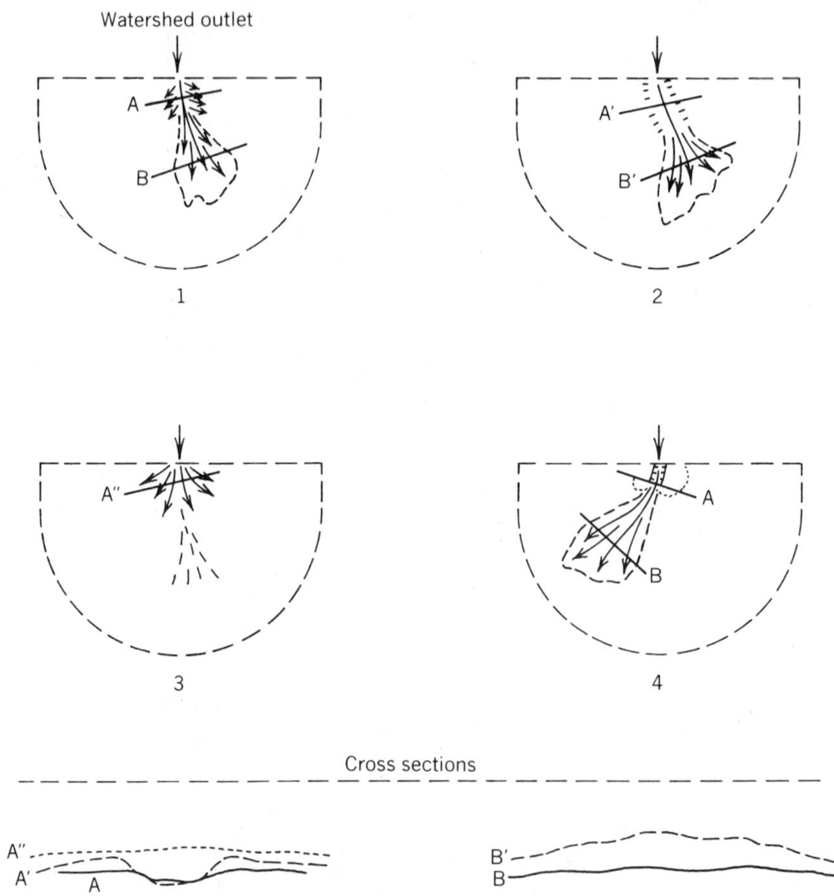

FIGURE 9.38. Simplified growth model for alluvial fans formed by episodic mudflow activity. Fluvial activity is comparatively unimportant. See text for explanation (from Weaver, 1984).

The long-term results of this avulsively shifting pattern of deposition is shown by the depositional maps of the mudflow fan (Fig. 9.40). During run 10, deposition was concentrated at the extreme right sector of the fan. Continued accumulation of sediment in this area could only lead to decreasing gradients above the sector receiving heavy deposition (e.g., along line AB of run 10, Fig. 9.40) and increasing gradients over the remainder of the alluvial fan (e.g., along line AC of run 10, Fig. 9.40).

During the following series of events (run 11), deposition continued in the right sector with one center of deposition shifting laterally towards the axis of the fan and another moving towards the apex. Midway through run 11, deposition at the apex obliterated the old channel path and subsequent flows were then directed down the steep fan surface to the left of the centerline. Runs 12 and 13 were characterized

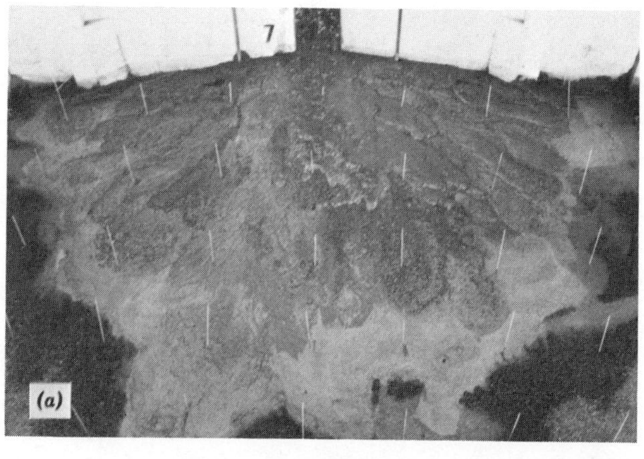

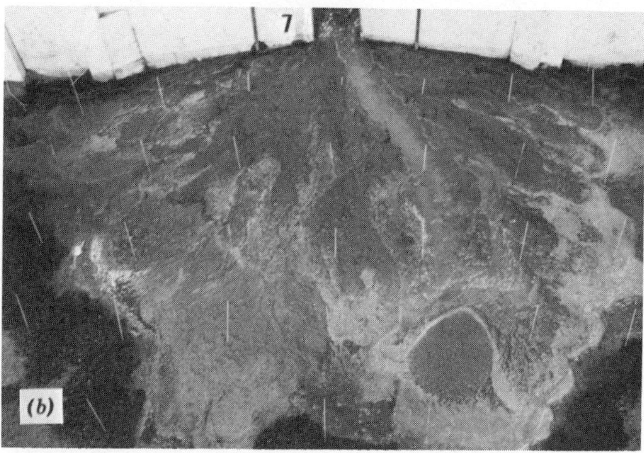

FIGURE 9.39. (*A*) Viscous mudflows early in run 7 were spread over the proximal region of the fan and rapidly built up apex slopes (Fig. 9.38, step 3). Measurement pins are spaced 30 cm apart. (*B*) More fluid flows channeled the apex and directed runoff to the topographically lower part of the fan (Fig. 9.38, step 4) (from Weaver, 1984).

by continued deposition in the left sector as topographically low areas were systematically built up by repeated mudflow events. At the same time, there was virtually no deposition on the right side.

By run 14 (Fig. 9.40) the left portion of the fan surface was topographically higher and deposition once again abruptly shifted back to the right half of the fan. Although active deposition was again restricted to one side of the fan, continued accumulation and buildup of the distal areas caused the depositional centers to migrate laterally as well as up and down the fan surface. The development of topographic highs and lows on the surface of the fan as well as their subsequent abandonment and reversal varied regularly from one extreme to another (Fig. 9.41).

Mudflows created large-scale topographic anomalies on the fan surface. They

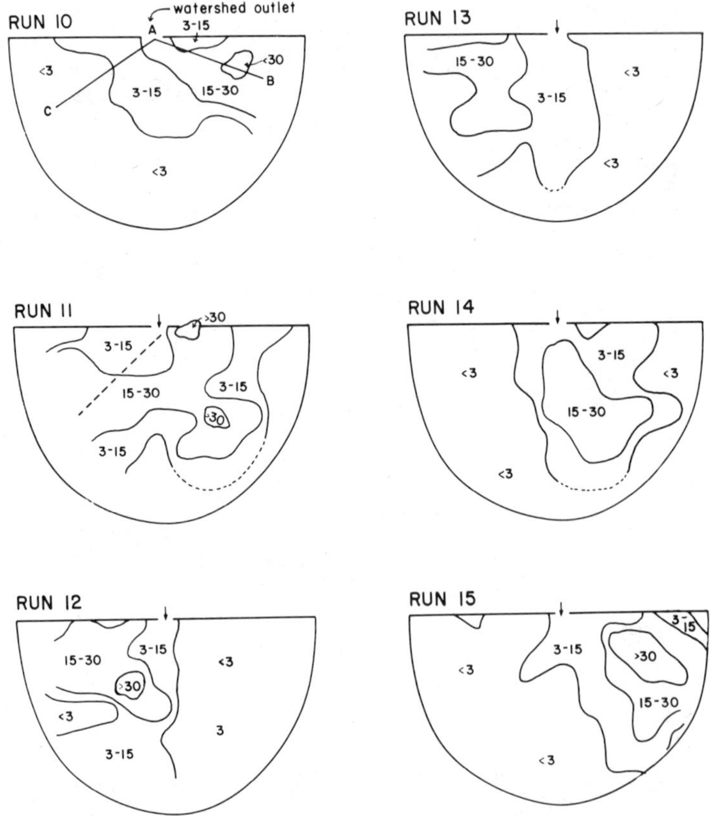

FIGURE 9.40. Deposition on the mudflow fan, runs 10–15. See text for discussion. Depths are in mm $\times$ 10^{-3} (from Weaver, 1984).

were responsible for decreasing fan slope above areas of mudflow deposition. Due to the viscous nature of mudflows, the decreased slopes caused the locus of deposition to move towards the apex. At the same time, continued apex aggradation together with virtually no deposition on the other portions of the fan surface caused a rapid increase in overall fan gradients.

When viewed over short periods, deposition on the mudflow fan was highly localized and episodic. However, over the long term, rates of vertical accumulation were relatively uniform, the overall fan surface slopes were maintained, and the rate of areal growth was relatively constant. As described above, these results were accomplished by the progressive shifting of deposition up and down the fan surface as well as laterally (Fig. 9.42).

As on the fluvial fan, patterns of deposition on the mudflow fan also exhibited some degree of regularity and predictability (Table 9.10). However, the variability

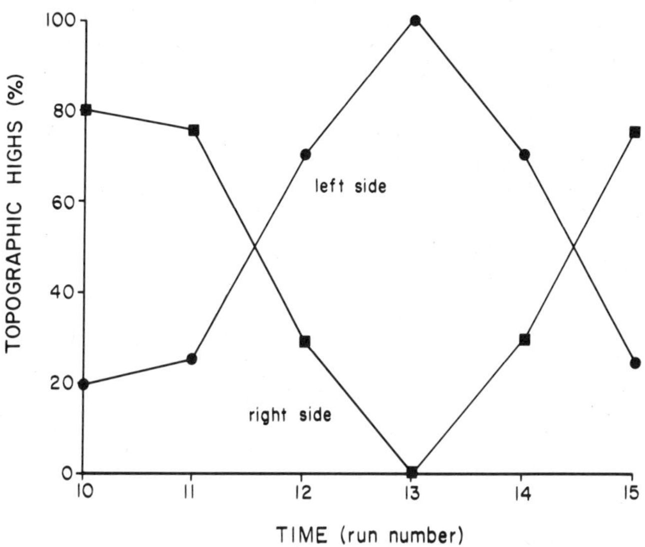

FIGURE 9.41. Distribution of topographically high points on each side of the mudflow fan as a percentage of the total high points on the fan during runs 10–15. Low values indicate periods when one side of the fan was lower than the other (from Weaver, 1984).

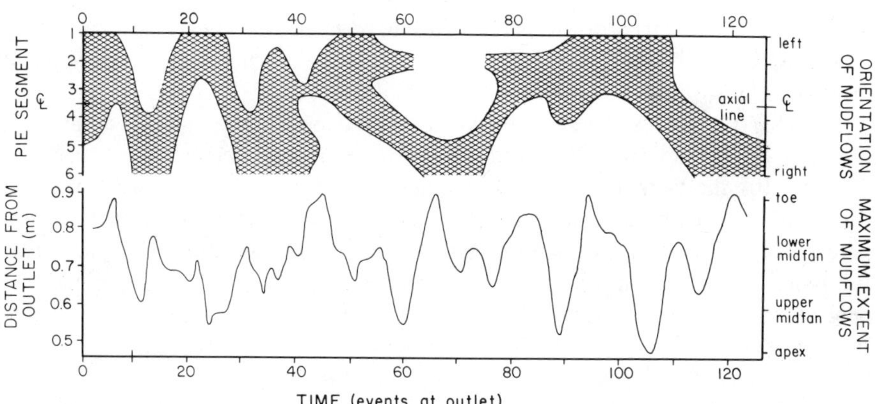

FIGURE 9.42. Lateral (side to side) and longitudinal (up- and downfan) patterns of deposition on the mudflow fan. In the top diagram, the fan was divided into six pie-shaped segments from left to right. The location of deposition is shown by the striped pattern, as recorded for each mudflow event that reached the surface of the fan. The locus of deposition migrates across the fan in an irregular fashion. In the lower diagram, the maximum downfan extent of each flow is plotted as a moving mean of every five measurements, which shows that similar shifts in depositional patterns occur longitudinally (from Weaver, 1984).

TABLE 9.10 Forward Transition Probability Matrix for the Mudflow Fan[a].

	Following State						
Former State	Erosion	No Change	Moderate Deposition	Moderately Heavy Deposition	Heavy Depositon	Percentage of total events	*n*
Erosion	0	0.50	0.17	0.33	0	2	6
No Change	0.01	0.52	0.27	0.17	0.03	45	172
Moderate deposition	0.01	0.31	0.36	0.27	0.04	30	113
Moderately heavy deposition	0.01	0.35	0.38	0.22	0.04	21	79
Heavy deposition	0	0.64	0.27	0.09	0	3	11
Percentage of total events	1	43	32	21	3		
n	3	162	122	81	13		381

[a]This matrix incorporates data from the entire fan surface. It predicts the probability of various erosional and depositional events occurring given the known magnitude of previous events at any point on the fan's surface. Event magnitudes are as follows; erosion: < −0.3 cm/run; no change: −0.3-0.3 cm/run; moderate deposition: >0.3-1.5 cm/run; moderately heavy deposition >1.5-3.0 cm/run; heavy deposition: >3.0 cm/run.

induced by large ranges of discharge concentrations (viscosities) and magnitudes (volumes) for each runoff event added to the random, nonpredictable component of alluvial fan evolution under mudflow conditions.

Probability matrices were developed based on deposition measured on the mudflow fan surface following each run. A matrix evaluated for the fan as a whole (Table 9.10) shows that the most common result of any given run was for some degree of significant deposition at any location (56% chance of occurrence) regardless of the magnitude of deposition that occurred during a preceding run. Areas receiving heavy deposition during one event were most likely to experience no deposition during the following event ($p = 0.64$). In contrast, areas of the mudflow fan that received moderate or moderately heavy deposition during any given run were most apt to experience continued deposition in succeeding events.

Although average deposition rates were high when compared to the fluvial fan, many locations along the mudflow fan margin were not subject to frequent flooding or sediment deposition. When the fan apex was analyzed separately (Table 9.11), the probability of any area experiencing moderate or moderately high rates of deposition rose to 84%, an increase of 31%. Regardless of the area analyzed, the frequency matrix indicates that the most common occurrence was for moderate or moderately heavy apex deposition during any given run. The relatively small probability of events resulting in no significant deposition at the apex ($p = 0.11$) as

TABLE 9.11 Forward Transition Probability Matrix for the Apex Area of the Mudflow Fan.[a]

	Following State						
Former State	Erosion	No Change	Moderate Deposition	Moderately Heavy Deposition	Heavy Deposition	Percentage of Total events	*n*
Erosion	0	0	1.00	0	0	1	1
No change	0	0.25	0.38	0.25	0.13	11	8
Moderate deposition	0	0.05	0.61	0.32	0.03	50	38
Moderately heavy deposition	0.04	0.12	0.52	0.28	0.04	32	25
Heavy deposition	0	0.25	0.50	0.25	0	5	4
Percentage of total	1	11	55	29	4		
n	1	8	42	22	3		76

[a]Note the increased probability associated with deposition at a point on the apex (88% likelihood) as compared to the entire fan surface (56% chance, Table 9.10). Event magnitudes are the same as defined in Table 9.10.

compared to the fan as a whole ($p = 0.43$, Table 9.10), indicates that deposition at the fan apex had a greater chance of being followed by continued aggradation than anywhere else on the fan surface.

DISCUSSION

Models of Fan Growth

A successful model of landscape evolution must first describe landform development under constant controlling conditions, and then the effects of varying conditions can be considered. Because varying environmental conditions are so common in areas that display active alluvial fans, they are commonly invoked as explanations for alluvial fan morphology and behavior. However, the conceptual model, which is based on the hypothesis that aggradation must eventually lead to oversteepening, instability, and incision, suggests that fans should develop in an episodic manner, even under constant conditions (Fig. 9.43). This model was confirmed by observation of four types of experimental alluvial fans that ranged from fluvial to mudflow in composition (Fig. 9.44).

The fluvial fans, developed by streamflow under conditions of high sediment yield, evolved through a regular cycle of events that resulted in long-term rates of uniform deposition and a characteristic pattern of horizontal and vertical growth (Fig. 9.43). For fans with constant water and sediment supply, and in the absence

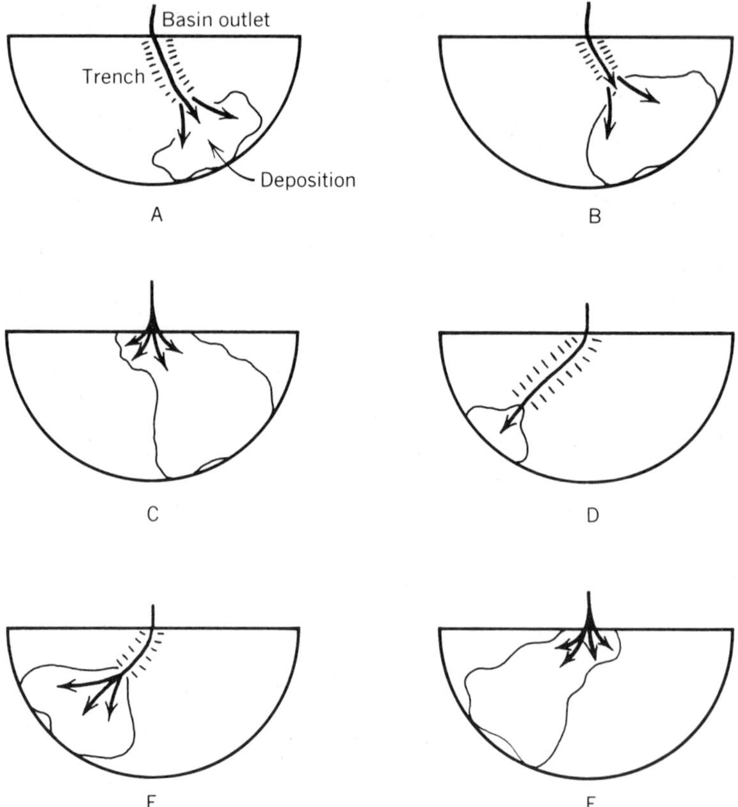

FIGURE 9.43. Schematic portrayal of the cyclic pattern of erosional and depositional events that characterized the evolution of the experimental alluvial fans. (*A*) Apex incision results in deposition near the toe. (*B*) Channel backfills and the locus of sedimentation (intersection point) moves up to the midfan region. (*C*) Channel is totally obliterated and flows emerging from the drainage basin now spread in all directions, depositing sediment at the apex and increasing local slopes. (*D*) Oversteepening of the apex continues until the inherent stability threshold is exceeded, downcutting is initiated, and the sediment is again directed towards the toe. (*E*) Channels in the topographically low area, as compared to the right side of the fan, slowly backfill and raise the relative elevation of this sector until (*F*) deposition again returns to the fanhead.

of baselevel change, repeated fanhead trenching and related episodic adjustments were a part of fan growth. Deposition at the fan apex oversteepened the fanhead until a critical, threshold slope was attained. Rapid channel incision at the apex and upper midfan flushed stored sediment towards the toe, the channels were backfilled until flow was once more spread over the surface, and deposition was again concentrated at the apex, which renewed the cycle. In accordance with this growth model, vertical and lateral growth was not continuous, even with constant discharge.

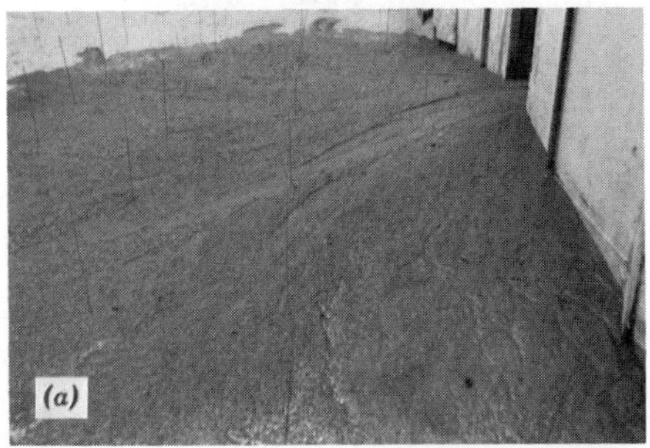

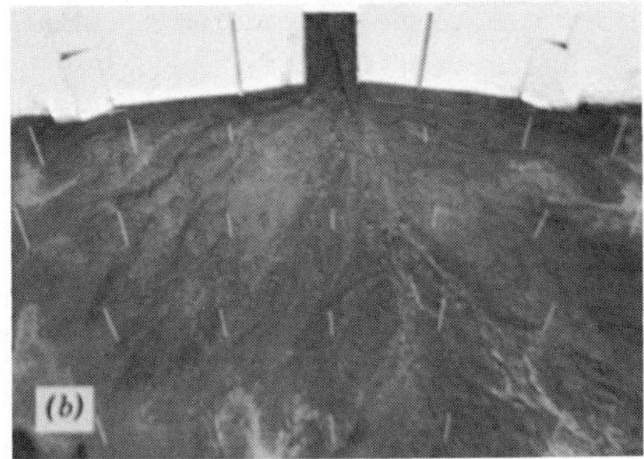

FIGURE 9.44. Comparison between the surface morphologies of the (*A*) episodic-fluvial fan, (*B*) mixed-mode fan, and (*C*) mudflow fan. Note especially the increase in surface roughness and the decrease in fluvial channels that accompanied the growing importance of mudflow activity (from Weaver, 1984).

There were important differences in the mode of development and the form of the fans formed by streamflow, mudflows, and combined streamflow–mudflow (Fig. 9.44), as a result of the erosion processes in the source area and the resulting sediment yields (Tables 9.12, 9.13). The rate of areal expansion and volumetric growth of the fans varied widely, but growth rates increased with sediment yield and sediment concentrations. As mudflows became increasingly dominant, volumetric growth rates increased until, during the mudflow-fan experiment, the same 10-sec burst of precipitation produced 14 times the sediment produced during the episodic-fluvial-fan experiment (Table 9.14).

Vertical growth rates were controlled by sediment yields and the surface area over which the sediment was deposited. Fans with more limited surface areas had greater vertical growth rates. Growth rates were also controlled by sediment concentration (which affected viscosity, distance of travel, and lateral extent of fans) and by frequency of events. Fan slope was also influenced by these factors. The

TABLE 9.12 Physical Characteristics of the Sediment Source Areas

Type of Fan	Source Area (m^2)	Mean Basin Slope	Stable Mean Channel Slope	Steepest Hillslope Gradients	Channel Length (m)
Fluvial	56	0.130	0.146	0.367	23.1
Episodic fluvial	56	0.097	0.159	0.474	16.0
Mixed mode	56	0.200	0.170	0.624	18.2
Mudflow	17	0.340	0.193	0.891	9.1

slope of the episodic-fluvial fan was 2.5 times greater than that of the fluvial fan, and the slope of the mudflow fan was 1.65 times greater than that of the episodic-fluvial fan (Fig. 9.45). Higher water discharges, lower sediment concentrations, and constant flow permitted sediment reworking and redistribution on the fluvial fan, which maintained a relatively low gradient, but this was markedly restricted on the episodic-fluvial, mixed-mode, and mudflow fans.

In spite of the differences between the fluvial and episodic-fluvial fans, fluvial fans formed by episodic runoff events of comparatively high sediment concentration display nearly all the important evolutionary characteristics of fans produced by continuous streamflow. In addition to the similarities derived from cyclic trenching of apex deposits, experimental evidence also suggests that the longer-term migration of depositional centers in response to the development of topographic highs and lows is a phenomenon common to alluvial fans formed by both episodic and continous streamflow processes.

Primary sedimentation on mixed-mode alluvial fans was overwhelmingly dominated by viscous mudflows, which accounted for up to 80–90% of the total fan deposition while comprising as few as 20% of the runoff events by number.

Experimental results suggest that even when mudflows dominate all other processes, one of the most important factors controlling the evolution of mudflow fans is the concentration of sediment or viscosity of the flows. While stream runoff may occur on such fans, it is neither frequent enough nor of great enough magnitude to alter significantly the depositional patterns of succeeding flows or to modify the steep, coarse-textured surface morphology of the fan (Fig. 9.44*C*). Major fanhead trenches were less common as sediment concentrations and the incidence of mudflows increased.

Mudflow fans evolved through a systematic series of stages that resulted in substantially the same long-term cyclic sedimentation patterns that were described for fluvial and mixed-mode alluvial fans (Fig. 9.43). Repeated deposition on one portion of the fan produced a relative topographic high, and mudflow deposits produced especially steep slopes in the apex area, which were unstable under more fluid runoff conditions.

TABLE 9.13 Hydrologic Characteristics of Experimental Fans

Type of Fan	Precipitation Characteristics	Type of Event (%)		Runoff (% by volume)		Sediment Yield (cm^3 per event)		Water Content (% by weight)	
		Mudflow	Fluvial	Mudflow	Fluvial	Mudflow	Fluvial	Range	Mean
Fluvial	Uniform constant 50 mm/hour	0	100	0	100	—	1.7[a]	91–98	96
Episodic fluvial	10-sec bursts over 8 areas	0	100	0	100	—	2.3	81–91	85
Mixed mode	12-, 15- and 18-sec bursts over 10 areas; surface etched	30	70	75	25	29.8	3.7	37–67	42
Mudflow	10-sec bursts over 8 areas; surface etched	80	20	97	3	31.2	2.8	10–50[b]	30[b]

[a]per 10 sec of precipitation and runoff.
[b]Estimated.

TABLE 9.14 Growth Characteristics of Experimental Alluvial Fans

Type of Fan	Areal Growth Rate (m^2/ event)	Lateral–Vertical Growth Rate Ratio[b]	Volumetric Growth (cm^3/10 sec)	Vertical Growth Rate[c] (m/ 10 sec) ($\times 10^{-5}$)	Fan Slope	Side to Axial Slope
Fluvial	0.004[a]	17	2.0[b]	0.9	0.043[e]	1.42[e]
Episodic fluvial	0.007	7	2.3	7.3	0.109	1.46
Mixed mode	0.020	11	8.5	13.7	0.130	1.43
Mudflow	0.036	5	25.2[d]	42.7	0.180	1.20

[a]Per 10 sec of precipitation and runoff.
[b]Lateral growth along fan axis divided by mean vertical growth measured along same line.
[c]Using last fan area recorded for each experiment, 20.3, 3.1, 5.9, 5.8 m^2, respectively.
[d]Following run 4, if only mudflow events are used (96% by volume), growth rate increases to 32 cm^3/event.
[e]Measured at end of run 10C.

Fanhead Trenching

During periods of fanhead trenching, the site of maximum deposition was primarily influenced by the angle of exit from the source area and the topographic irregularities on the fan surface. These and other influences combined with the repeated cycle of fanhead trenching to produce a pattern of deposition that was generally predictable (Tables 9.4–9.6).

Three types of fan trenches were observed. One type was small channels that formed on locally steep slopes on the midfan. These were short-lived channels that

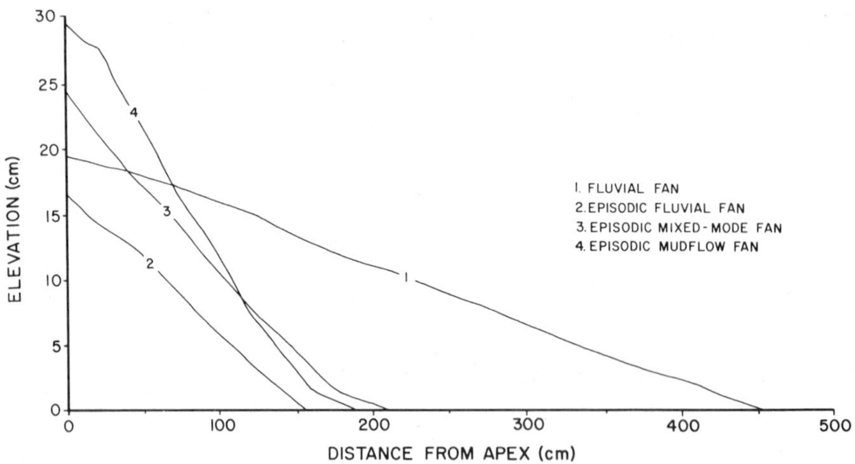

FIGURE 9.45. Longitudinal profiles of experimental alluvial fans (from Weaver, 1984).

rarely captured all of the flow from the source area. They were responsible for less than 5% of the episodes of fanhead trenching. The second type of trench developed after the critical slope threshold at the apex was exceeded (Fig. 9.7). This second type was more likely to capture the bulk of streamflow because of its location at the apex. The third type was a large midfan trench that formed as sediment yields were reduced. This channel reflected a major change in external conditions and it was the most significant incision observed (Fig. 9.25).

The initiation of fanhead trenching was dependent not only on general aggradation of the fan apex but, more specifically, on the formation of locally oversteepened slopes that led to major entrenchment (Fig. 9.46). The critical slope for entrenchment on the experimental fluvial fan was 2.6–2.75°; it would of course be different under other conditions, depending on a wide variety of climatic, hydrologic, and sedimentologic factors. Changes in these factors might change the critical slope for entrenchment or modify the landform so that its stability threshold is changed. Thus, for example, tectonics may increase slope beyond its threshold value and initiate trenching (Fig. 9.47). Tilting might also reduce slope, increasing stability and promoting apex aggradation. On the other hand, climatic change may raise or lower the threshold of stability by changing sediment yields, flood magnitudes, vegetational characteristics, and overall erodibility, so that a new threshold

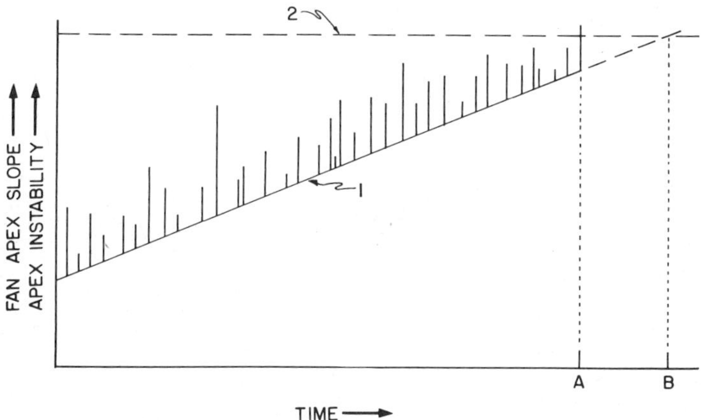

FIGURE 9.46. Relation through time between gradient at the fanhead and apex instability. Line 1 portrays the gradually increasing slope of the fanhead. When the ascending line of fanhead slope intersects line 2, which represents the maximum slope at which the apex is stable, trenching will occur at time B. Superimposed on line 1 are vertical lines representing changes in fanhead instability that are related to high-magnitude runoff events or longer-term climatic fluctuations. Normally, the operation of these processes has little significant morphologic effect on the alluvial fan. However, when the fan slope and apex instability are high, trenching will occur sooner than expected (at time A) when a large-magnitude event exceeds the stability threshold (line 2). In reality, the event merely precipitated the eventual incision at time A rather than at time B (after Schumm and Hadley, 1957).

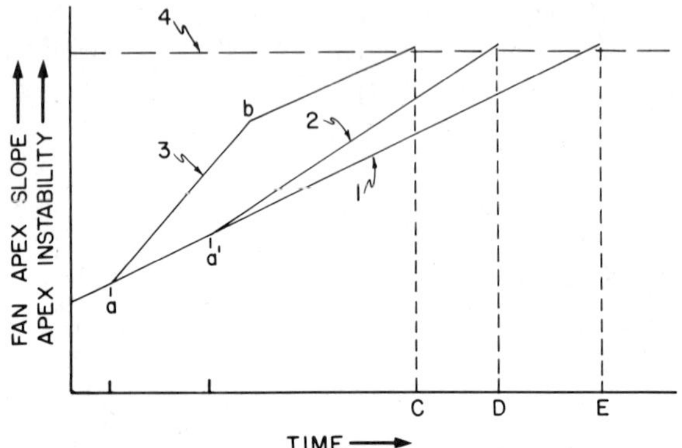

FIGURE 9.47. Effect of tectonics on the timing of morphologic response. Under unvarying external conditions, apex slopes increase (line 1) until they exceed the threshold (line 4) at time E. When a fan is steepened by tectonic tilting, beginning at a and ending at b, and then normal depositional steepening follows, during a tectonically quiescent period, trenching will occur at time C (line 3). Constant, uniform tectonic steepening initiated at a′ will similarly cause incision at time D (line 2). If the fan is tilted in the other direction and gradients are reduced (or steepened more slowly), the initiation of trenching will occur sometime following time E (from Weaver, 1984).

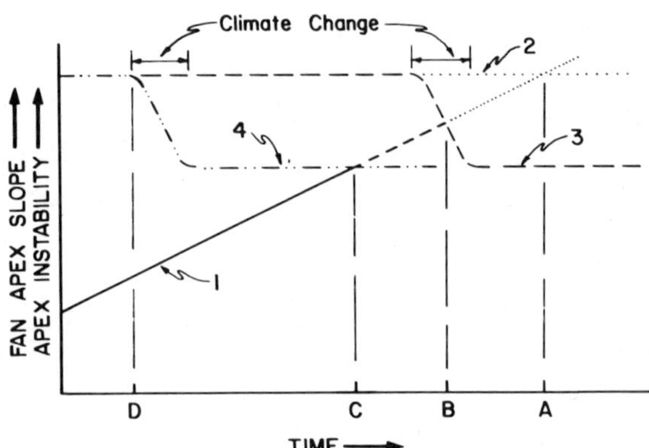

FIGURE 9.48. Effect of climate change on the timing of morphologic response. Line 1 depicts the increase of apex slopes with time. Under unchanging threshold conditions (line 2) fanhead slopes exceed their stability threshold at time A and incision commences. Climate change, which in this case establishes a new, lower threshold of stability (line 3), triggers incision at time B. If the same change was initiated much earlier, at time D, when the fan displayed greater inherent stability, then fanhead trenching at time C would substantially lag and appear essentially unrelated to the climate change depicted by line 4 (from Weaver, 1984).

slope is established that may be either closer to or farther from the actual value of slope on a fan apex (Fig. 9.48).

Recognition of the importance of slope thresholds permits the development of explanations of alluvial fan behavior that do not rely on changes of external conditions. This is not to deny the significance of such factors as tectonism or climatic change, but it also permits an explanation of differences in the timing, magnitude, and direction of responses of alluvial fans to the same external change, in terms of their relative stability (Fig. 9.49). Therefore, according to this model, a group of alluvial fans that are not subject to changing external conditions may be expected to display a variety of morphologies and histories (Harvey, 1984a). The experimental results indicate that fan behavior may be directly explicable in terms of a threshold model. However, changes of external factors may modify the threshold values.

Alluvial Fans of Southeast Idaho: A Field Test

In order to test the conclusions reached during the experimental alluvial fan study, Weaver (1984) selected an area in east-central Idaho where the results of experimental studies could be evaluated (Fig. 9.50). The Pahsimeroi and Big Lost river valleys contain a large number of alluvial fans that are in close proximity to each other and display a wide variety of morphologic conditions (from deeply entrenched to unincised). Severe and perhaps multiple changes in the climate occurred within this area during the late Quaternary (Dort, 1962, 1965; Funk, 1976; Knoll, 1973).

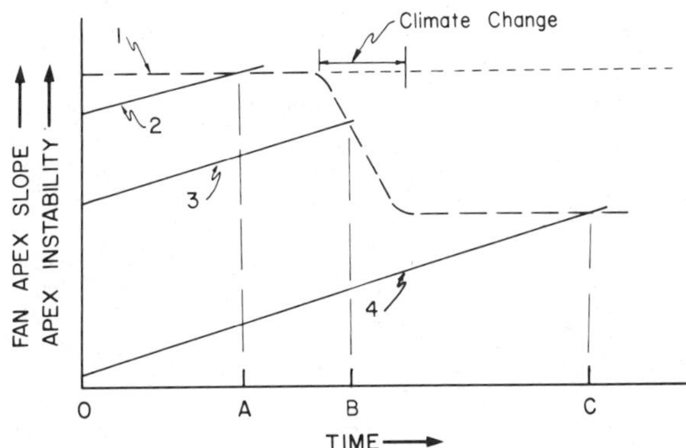

FIGURE 9.49. Effect of climate change on a group of alluvial fans. At time 0, the alluvial fans of an area display contrasting degrees of inherent stability under a specific climatic and tectonic setting. One fan, represented by line 2, exceeded its stability threshold without external change and fanhead trenching occurred at time A. Climate change prematurely triggered incision at time B on another fan (line 3), which would have been stable at time B had conditions remained unchanged. Finally, the fan displaying the greatest relative stability at time 0 showed considerable lag by responding to the climate change at time C, well after the establishment of new conditions (line 4) (from Weaver, 1984).

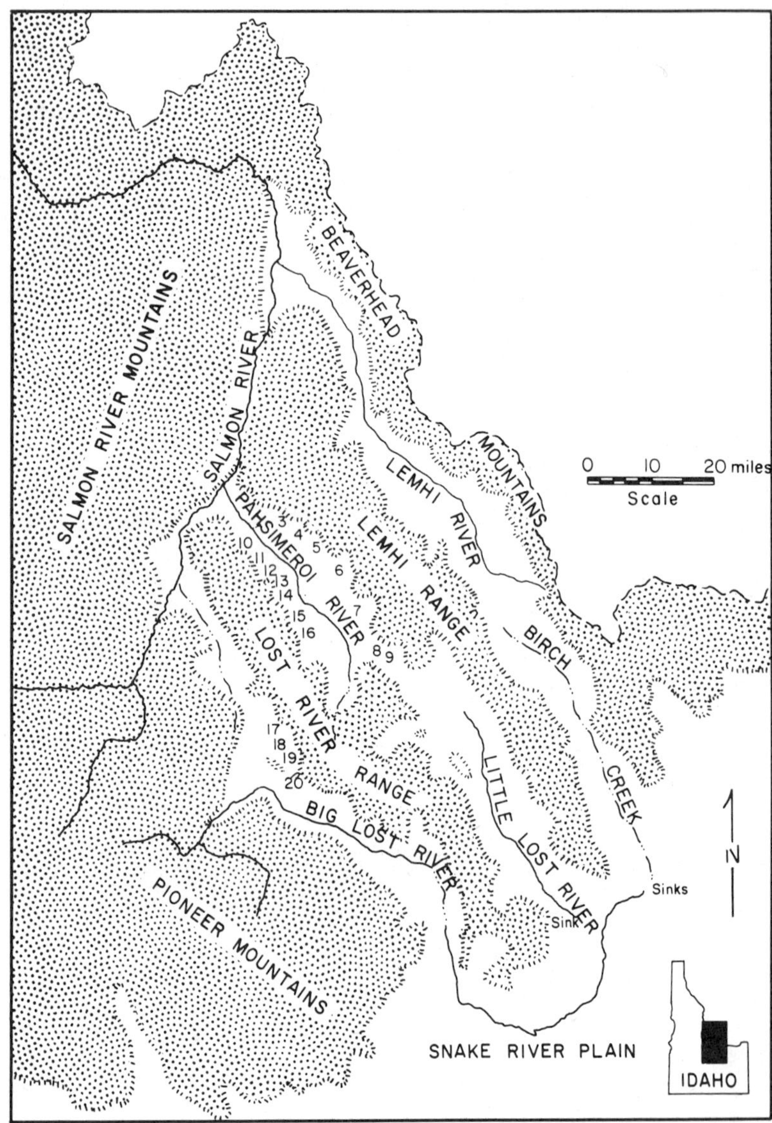

FIGURE 9.50. Index map showing location of alluvial fans in Pahsimeroi and Big Lost river valleys (from Weaver, 1984). Fans are identified as follows: (3) Tater, (4) Morse, (5) Falls, (6) Patterson, (7) Big Creek, (8) Goldberg, (9) Ditch, (10) Lawson, (11) Trail, (12) Antelope, (13) Sulpher, (14) Grouse, (15) Dead Cat, (16) Rock Spring, (17) Willow, (18) Rock Creek, (19) Cedar, (20) Elkhorn. For location of Ennis (1) and Morgan (2) fans see Fig. 9.51.

Sediment delivery to the fans was high during glacial stages, but the glacial-derived sediments have not been produced and delivered to the fans in the Holocene. In addition, it is generally accepted that crustal warping, which caused earlier drainage reversals, continued on a line transverse to the trend of the major mountain ranges through late Pleistocene time (Funk, 1976; Rupple, 1964, 1967).

Although the evidence for longitudinal tilting of the Pahsimeroi and adjacent valleys is good (Funk, 1976; Ruppel, 1964), the overall configuration and orientation of alluvial fan deposits and observations regarding the distribution, degree, and style of channel incision suggest that such tectonic activity, if it occurred during the most recent period of alluvial fan growth, played a negligible role in determining fan morphology (Funk, 1976).

Because the local climate varied from an extreme of glacial and periglacial conditions to interglacial semiaridity, it is reasonable to assume that climate change profoundly affected alluvial fan growth and morphology in these valleys. However, the geomorphic effects of climate change on alluvial fans should be similar in nature and widespread in extent (Bull, 1964b, p. 105; Eckis, 1928, p. 240; Lustig, 1965, p. 184), and where fans have evolved in similar and adjacent drainages, their response to climate change should be similar (Bull, 1964a). It is therefore surprising to find the broad array of alluvial fan morphologies that exist in the study area and to discover that, almost without exception, nearly identical and adjacent drainage basins and alluvial fans responded in very different ways to the same changes (Figs. 9.51–9.53).

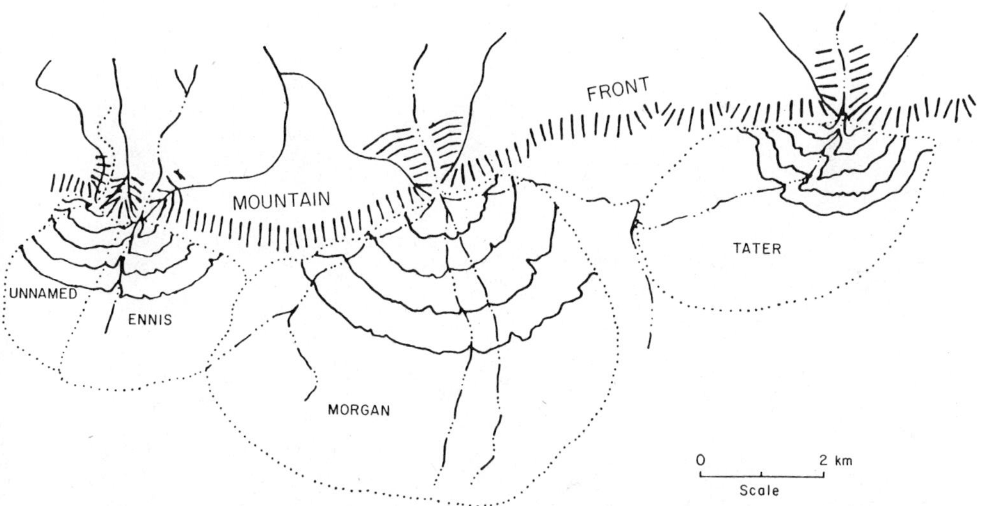

FIGURE 9.51. Map showing different topographic characteristics of unnamed (left), Ennis, Morgan, and Tater Creek alluvial fans. Dotted lines marked fan margins. Solid lines on fan surface are contour lines traced from U.S.G.S. topographic map (CI = 80 ft). Note that the Morgan Creek alluvial fan is not trenched, in comparison to the trenched Ennis and Tater fans (from Weaver, 1984).

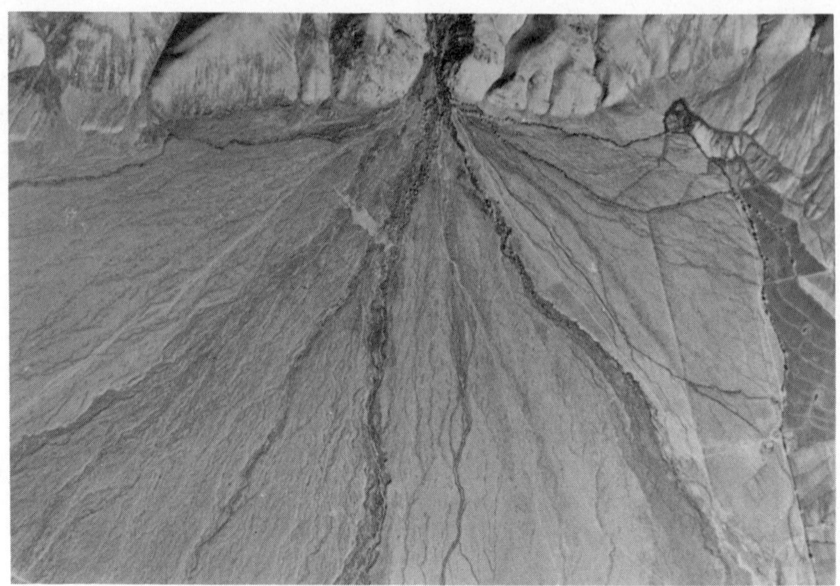

FIGURE 9.52. The Morgan Creek alluvial fan has no well-defined fanhead trench. Flow spreads over the fanhead at the mountain front and moves down the fan in a number of distributary channels (from Weaver, 1984).

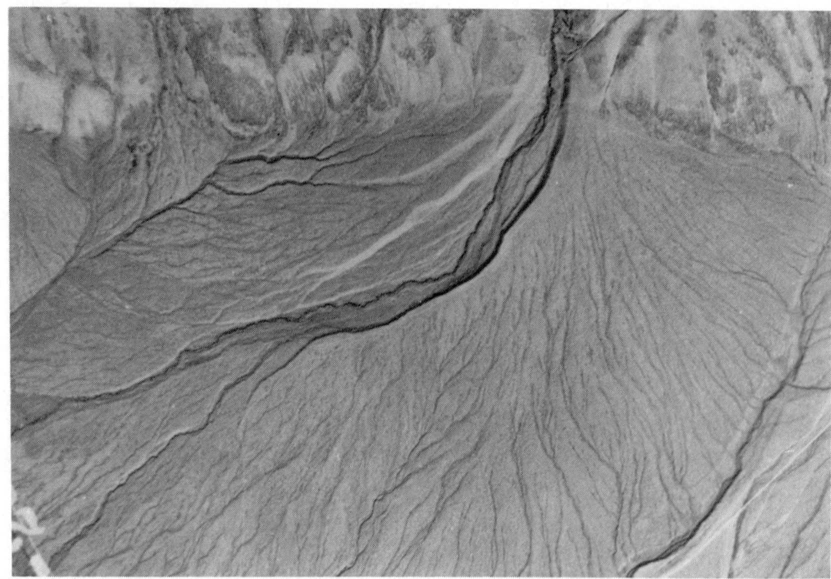

FIGURE 9.53. Deeply trenched Tater Creek alluvial fan. Note the multiple terrace levels along the left side of the channel and the moderately dissected fan surface that was left stranded by the entrenchment (from Weaver, 1984).

There appears to be little pattern to the distribution, magnitude, or timing of erosional responses to Quaternary climate change. Some fans have fanhead trenches that are morphologically young; that is, they are narrow, steep-walled, V-shaped incisions that indicate rapid rates of downcutting (Fig. 9.54*A*). Other streams developed boxlike, flat-bottom trenches, which presumably indicate downcutting followed by some period of stability during which channel-widening processes predominated (Fig. 9.54*B*). Although some streams display simple rectangular trenches, others have developed complicated sets of paired and unpaired inset fan segments, or terraces, with up to seven identifiable surfaces (Fig. 9.54*C*). Still

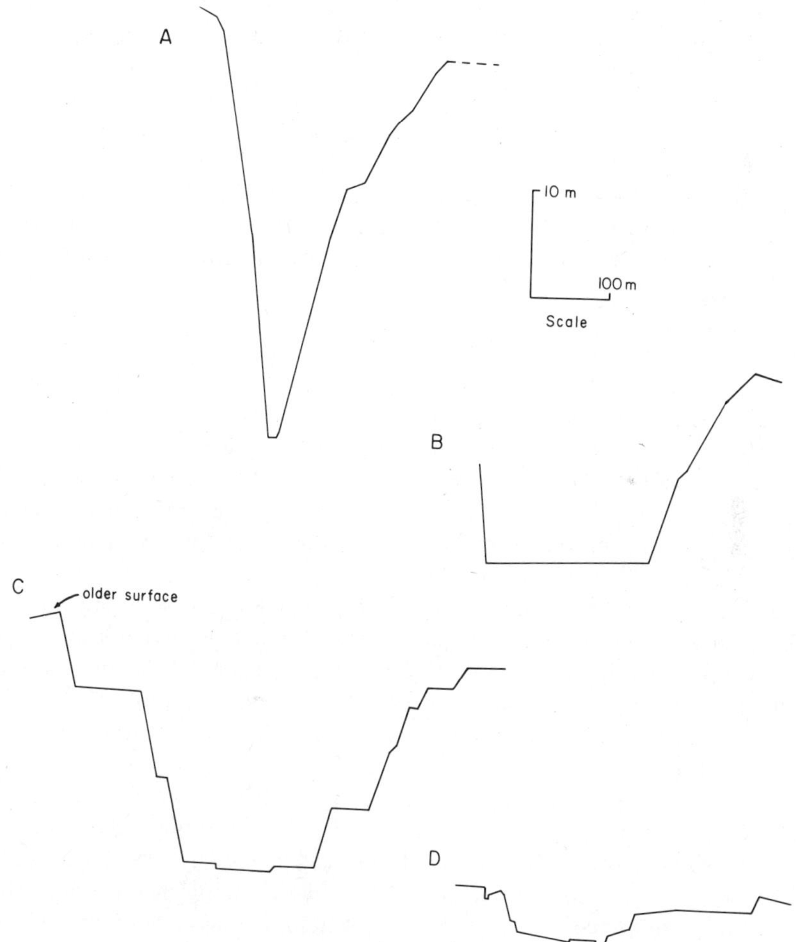

FIGURE 9.54. Cross sections at apex of (*A*) Dead Cat Canyon, (*B*) Rock Springs, (*C*) Grouse Creek, and (*D*) Falls Creek alluvial fans (from Weaver, 1984).

others display little or no channelization at all (Figs. 9.52, 9.54*D*). There are no obvious external controls that dictate the timing and rates of trench development.

If climate change is presumed to have been regional in extent, one might expect that trenches developed by streams draining comparable or adjacent areas should be grossly comparable in depth and extent, since trench development would have been primarily controlled by the volume of meltwater available and the duration of its occurrence (Funk, 1976, p. 218). However, paired basins in the area show no such similarity in erosional response. One might also expect that the largest and highest drainage basins (i.e., ones containing the most extensive alpine glaciers) would display the most well-developed, deeply incised fans, as Funk (1976, p. 223) found to be the case in the nearby Birch Creek Valley. Yet, the existence of large glaciated drainages, which feed unincised fans, and the number of small, less intensely glaciated basins, which are associated with fans with immense trenches, implies the existence of additional controls on alluvial fan behavior.

The different responses of the alluvial fans to regional climate change imply that, at any one time, each fan displayed a certain level or degree of inherent stability or susceptibility to erosion. When this threshold of stability was exceeded, either as a result of the cyclic, natural progression of events, as observed in the laboratory situation, or by a change in external conditions (climate change, tectonism), fans responded by trenching through previously stable fanhead sediments, which reduced stream gradients in the apex and midfan areas. If one or more of the fans in a region were close to the stability threshold, a comparatively small change in one of the controlling external variables would have triggered a dramatic erosional response. On the other hand, alluvial fans that were at or near their point of maximum stability would not have responded to even major climatic or tectonic changes unless such changes rendered them unstable under the new conditions.

Since each alluvial fan in the area developed under somewhat different conditions, owing to differences in drainage basin morphology, bedrock, glacial activity, erosion rates, sediment yields, and streamflow discharges, no two fans will have the same inherent susceptibility to erosion.

Stability is dependent on how closely fanhead slopes approach the critical threshold gradient beyond which incision will occur under prevailing climatic and tectonic conditions (Fig. 9.55). The greater the difference between the existing apex slope and the threshold slope, the more stable will be the fan. Since each alluvial fan responds to its own unique threshold slope, even adjacent fans are likely to be at different stages of geomorphic development.

Climate change and tectonism will affect the stability of alluvial fans. For example, under a given set of climatic conditions tectonic tilting may increase the slope at the fan apex beyond its threshold value and thereby initiate channel incision (Fig. 9.47). Tilting in the opposite direction could similarly increase the stability of an alluvial fan by reducing surface slopes and promoting deposition at the apex (Fig. 9.47). In these examples, the physical alteration of the fans by tectonics modified their stability in relation to a fixed or unchanging threshold value. However, unlike tectonism, climate change typically lowers or raises the threshold

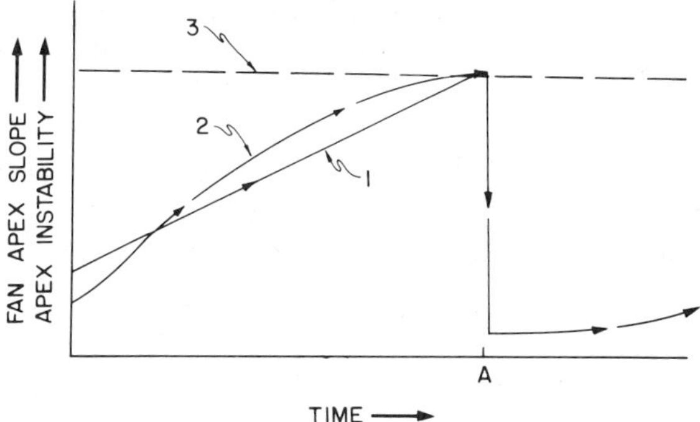

FIGURE 9.55. Hypothetical relation between fan apex slope and fanhead stability with time. Curve 1 shows slopes steadily increasing with time until they exceed the maximum gradient that is stable under existing water and sediment discharge conditions (line 3). Once the threshold is exceeded (time A), trenching rapidly reduces channel gradients and the cyclic growth process is reinitiated. Line 2 depicts actual shape of curve identified in experimental fluvial fan study (from Weaver, 1984).

values themselves (Fig. 9.48). Hence a climatic change resulting in increased streamflow and/or decreased sediment yield (i.e., increased Q_w/Q_s) would lower the maximum slope at which an alluvial fan could remain stable. If such a change were substantial, the threshold value could be lowered enough to cause alluvial fans to be rendered unstable and trenched. Conversely, if sediment concentrations were increased, slopes previously near the point of instability could continue to aggrade to a new, steeper threshold slope, which represents the maximum stable configuration under the new conditions.

Most landform change in the semiarid southwestern United States has been attributed to complex patterns of environmental change (tectonism, climate change, land use). Although partially masked by the external conditions imposed on the system, the frequently nonuniform reaction of landforms in such areas can be explained by the existence and operation of geomorphic thresholds (Fig. 9.55). Since landforms are generally out of phase in their geomorphic development and because each displays a unique level of stability at any one time, it can be expected that not all of them will respond to even severe external changes. The ones that react do so at different times depending on when they became unstable relative to the changing conditions of sediment and water discharge (Fig. 9.49). Field and experimental evidence indicates that instead of controlling landform behavior, external stimuli act within a cyclic framework of landform development to merely hasten or delay the conclusion of one cycle and the initiation of the next.

In conclusion, the cyclic model of alluvial fan evolution proposed in this study (Fig. 9.55) is not intended to supplant the existing, well-documented explanations provided by the climatic, tectonic, or erosion cycle theories summarized by Lustig (1974). However, the concepts outlined here do provide a framework from which to view other models and to determine the conditions within which they operate. The variable and frequently contrasting responses of alluvial fans in a given area strongly suggests that external factors, such as climatic change and tectonism, operate within, but do not control, the most fundamental cyclic growth pattern of alluvial fans. Therefore, it is not surprising that the alluvial fans studied by Weaver (1984) in east-central Idaho should show such dramatically contrasting geomorphic responses to regional climate change.

10 Fan Deltas

A fan delta is "an alluvial fan that progrades into a body of water from an adjacent highland" (McGowen, 1971). In contrast to alluvial fans, fan delta sediments can be identified as having formed in two environments: (1) the subaerial fan, affected primarily by fluvial processes; and (2) the subaqueous fan, affected primarily by basin configuration and marine or lacustrine processes. A fan delta has many of the surface characteristics of an alluvial fan and many of the subaqueous characteristics of a delta. The fan delta is influenced by both environments and obtains its character from the interaction between subaerial flow, water levels, and depositional basin configuration.

Fan deltas form in climates that range from arid to humid. Sediment is transported by mudflows and ephemeral streams under arid conditions and by braided streams under humid conditions (McGowen, 1971; McGowen and Scott, 1974). Climate-related fluctuations in water level and storm intensity may also affect fan delta characteristics, as does the nature of the receiving basin, which can be shallow or deep, lacustrine, lagoonal, or marine (McGowen and Scott, 1974). Fan deltas were first mentioned by Holmes (1965), who observed the fan-shaped deposit of sediment at the mouth of the Lyn River on the south coast of the Bristol Channel, Devon, England. Holmes published only cursory observations, and little appeared on the subject until McGowen's (1971) detailed study of the Gum Hollow fan delta on the Texas coast. Modern fan deltas have also been studied along the Gulf of California, the Red Sea, the southeast coast of Alaska, the coasts of Honshu Island in Japan, the south coast of Puerto Rico, and the southeast coast of Jamaica (Wescott and Ethridge, 1980).

Wescott and Ethridge (1980) named fewer than 10 authors who referred to fan deltas as such in the rock record. However, many sedimentary sequences described in the literature have since been identified as fan deltas, and the term is in common use today.

EXPERIMENTAL STUDY

Experimental work was undertaken by Jackson (1981) to observe the depositional and erosional processes active on a fan delta and to examine the morphology and sedimentary facies resulting from these processes. Figure 10.1 illustrates the terminology used to describe the experimental fan deltas. In many respects, Jackson's results are similar to and compliment those of Chang (1967) and Chang (1982), who studied the hydraulics of fan delta growth.

Experimental Facility and Procedure

Jackson's (1981) fan delta experiments were performed in a rectangular wooden box 4.2 m wide, 8.5 m long, and 0.53 m deep, across which a partition was constructed to create a receiving basin. There was an embayment 1.8 m wide and 0.7 m deep in the center of the partition, which received water and sediment from a steel flume 2.4 m long and 0.2 m wide (Figs. 10.1, 10.2). The flume could be raised or lowered to vary slope, and sand was cemented to its floor to create a rough bed. Enough sediment was always kept in the flume to permit scour. A Syntron apparatus fed sand to the flume at an average rate of 16 g/sec, except during runs at the end of the fourth experiment, when floods were simulated (Table 10.1). Water level in the receiving basin was controlled by a standpipe.

Four fans deltas were formed during the study, each under a different set of conditions. Fan delta 1 was developed as a control, and for the other three fan deltas one or more of the following variables were altered (Table 10.1): (1) depth of basin water, (2) grain size of the sediment (Fig. 10.3), and (3) water and sediment discharge into the flume.

FAN DELTA MORPHOLOGY

Fan delta 1 will be described in some detail, as it is generally representative of the morphology and mode of growth of all four experimental fan deltas.

Fan Delta Profiles

The longitudinal profiles measured during the formation of fan delta 1 provide a record of fan delta growth (Fig. 10.4). The gradient of the fan plain averaged 2^0, with a range of $1–4^0$, and remained approximately constant with time. The gradient of the fan delta slope averaged 26^0, with a range of $18–34^0$. Some of the variation was due to differences in sediment character, with steeper gradients forming when there was less channel sand deposited at the edge of the fan delta.

The longitudinal profile (Fig. 10.4) of the fan plain was usually concave at the

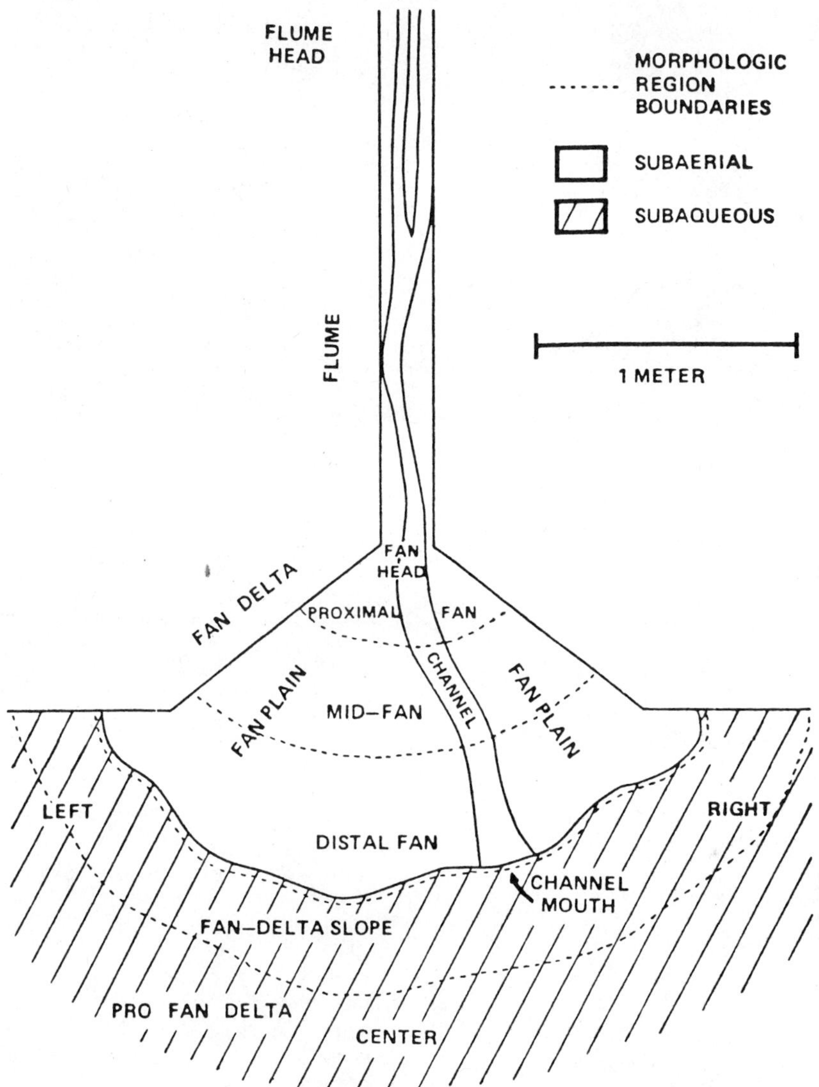

FIGURE 10.1. Fan delta terminology and plan of experimental fan delta (from Jackson, 1981).

midfan and convex near the waterline. This shape of the longitudinal profile was generally due to fanhead deposition and a decrease of sediment load downstream coupled with the effect of the water level. Deposition above the water level produced a convex lower fan profile. The result was a concave–convex longitudinal profile. The steep fan delta slope that formed below the waterline was a major difference between the alluvial fans (Chapter 9) and the fan delta.

FIGURE 10.2. Photograph of fan delta experimental facility.

TABLE 10.1 Summary of Experimental Conditions

	Basin Water Depth	Sediment Type (Fig. 10.3)	Sediment Load (g/sec)	Water Discharge (L/sec)
Fan delta 1, runs 1–9	Constant, 23 cm	Slightly gravelly muddy sand	16	0.14
Fan delta 1, runs 10–12, baselevel rise	Rising	Slightly gravelly muddy sand	16	0.14
Fan delta 2, runs 1–5, 8–11	Constant, 27 cm	Slightly gravelly muddy sand	16	0.14
Fan delta 2, runs 6–7, baselevel drop	Falling	Slightly gravelly muddy sand	16	0.14
Fan delta 3	Constant, 20 cm	Sandy gravel	16	0.14
Fan delta 4, runs 1–2	Constant, 8 cm	Slightly gravelly muddy sand	16	0.14
Fan delta 4, runs 3–4, flood discharges	Constant, 8 cm	Slightly gravelly muddy sand	16	Floods of 5.5 L

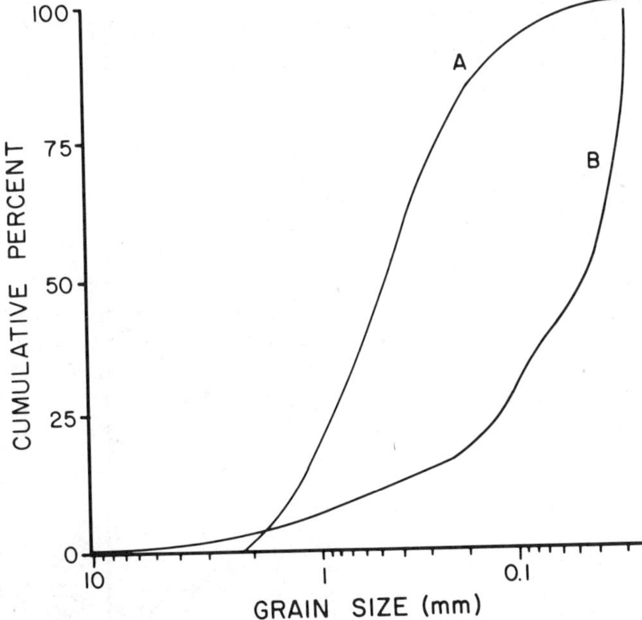

FIGURE 10.3. Sediment used for experimental fan deltas (from Jackson, 1981): A, muddy sand; B, washed sand.

Much of the deposition on the proximal fan was along the axis of the fan delta, making it higher than the surface on either side (Fig. 10.5). As a result, the transverse profile of the fan delta was convex (Fig. 10.5), like that of an alluvial fan, but the transverse profiles were often asymmetrical due to uneven deposition (Fig. 10.5).

The channel flowed on the right side of fan delta 1 26% of the time, on the left side 31% of the time, and in a central location 43% of the time. The higher percentage of time that the channel was flowing in the center of the fan delta reflects the influence of flow momentum from the flume, and it enhanced the convexity of the transverse profile. Depressions or concavities on the transverse profiles, and in some instances on the longitudinal profiles, were the result of channel incision.

Cycles of erosion and deposition were noted during the first fan delta experiment (Figs. 10.4, 10.6). Profile 1, measured after run 1, shows a gently convex fan-plain slope that is regular, in contrast to the concave–convex profiles of the later runs. Flow during the first run was almost totally sheetflow.

Both progradation and aggradation increased fan delta size during run 2. Sheetflow was predominant during run 2, and deposition formed a convexity on the distal fan delta, and it steepened the fanhead (Fig. 10.4). During run 3, which was the same length as run 2, erosion restored regular profile (Fig. 10.4).

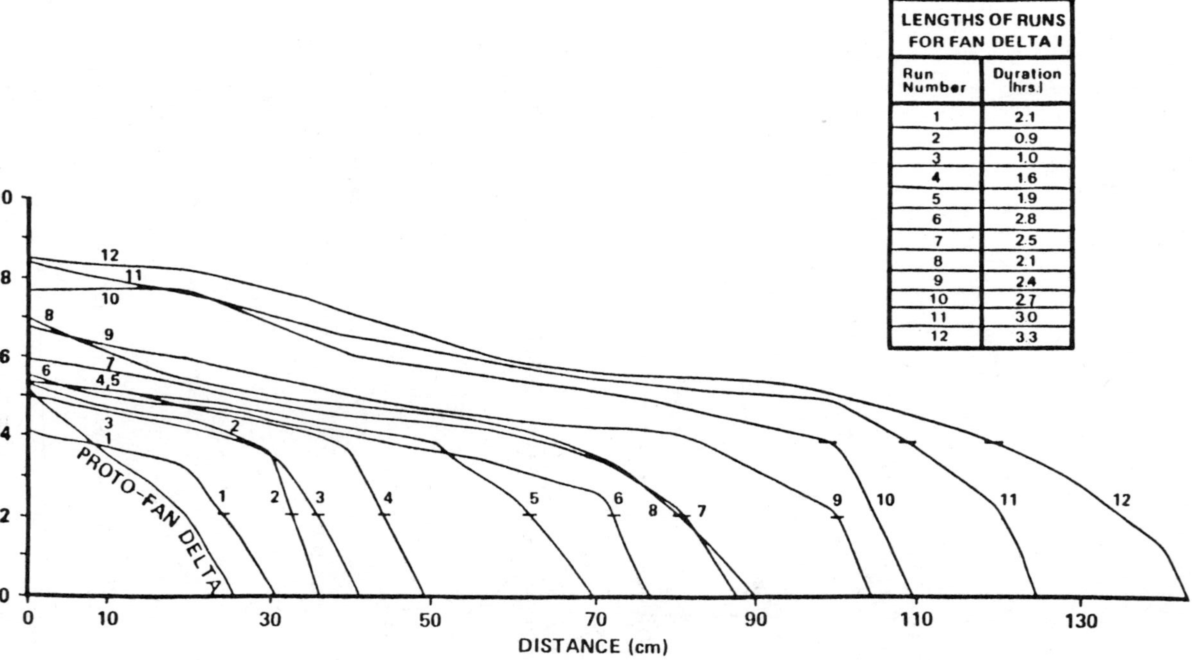

LENGTHS OF RUNS FOR FAN DELTA I

Run Number	Duration (hrs.)
1	2.1
2	0.9
3	1.0
4	1.6
5	1.9
6	2.8
7	2.5
8	2.1
9	2.4
10	2.7
11	3.0
12	3.3

FIGURE 10.4. Longitudinal profiles of fan delta 1, measured down the central axis. Tick mark on each profile indicates water level. Runs 10–12 were at higher baselevel (from Jackson, 1981).

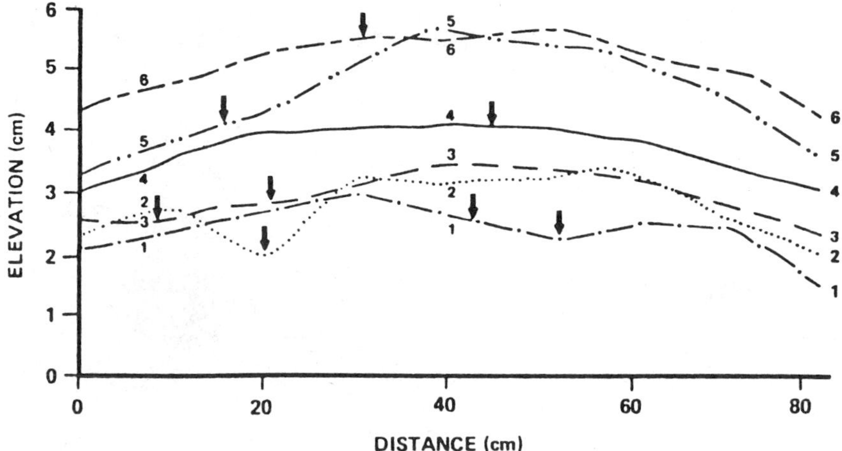

FIGURE 10.5. Transverse profiles of fan delta 1. Each profile represents one run. Approximate channel centers are marked with an arrow. Vertical exaggeration is 6.6 ×. Distance is measured from right side of fan looking downfan (from Jackson, 1981).

Aggradation and progradation continued during run 4, but a concavity at 20 cm (Fig. 10.4) was formed by headcut erosion close to the fan axis. During the long run 5, the fan and proximal fan slope were largely unchanged, but there was major progradation along the axis of the fan (Fig. 10.6, run 5), which produced a pronounced irregularity in the transverse profile (Fig. 10.5).

During run 6, the mid- and distal-fan surface was lowered by incision caused by a migrating nickpoint and the transverse profile was smoothed (Figs. 10.4–10.6). There was uniform vertical accretion and progradation during run 7 (Figs. 10.4–10.6) following the gradient change of run 6 (Table 10.1). Progradation was insignificant during run 8, although this run was almost as long as run 7. Instead of being transported, sediment was stored at the steepening fanhead (Fig. 10.6), which formed a smooth concave–convex profile (Fig. 10.4). With continued deposition, the fanhead incised and was lowered during run 9 (Fig. 10.4).

Before run 10, the water level was raised 2 cm and as a result there was major deposition on the fan plain during runs 10–12 (Figs. 10.4, 10.6).

The coastal margin of the experimental fan deltas was arcuate. Channel lobes formed protrusions that coincided with active and abandoned channel mouths. Slight differences in the extent of the protrusion of channel-mouth lobes were due to sediment size differences and basin water depth. Channels of fan delta 3 moved laterally at a much faster rate than the channels of the other fan deltas because the sandy sediment was easily eroded (Table 10.1). The coastal margin was therefore smoother and more regular than that of the other fan deltas. In contrast, channels transporting muddy sand (Table 10.1) remained relatively stationary during incision because the mud was cohesive, and distinct lobes formed at the channel mouths.

Fan deltas prograded into basins of both high (27 cm) and low (8 cm) water

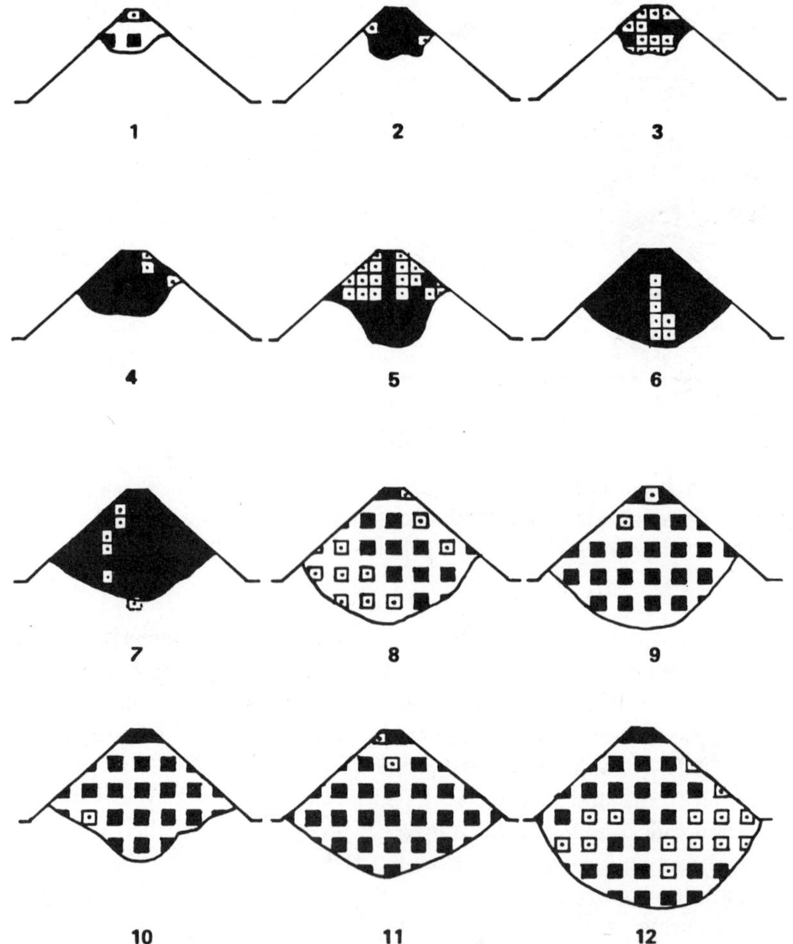

FIGURE 10.6. Depositional patterns for fan delta 1, runs 1–12 (plan views). Each black square represents deposition or nonerosion on a 100-cm^2 area of the fan delta. Squares with dots represent erosion. Empty spaces indicate no measurement was taken for that square (from Jackson, 1981).

depths (Table 10.1), and this caused a visible difference in the fan delta coastal profiles. The first two fan deltas showed small channel lobes in view, but channel lobes were much more prominent on the shallow-water fan delta 4.

Fan Delta Dynamics

Cyclic Processes

Fan delta growth under conditions of constant discharge, sediment input, and flume slope was the result of many alternating erosional and depositional episodes. A typical erosional–depositional cycle began with a single straight channel eroding at the edge of the fan plain (Fig. 10.7*a*). Scouring of the channel bed caused a relatively large amount of sediment to move off the edge of the fan plain. This sediment piled up at the top of the fan delta slope and frequently (every 5–10 sec) slumped, causing progradation in the form of a lobe at the channel mouth, which lengthened the channel (Fig. 10.7*b*).

When the distal part of the fan was steepened by deposition near the waterline, gradient adjustment was accomplished by migration of a nickpoint upstream. Then channel backfilling at the mouth initiated the depositional stage of the cycle (Fig. 10.7*b*). Increased roughness and decreased slope created by the channel backfill

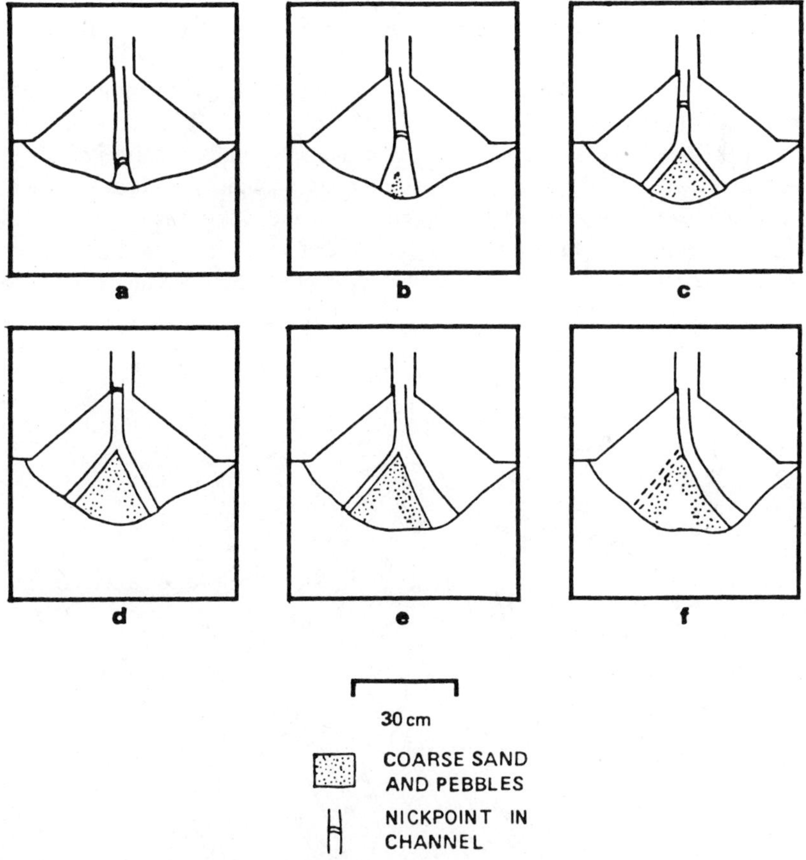

FIGURE 10.7. Simple six-stage erosional–depositional cycle (from Jackson, 1981). (*a*) Nickpoint at channel mouth. (*b*) Nickpoint migrates into midfan. (*c*) Nickpoint migrates into proximal fan. Mid-channel bar forms and channel divides. (*d*) Nickpoint disappears at fanhead. Mid-channel bar accretes at upstream end. (*e*) One channel becomes dominant. (*f*) All of the flow gradually shifts to the new channel.

deposit caused further deposition, and a mid-channel bar formed that continued to grow at its upstream end (Fig. 10.7*c, d*). The nickpoint eventually disappeared at the head of the fan delta. At this stage, one of the two channels flanking the mid-channel bar enlarged (Fig. 10.7*e*) as it steadily captured flow from the other branch, which was eventually abandoned (Fig. 10.7*f*). Progradation occurred at the new channel mouth, and after a few minutes a new erosional–depositional cycle began.

The erosion began with either a single nickpoint, multiple nickpoints, or simultaneous nickpoints in several channels. Relief at the nickpoint was greatest (1–1.5 cm) at the channel mouth, but it gradually lessened as it migrated upstream. Single nickpoints did not advance upstream into the confined flume channel. Multiple nickpoints resulted from the development of sucessive nickpoints in the same channel. The first nickpoint was comparatively large and, in contrast to the single nickpoint, almost always migrated past the fanhead into the flume. One or two secondary nickpoints followed and completed the channel incision and lowering of gradient.

When more than two channels formed as a result of the construction of multiple mid-channel bars, simultaneous nickpoints developed (Fig. 10.8). Up to five channels could incise at the same time, but after a few seconds usually one channel succeeded in capturing all of the flow and creating a single new main channel.

Nickpoints usually formed at the shoreline. An exception was erosion in small overflow channels at the fanhead and in the midfan area where bed load remained in the main channel when overbank flow took place. The overbank flow carried

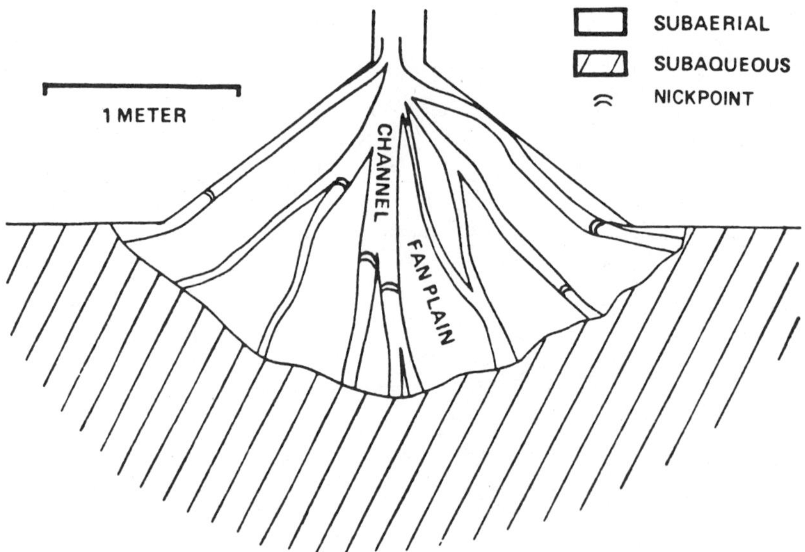

FIGURE 10.8. Subaerial fan delta showing channel pattern and simultaneous nickpoints (from Jackson, 1981).

little sediment, which, when combined with a steeper fan-plain gradient, caused small headcuts to form in the overbank channels.

Backfilling at the channel mouth began during the later stage of nickpoint migration and backfilling of the channel completed each episode of fan growth. After several erosional–depositional cycles had built up the central area of the fan, active deposition moved to either the right or left side of the fan delta and the erosional–depositional cycle began again.

Fan delta 3 formed under nearly the same conditions as fan delta 1, but the sediment used was better sorted and contained less silt and clay. The low cohesiveness of the sand caused bank instability and a substantial increase of lateral accretion. Overbank flow usually did not result in channel avulsion because the sediment was not cohesive enough to allow continuous channels to form. The water simply flowed over the fan delta surface, creating an irregular surface with small scours.

Baselevel Changes

The baselevels of fan deltas 1 and 2 were raised and lowered to observe the geomorphic and sedimentologic responses to these changes (Table 10.1).

Baselevel Rise

The adjustment of the gradient of fan delta 1, after a small rise in baselevel, took place in several steps (runs 10–12, Fig. 10.4). There was considerable vertical accretion near the fan apex during run 10, and the entire fan plain aggraded 1 cm in response to the rise of baselevel. Progradation and erosion during run 10 were minimal, as deposition was concentrated on the surface of the fan delta plain. During run 11, further aggradation completed gradient adjustment and renewed progradation was significant. A nickpoint at 60 cm on profile 12 indicated the renewal of erosional–depositional processes on the fan delta.

Baselevel Lowering

Near the end of the development of fan delta 2, baselevel was lowered slowly and, as a result, nickpoints formed in all channels. Larger channels incised faster and adjusted more quickly to the change in baselevel. Smaller channels were abandoned, and eventually all of the surface water was confined to one central channel.

The cyclic patterns of geomorphic change on the fan delta were temporarily interrupted during the drop in baselevel. The single straight channel developed quickly by nickpoint migration, and mid-channel bar formation ceased. A small central portion of the fan delta surface was eroded, and this sediment was redeposited in front of the fan delta to form a small fan delta at the lower baselevel (Figure 10.9*A*). Lateral migration of the channel removed the topset beds of the main fan delta and truncated the upper portions of the foreset beds. Eventually almost all of the upper fan plain was eroded to the new level. The central third of the fan delta was eroded relatively quickly, whereas the sides were eroded more

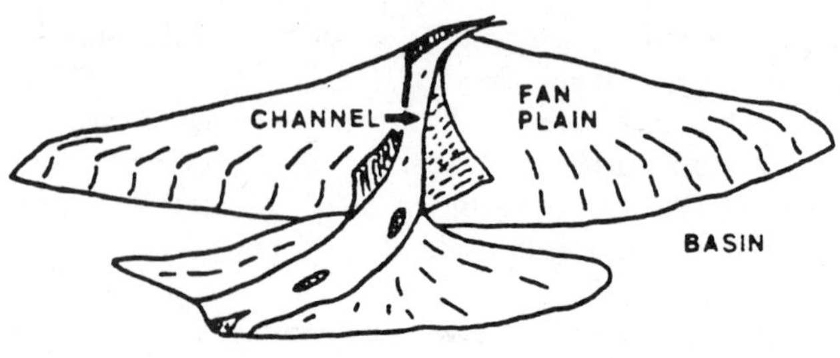

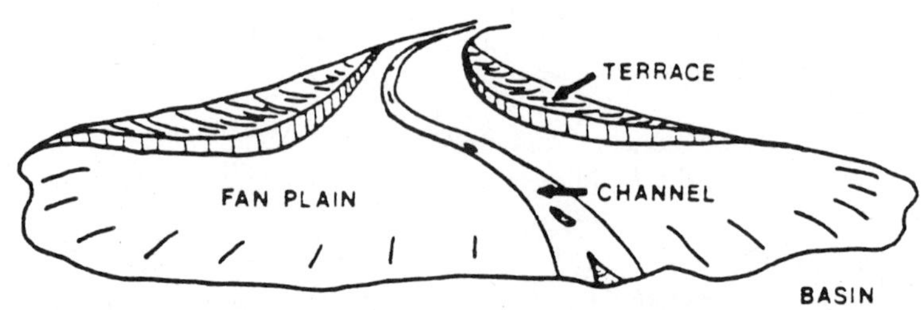

FIGURE 10.9. Fan delta morphology during a lowering of baselevel (from Jackson, 1981). (*A*) During the early stage of incision, a small fan delta is deposited in front of the older fan delta. (*B*) During later stages, a new fan delta is produced at a lower level. The old fan delta surface is preserved as terraces.

slowly, leaving a terrace at the far left and far right sides of the fan delta (Figure 10.9*B*). Finally, deposition resumed after the channel had lengthened and the original gradient had been reestablished, and a new fan delta was created at the lower baselevel. This sequence of events was similar to that which occured on Gum Hollow fan delta as a result of a baselevel drop during the winter of 1967 (McGowen, 1971); similar changes have been reported by Sneh (1979).

Progradation

An inverse relationship existed between progradation rate and baselevel change on the fan deltas. A rise in baselevel decreased, and a drop in baselevel increased, progradation rates. The effect of baselevel rise on progradation is shown on Fig. 10.10. The fan delta 2 experiment was continued for 10 hr longer than fan delta 1, yet both prograded the same distance. Discharge and sediment load were the same, but baselevel was higher for fan delta 2, so that more sediment had to be deposited for a given amount of progradation than was necessary for fan delta 1. Therefore, fan delta 2 took longer to prograde than fan delta 1.

The increased rate of progradation resulting from a lowering of baselevel is

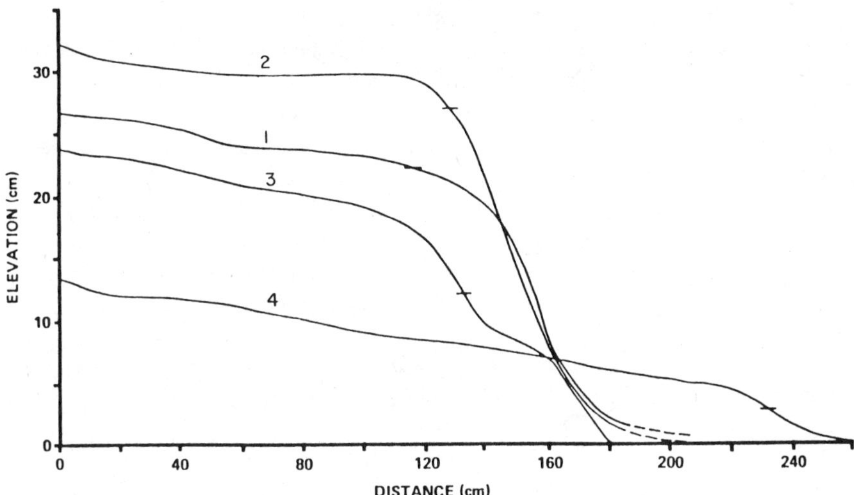

FIGURE 10.10. Final profiles for fan deltas 1–4. Tick on profile shows water level. Irregular slope of fan delta 3 at 140–160 cm is due to slumping (from Jackson, 1981).

illustrated by fan delta 4 (Fig. 10.10). It was built in 14 hr, 5 hr less than any of the other fan deltas, and yet it prograded 30% farther. This is simply a reverse of the situation that existed at the higher baselevels.

Chang (1982; see also Schumm, 1977, p. 309–313) also documented the response of a model fan delta to baselevel change as well as to changes in water discharge and sediment load. A lowering of baselevel, an increase of discharge, or a decrease of sediment load induced channel incision, which in turn caused delta progradation and a relative elongation of fan delta shape. Opposite changes in these controlling variables induced aggradation, the creation of distributary channels, and a relative widening of the delta.

Progradation rates are also controlled by water depth in both experimental and natural conditions. Chang (1982) found that rates of progradation increased with a decrease in water depth, and McGowen (1971) recorded relatively rapid progradation in his study of the Gum Hollow fan delta, which was built into a maximum of 0.9 m of water. During the spring of 1966 there were three heavy rains, and the fan delta prograded 300 m. Increased prominence of channel-mouth lobes was another morphologic response to low water depth, and the coastline of fan delta 4 was more irregular than the margins of the other three.

DISCUSSION

Episodic, cyclic erosion and deposition characterized the formation of Jackson's (1981) fan deltas, much as they did the experimental alluvial fans (Chapter 9). Erosion at the fanhead was contemporaneous with deposition downstream. Erosion was accomplished by single, multiple, and simultaneous headcuts, which originated on the distal fan and migrated toward the apex. Deposition was accomplished by mid-channel bar construction, channel bifurcation, lateral accretion, and sheet-flooding over the fan surface.

The major difference between the fluvial fan and the fan delta was the development of headcuts and channel incision at the margin of the fan delta. This was the result of deposition of sediment above water level, which produced a convexity on the distal fan plain. Incision started on the steeper downfan part of the convexity and then migrated to the fanhead. Deposition at the fanhead was a process common to both types of fans and headcuts often migrated through these steepened proximal deposits on the fan-delta.

Five depositional environments were present on the fan deltas: (1) channels, (2) abandoned channels, (3) the fan delta plain, (4) the fan delta slope, and (5) the

TABLE 10.2 Characteristics of the Experimental Fan Deltas as Compared to the Yallahs and Gum Hollow Fan Deltas.[a]

	Similarity of Natural Fan Deltas	
Characteristic of Model	Yallahs	Gum Hollow
Morphology		
Similar to an alluvial fan	X	X
Has an escarpment	X	X
Has only one apex	X	X
Process		
Fluvial deposition by a shifting braided stream	X	X
Subaqueous slumps	X	None recorded
Geometry		
Wedge-shaped, thinning seaward from the coast and landward from the midfan	X	X
Sedimentology		
Sedimentary structures and microdelta foreset laminae	No	No
Coarse sediment predominates in channels and on upper midfan	In channels only	X
Environments of deposition		
Channels	X	X
Abandoned channels	No, not silted in	X
Fan delta plain	X	X
Fan delta slope	No	No, bay too shallow
Pro-fan delta	No, truncated by canyons	X

[a] From Jackson (1981). X indicates the fan delta has the same or a very similar characteristic.

pro-fan delta (Fig. 10.1). Each was characterized by an individual shape, extent, sediment size distribution, and set of sedimentary structures.

A prototype example is useful to confirm the validity of the fan delta model, but no single natural fan can be expected to exhibit all the characteristics of the models. The modern Yallahs fan delta in Jamaica and the Gum Hollow fan delta on the Texas coast do, however, show the most important characteristics of the model, with the exception of some sedimentary structures and additional coastal environments of deposition (Table 10.2). In addition, the freshwater Gilbert deltas that formed along the margins of Pleistocene-age Lake Bonneville (Gilbert, 1890) are similar in morphlogy and stratigraphy to the experimental fan deltas.

11 Alluvial Fan Sedimentology and Stratigraphy

Alluvial fans form an important part of the geologic record. Those of arid and semiarid regions are the most obvious and have long been a topic of geomorphic study (Blissenbach, 1952, 1954; Bull, 1964a, Denny, 1965, 1967; Hooke, 1974). Within the past few years the fluvial fan model has been described, largely due to work conducted in South Africa on the Witwatersrand deposits (Pretorious, 1975). Other fluvial fan deposits include the quartz–pebble conglomerate deposits of the Banket formation of Ghana (Sestini, 1973), the Huronian Elliot Lake deposits of Canada, and other, similar deposits in South American and east-central Asia. McGowen and Groat (1971) have applied the fluvial fan model to explain the Van Horn sandstone of Texas.

Large modern analogues of fluvial fans are difficult to find, and the body of literature available is still small in comparison to that dealing with mixed-mode and mudflow fans. Two possibly analogous deposits have received some attention in the literature. The first is the glacial outwash fan (sandur) studied primarily in Iceland and Alaska by various workers (Boothroyd, 1970; Gustaufson, 1974; Krigstrom, 1962). The second is the braided outwash fans of the Himalayas, Lesser Himalayas, and Siwalik Range of northern India (Nossin, 1971). For example, the Kosi River fan near Bihar, India, is a braided alluvial fan about 210 km long and nearly 95 km wide.

For the most part, the sedimentology of large modern fluvial fans has not been studied in great detail. Little is known about the variability of their internal composition, although the internal characteristics of presumably comparable rock strata in the geologic record have received considerable attention.

EXPERIMENTAL STUDY

During Weaver's (1984) experiments, four alluvial fans were produced to evaluate models of alluvial fan growth and evolution (Chapter 9). Macke (1977) analyzed two of the fans, paying particular attention to grain-size distributions, heavy-mineral concentrations, and internal stratigraphy. The first fan is analogous to fluvial fans developed in humid regions prior to colonization of the land surface by plants. The second fan is analogous to mudflow fans common in arid and semiarid regions. The two fans represent the end members of fan-building processes (Blissenbach, 1954). Macke's objective was to integrate the geomorphic, stratigraphic, and sedimentologic characteristics of experimental alluvial fans in order to help explain related deposits in the rock record.

Experimental Procedure

The fluvial fan (see Chapter 9) was the result of the application of constant, uniform precipitation from the overhead sprinkler system over an initially smooth, slightly concave 46-m^2 sediment source area (Fig. 9.1). This resulted in a continuous water and sediment discharge of 0.8 L/sec to the depositional area.

The mudflow fan (Fig. 9.44*C*) was produced experimentally using a 17-m^2 source area with steep V-shaped valley walls and steep gradient channels (Fig. 9.28) that was designed to produce maximum sediment production and the highest possible sediment concentrations. A 38-mm-diameter cylinder was used to obtain cores 20 mm long. Magnetite, which comprised 0.5% of the sediment in the source area, was separated from the other sediment in order to study heavy-mineral distribution within the two fans.

SEDIMENTOLOGY AND STRATIGRAPHY

Fluvial Fan

As discussed in Chapter 9, the processes of deposition on the fluvial fan were complex and variable. Water leaving the source area spread out over the fan apex, depositing sediment, raising the level of the apex, and steepening its gradient. Periodically, trenching of the apex was initiated when the gradient exceeded a threshold of stability, and the dominant locus of deposition shifted downfan to below an intersection point, where the fanhead trench emerged onto the fan surface (Fig. 11.1). In effect, the trench extended the main stream of the source area onto and down the fan. Incision was followed by backfilling as deposition, which began below the intersection point, migrated headward. During backfilling, the point of

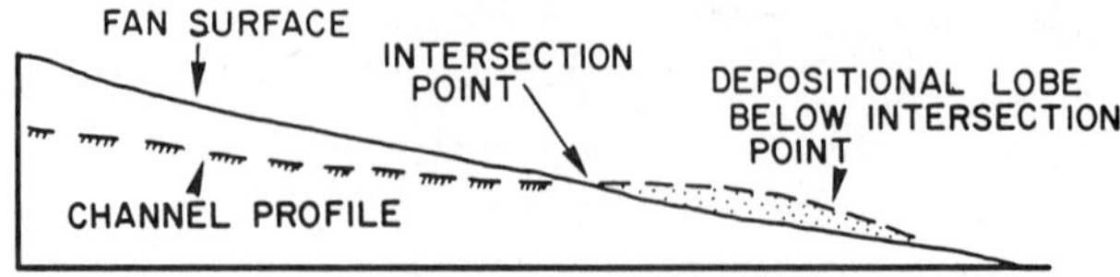

FIGURE 11.1. Idealized longitudinal profile of the fluvial fan during fanhead trenching.

divergence of flow moved upfan until, at the end of the cycle, deposition was again concentrated at the apex.

Sedimentology

The normal grain-size relationships on the surface of arid region alluvial fans is one of decreasing grain size downfan (Blissenbach, 1954; Bull, 1964a,b). On the experimental fluvial fan, grain size increased downfan (Fig. 11.2). Explanation of this anomalous pattern requires consideration of three factors; (1) the loss of fines from the fan, (2) sampling procedures, and (3) fanhead trenching.

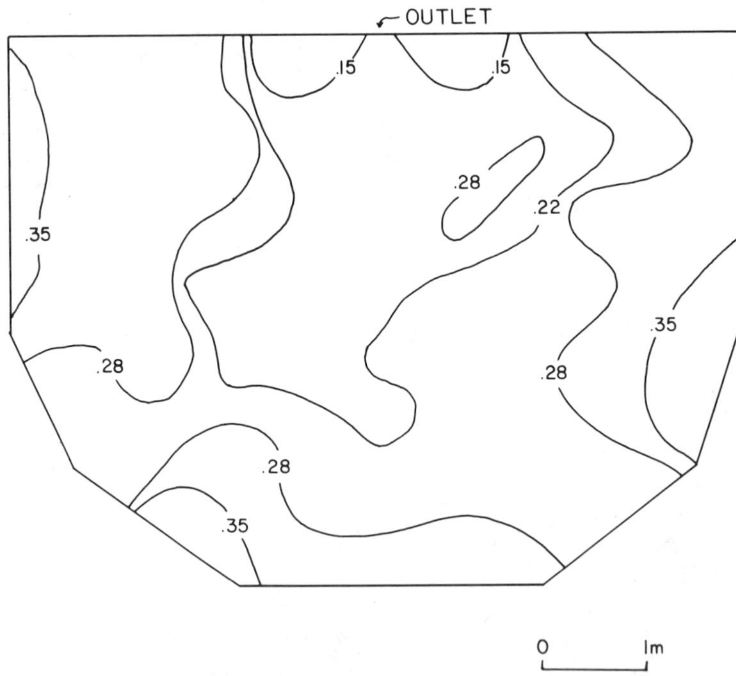

FIGURE 11.2. Grain-size distribution on the fluvial fan (isopleths in millimeters) (from Macke, 1977).

First, the source area and fan are not a closed system with respect to fines, because most of the suspended load was carried through the gravel barrier at the toe of the fan and out of the depositional area. Hence the fluvial fan, as studied, is probably only analogous to the proximal and midfan regions of a large natural fan. If there had been a closed depositional basin in the REF, the distal segment would be fine grained and an overall decrease would become apparent.

Second, the vertical sampling of the fan was done in 20-mm increments, which provided bulk sample data but no information on changes within single sedimentary layers. Although many of the layers might have become finer downfan, the layering of the deposit was too fine to be sampled in more detail. For this reason, the bulk sampling reflects the relative proportion of coarse-grained sediment within the entire sample rather than the size distribution of any individual layer. Actually, the distribution of sediment within a vertical sequence that was developed entirely during an interval of untrenched flow should show a relatively uniform decrease in sediment-size distribution downfan.

Third, the trenching at the fanhead was the primary cause for the anomalous grain-size distribution of the fluvial fan. Trenching caused the intersection point (Fig. 11.1) at which channelized flow was dispersed across the fan surface to move downfan. The fanhead trench reworked and resorted previously deposited sediment at the fanhead and increased the size of sediment that could be transported by a given discharge because the incised channel was more competent than the spread flow that originally formed the deposits (Fig. 9.20*C*). The trench eroded and transported sediment previously deposited at the fanhead as well as coarse-grained sediment of the source area channels, which were then deposited as coarse-grained wedges of sediment downstream of the intersection point (Fig. 11.1).

When the fanhead was not trenched, relatively fine-grained sediments that were eroded from drainage basin hillslopes and channels were deposited on the fan surface and grain size decreased from the fan apex downstream. During trenching, this sediment was mixed with coarser-grained sediment that was produced as nickpoints migrated through the fan apex and the channel of the lower source area. This coarse sediment was deposited on the middle and distal parts of the fan with a grain size decrease downfan from the new intersection point.

The sedimentology of the alluvial deposit is therefore extremely complex due to the several processes involved in building the fan. To a large extent, this complexity represents the effects of fanhead trenching and the complex response of the drainage basin, which caused increased variability of both the sedimentologic and stratigraphic aspects of the fan.

The geometry and sedimentology of the sediment deposited on the experimental fan was similar to that of the conglomerates of the ancient Tarkwaian fan in Ghana (Sestini, 1973), which increase in thickness in a downfan direction. Fanhead trenching allowed coarse sediments to bypass the fanhead and be deposited downfan. This formed wedge-shaped deposits that thickened downfan. Therefore, although mean grain size within an individual layer may decrease downfan, as suggested earlier, bulk sampling of the deposit reflects an increase in relative importance of these coarse-grained layers, and therefore there is an overall increase in grain size in a downfan direction.

Stratigraphy

Depositional units of the fluvial fan were generally much more continuous longitudinally than laterally. While deposition during a single episode might take place over an entire longitudinal segment of the fan, lateral distribution of sediment was usually more restricted. Broad areas of the fan might undergo long periods of nondeposition while sediment was being deposited in adjacent regions. Also, deposition did not migrate over the fan at a constant rate, and the changes in the sites of deposition were frequently avulsive. Therefore, deposits containing sediments of similar age could be located relatively far apart, especially in the distal portions of the fan, making stratigraphic time lines within the deposit laterally discontinuous.

At the toe of the experimental fluvial fan, coarse-grained layers interfingered with fine-grained sediment deposited as the result of some ponding of water by the gravel barrier that was placed around the fan margin. The fine-grained layers were deposited during periods when no coarse sediment reached that location. An analogous situation is encountered in nature when fans on opposite sides of a valley impinge on one another or on a shallow playa lake. Another example is provided by the distal facies of the Witwatersrand fluvial fan (Pretorius, 1974), where discontinuous sheets of fine-grained bedload sediment are interbedded with finer-grained, suspended-load sediments. The coarse-grained units are the result of channelized flow extending to the toe of the fan when flow is directed over that region (Minter, 1976). Maximum development of these layers occurs when the fanhead is trenched and the channels have attained their maximum length downfan.

Mudflow Fan

Sediment moved from the source area onto the surface of the mudflow fan as discrete mudflows (Fig. 9.39*A*). Tongues of sediment moved with sharply defined fronts, which remained distinct after movement ceased. The trailing portion of an individual lobe was often poorly defined. After the mudflow stopped, water, which was derived from the dewatering of sediment on the fan surface and in the source area channels, flowed over the fan carrying high concentrations of suspended load.

Lobate sedimentation described by Hooke (1974) is attributed to sieve deposition, that is, deposition due to loss of water from the flow by infiltration. This type of deposition was not observed on the mudflow fan because fine sediment within the deposit and the short time between successive depositional events caused the fan to remain saturated, which prevented dewatering of the subsequent flows by infiltration. Nevertheless, lobate deposits were formed and dewatering did occur as mudflows spread over the fan surface and lost velocity. Water derived from the dewatering of the frontal lobe of each mudflow generally proceeded to the toe of the fan, where a thin sheet of clay-rich sediment was deposited. This formed a low-permeability basal layer over which the fan prograded.

Mudflows that did not reach the fan apex dewatered in the source area valley and gave rise to flows that incised the fanhead. This caused minor reworking at the fanhead and the production of better-sorted sediment than was originally deposited by the mudflows. This minor incision of the fanhead by streamflow was a common occurrence and was different from the fanhead trenching induced by climatic and tectonic adjustments in nature (Hooke, 1974).

Areal growth of the fluvial fan was accomplished by fanhead trenching and downfan shift of sediment, but areal growth of the mudflow fan was accomplished primarily by high-velocity, low-viscosity flows that reached the toe of the fan. The distance traveled by an individual mudflow in a downfan direction was dependent on both the size and viscosity of the flow.

The streamflow that followed a mudflow event greatly influenced the lateral and longitudinal extent of the following mudflow event. If the fanhead was deeply incised as the result of a streamflow or fluid mudflow event, the following mudflow was largely channelized at the fanhead, and it did not spread until it had traveled farther downfan. Overbank deposition by mudflows at the head of the fan occurred frequently.

Sedimentology and Stratigraphy

A high degree of stratification was preserved within the mudflow fan. Internal organization of the coarse-grained layers was poor, but individual layers were distinct. This was due to the presence of layers of fine-grained sediment that were deposited by streamflow after each mudflow event. Unlike the fluvial fan, no distinct concentrations of heavy minerals could be observed within the layering.

Channel fill deposits were observed in cross sections as the result of incision by minor streamflows and low-viscosity mudflows. This was a simple cut-and-fill process without major reworking of fanhead sediments. The dramatic fanhead trenching sequence observed on the fluvial fan was not evident on the mudflow fan.

The internal organization of the mudflow fan did have several things in common with that of the fluvial fan. The continuity of layering in the deposit was much greater in a direction parallel to flow than perpendicular to flow. This, again, was due to deposition occurring over a restricted radius of the fan surface, but along nearly the entire length of the fan during many events. Depositional units at the fanhead were more continuous radially in the mudflow fan than in the fluvial fan because sediment was spread over a larger area of the fanhead due to the higher viscosity of the flows. Deposition from these flows formed individual layers of unsorted sediment. Mudflow deposition reached a relatively lower position on the mudflow fan than did untrenched fluvial fan deposition. However, the mudflow fan was much smaller than the fluvial fan and distance of sediment transport was much less. Also, fanhead trenching was less important on the mudflow fan, which favored preservation of these layers.

Figure 11.3 is a map of the grain-size distribution of the top 20 mm of the mudflow fan. The concentric contours at the toe of the fan represent the effect of the bulk sampling, which incorporated the basal clay layer. The central portion of the fan shows very little segregation of sediment by size, reflecting the unsorted nature of the mudflow deposits (Blissenbach, 1954).

Although sediment sorting by dewatering streamflow (afterflow) had a relatively minor effect on the overall grain-size distribution within the alluvial deposit, it was the dominant agent controlling the distribution of coarse-sediment concentrations and the basal clay layer. The afterflow was also the dominant agent controlling the distribution of heavy minerals within the fan and source area channels.

Fan Delta

Subaerial Zone

Grain-size analysis of sediments from the subaerial portion of fan delta 2 (Table 10.1) shows four significant areas of coarse-sediment deposition (Fig. 11.4). The first area to the right was the site of the most recently active main channel. It contained the coarse bed material, which is in contrast to finer-grained overbank deposits. Two other radial areas of coarse sediment reflected other major channel paths. Similar trends were found in the Witwatersrand basin in South Africa (Pretorius, 1974) and along the coastal profile of the Yallahs fan delta (Wescott, 1979).

The fourth significant area of coarse sediment was oriented transverse to the flow, on the upper midfan delta (Fig. 11.4). This trend marked the primary region

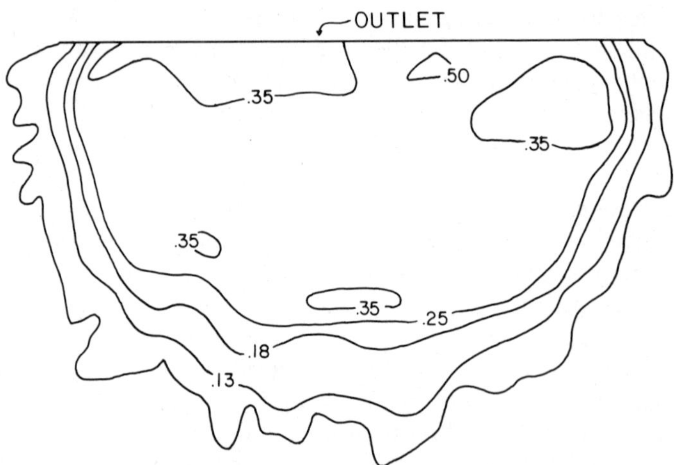

FIGURE 11.3. Grain-size distributioin of upper 20 mm of mudflow fan (isopleths in millimeters) (from Macke, 1977).

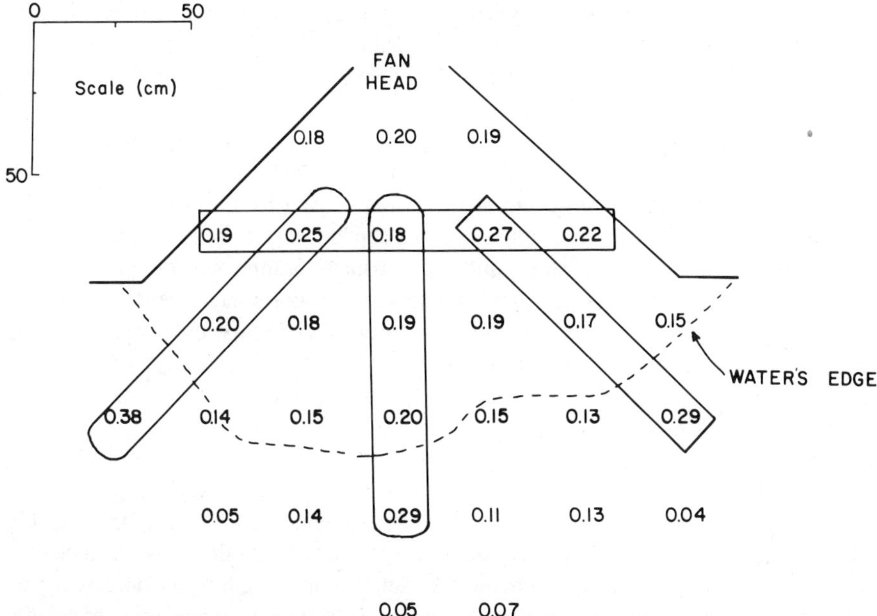

FIGURE 11.4. Grain-size distribution, in millimeters, on fan delta. Zones of coarsest sediment are outlined (from Jackson, 1981).

of coarse-sediment deposition. Much of the coarse sediment was initially dropped at the fan apex, because stream competency was lost as the flow spread over the fan delta. During fanhead incision by multiple nickpoints, coarse sand and granules were moved intermittently downfan, but because there was not sufficient energy to transport them across the fan, a coarse deposit remained on the upper midfan. Macke (1977) also documented a trend of increasing sediment size in a downfan direction (Fig. 11.2).

Subaqueous Zone

Fanhead trenching and resultant progradation at the channel mouth carried coarse sand to the fan delta slope. The coarse sediment was then covered by silt and clay delivered from small channels after the main channel shifted to another location. The three coarsest fan delta slope samples were channel sediments that were not buried.

Subaqueous channel slumps, fine sand stringers, and lobate mud slumps were apparent in fan delta 2 deposits. Channel slumps were formed when a headcut formed and an abundance of sediment was delivered to the delta margin. Coarse sand accumulated at the water's edge as erosion continued upstream. When the subaqueous angle of repose of the channel sand was exceeded, the sand slumped down the fan delta slope, making way for more deposition from above. The size of the slumps was proportional to the size of the channel feeding them. Fine sand stringers occurred in the vicinity of small channel mouths. They were about 1.5 cm wide and reached halfway or slightly less down the fan delta slope.

Mud slumps formed symmetrical, arcuate tongues of fine sediment at the base of the fan delta slope. Width and length of the slumps were about 15 cm and thickness, about 1 cm. Mud accumulation was caused by the prolonged absence of a channel at the location of the slump. Seepage may have dislodged the thin mud layer on the fan delta slope and caused mass failure, or perhaps the slope simply became too steep. Delta slope failures due to mass movement on slopes of about 1° have been described for the Mississippi Delta slope by Shepard (1955), who attributes them to gravity sliding.

Sedimentary Structures

The only visible sedimentary structures within the fan deltas were microdelta foreset laminae (Reineck and Singh, 1975), which are similar to Gilbert delta crossbeds and horizontal laminations. Because of the small scale of the fan delta channels, no fining or coarsening sequences could be detected within the channel deposits.

Sand and mud layers made up the fan delta. Topset sediments occupying the top few centimeters of the fan delta sequence were sandy, water-lain, horizontal deposits that were produced by relatively shallow water flow and small discharges. Foreset sediments were composed of alternating sequences of coarse channel sand that contained pebbles and fine sand and silt layers that were deposited by small channels. Bottomset sediments were horizontal layers of silt and clay. In a cross section parallel to flow, sand and mud laminations were continuous along the foreset slope. However, when baselevel was lowered, the foreset sediments were truncated at the top.

Channel sand deposits had a maximum thickness of 3 cm parallel to flow. Perpendicular to flow direction, channel sands were lens shaped and embedded in a matrix of silt and clay. Average width of the lenses was about 10 cm. In three dimensions, the fan delta deposit was a series of channel-sand tongues in a mud matrix.

Although very similar channel sands were present in both the topset and foreset laminae of the fan deltas, the mode of deposition for each position was different. The subaerial deposits were water-lain, whereas the subaqueous sand deposits resulted from slides or slumps at channel mouths. The differences between these deposits are easily distinguishable in nature. For example, conglomerate slope deposits on the Yallahs fan delta have abundant (greater than 15%) mud matrix and are matrix supported. Channel and beach deposits are grain supported (Wescott, 1979).

Heavy-Mineral Concentrations

Heavy-mineral placer deposits can be an economically significant aspect of alluvial fan sediments. Macke's (1977) study afforded a unique opportunity to correlate magnetite distributions in an alluvial fan with known processes acting at the time of deposition. The source of the magnetite was naturally occurring grains dispersed homogeneously throughout the sediment in the source area.

Magnetite concentrations within the fluvial fan show a generally lobate pattern (Fig. 11.5*A*), although the coarse sample spacing prevents accurate depiction of smaller-scale deviations. The contours show that the pattern of concentrations at a

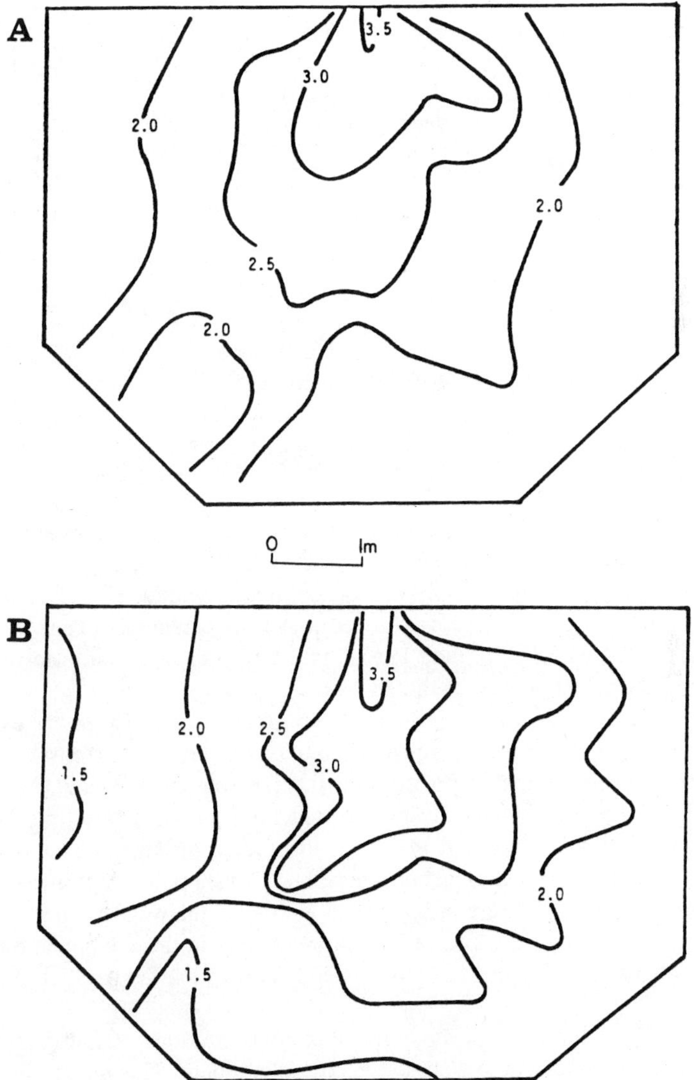

FIGURE 11.5. Isopleths of total magnetite concentration expressed as $\log_{10}$ of ratio of weight of magnetite to weight of sample × 10^6 (from Macke, 1977): (*A*) Upper 20 mm of fluvial fan; (*B*) upper 50 mm of fluvial fan; (*C*) 50–100-mm depth fluvial fan.

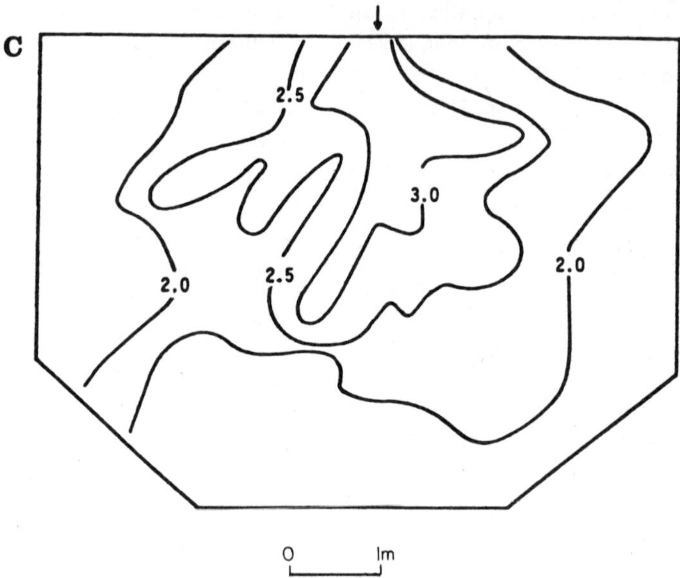

FIGURE 11.5. (*Continued*)

given distance from the fan apex is not comparable to the simple, smooth, semicircular form of topographic contours.

Sample depths of 0–50 mm and 50–100 mm were also analyzed separately (Fig. 11.5*B, C*). Due to the depth of material included in the sample interval, the curves produced represent average magnetite concentrations within several sedimentary layers. Examination of latex peels showed that heavy minerals occur in discrete stratigraphic horizons. The limited lateral extent of individual laminae normal to flow increases the complexity of the pattern.

Fanhead trenching decisively influenced the irregular distribution of heavy minerals in the fluvial fan. Some minor accumulations of heavy minerals occurred in the channels during normal deposition on the fan, but these were of secondary importance to those developed during the trenching phase of fan growth. It was the numerous trenching events at the apex, together with lateral channel shifting, that gave rise to the heavy-mineral concentration. Transport and deposition of heavy minerals during trenching were concentrated within the thalwegs of channels; these episodes of sediment sorting and redeposition resulted in the accumulation of thin bands or ribbons of heavy minerals that extended down the fan parallel to the flow direction.

Examination of the latex peels shows that concentrations of fine-grained dark minerals are often subjacent to well-sorted coarse sediment. This reflects backfilling of the trench and burial of a lag concentration of magnetite. After backfilling of the trench is complete, the normal spread condition of flow at the fan apex is reestablished. As a result, the heavy-mineral concentrations are isolated in space,

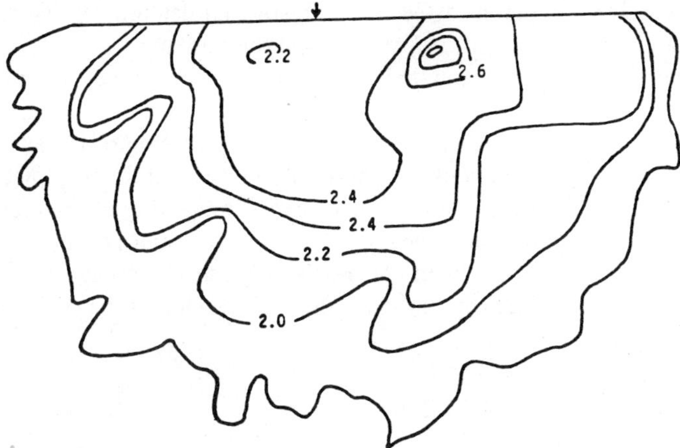

FIGURE 11.6. Isopleths of total magnetite concentration expressed as $\log_{10}$ ratio of weight of magnetite to weight of sample $\times$ 10^6 in upper 20 mm of mudflow fan (from Macke, 1977).

both vertically and laterally, due to the isolated nature of the trenching events during fan development.

Distribution of magnetite in the mudflow fans illustrates clearly the significance of reworking on heavy-mineral concentrations. The repeated reworking of the fluvial fan apex produced significant magnetite concentrations, whereas this distribution or concentration of heavy minerals did not occur in the mudflow fan (Fig. 11.6). The mudflow fan was not repeatedly reworked, and the magnetite was distributed essentially uniformly throughout this deposit.

DISCUSSION

The complexity of fluvial fan, mudflow fan, and fan delta growth as described in Chapters 2, 9, and 10 predicates that the sedimentology and stratigraphy will also be complex. The contrast between the fluvial and mudlfow fans is pronounced. The mudflow fan, because of significantly less fanhead incision and reworking, was characterized by a downfan decrease of grain size but no heavy-mineral concentrations. The downfan increase in grain size displayed by the fluvial fan and fan delta reflects the development of sediment-transport-efficient incised fanhead trenches, which provided a conduit from the fanhead and sediment source area to the midfan. Coarse sediments that normally would be deposited at the fanhead were moved farther downfan.

Just as the episodic movements of heavy minerals from the drainage basin is a function of the inherent workings of the fluvial system (Chapter 3), so the development of fluvial fanhead placers is a function of the repeated incision and reworking of the fanhead. Herail (1984, p. 142) sampled the gold content of alluvial fans on the Teleno piedmont of northwest Spain and determined that the fanhead contained four times the gold concentration of the midfan.

The heavy minerals concentrated in the fluvial fanhead may be third-or even fourth-generation placers. For example, heavy minerals are concentrated at preferred sites in the drainage basin (Chapter 3), but reworking of the sites shifts the heavy minerals downvalley, either to be moved out of the drainage network or to be stored again at a preferential site. Incision and reworking of the preferred sites delivers a high concentration of heavy minerals to the fanhead, and later fanhead incision further concentrates them. This development of rich placer deposits from a totally diffuse source reflects the geomorphic behavior of the drainage network and fanhead channels.

References

Abrahams, A.D. (1972) Drainage densities and sediment yields in eastern Australia, *Austr. Geogr. Studies* **10,** 19–41.

Abrahams, A.D. (1984) Channel networks: A geomorphological perspective, *Water Resources Res.* **20,** 161–188.

Abramowitz, M., and Stegun, I.A. (1964) Handbook of mathematical functions, U.S. Dept. Commerce, Bureau of Standards, Applied Math. Series, Vol. 55, 2046 pp.

Ackers, P. (1964). Experiments on small streams in alluvium, *J. Hydraulics Div., Am. Soc.* Civil Eng. **90,** 4, 1–37.

Ackers, P., and Charlton, F.G. (1970a). Dimensional analysis of alluvial channels with special reference to meander length, *J. Hydraulic Res.* **8,** 287–316.

Ackers, P. and Charlton, F.G. (1970b). The meandering of small streams in alluvium, Rept. 77, Hydraulics Research Station, Wallingfore, U.K., 78 pp.

Ackers, P., and Charlton, F.G. (1970c). The slope and resistance of small meandering channels, Inst. Civil Engineers Proc. Sup. XV, paper 73625, pp. 349–370.

Adams, J. (1980). Active tilting of the United States midcontinent: Geodetic and geomorphic evidence, *Geology* **8,** 442–446.

Adams, J., Zimpfer, G.L., and McLane, C., (1978). Basin dynamics, channel processes and placers, and placer formation: A model study, *Economic Geology* **73,** 416–426.

Ahnert, F. (1970). A comparison of theoretical slope models with slopes in the fields, *Zeit. Geomorph. Suppl.* **9,** 88–101.

Ahnert, F. (1976), Brief description of a comprehensive three-dimensional process response model of landform development, *Zeit. Geomorph. Suppl.* **25,** 29–49.

Albertson, M.L., Barton, J.R., and Simons, D.B. (1960). *Fluid Mechanics for Engineers,* Prentice-Hall, Englewood Cliffs, N.J., 561 pp.

Allen, J.R.L. (1971a). Bed forms due to mass transfer in turbulent flows: A kaleidoscope of phenomena, *J. Fluid Mech.* **49,** 49–63.

Allen, J.R.L. (1971b). Transverse erosional marks of mud and rock: Their physical basis and geological significance, *Sedimentary Geol.* **5,** 167–385.

Amerman, C.R. (1965). The use of unit-source watershed data for runoff prediction, *Water Resources Res.* **1,** 499–507.

Amorocho, J., and Hart, W.E. (1965). The use of laboratory catchments in the study of hydrologic systems, *J. Hydrology* **3,** 106–123.

Anderson, G.S., and Hussey, K.M. (1962). Alluvial fan development at Franklin Bluffs, Alaska, *Proc. Iowa Acad. Sci.* **69,** 310–322.

Andrews, E.D. (1979). "Hydraulic adjustment of the East Fork River, Wyoming, to the supply of sediment," in D.D. Rhodes and G.P. Williams, Eds., *Adjustment of the Fluvial System*, Kendall-Hunt, Dubuque, Iowa, pp. 69–94.

Antevs, E. (1952). Arroyo cutting and filling, *J. Geol.* **60,** 375–385.

Araya, T., and Higashi, S., (1983). Debris movement in torrential rivers of volcanic areas, Symposium on Erosion Control in Volcanic Areas, Technical Memorandum, Public Works Research Inst., Tokyo (Japan), pp. 5–30.

Armstrong, A.C. (1980). Simulated slope development sequences in a three-dimensional context, *Earth Surface Proc.* **5,** 265–270.

Ashida, K. and Sawai, K. (1976). Erosion and cross section on the cohesive stream bed, *Disaster Prevention Research Inst., Kyoto Univ., Bull.* **26,** 145–161.

Ashida, K., Takahashi, T., and Sawada, T. (1976). Sediment yield and transport on a mountainous small watershed, *Disaster Prevention Research Inst., Kyoto Univ., Bull.* **26,** 119–144.

Ashmore, P.E. (1982). Laboratory modelling of gravel braided stream morphology, *Earth Surface Proc. and Landforms* **7,** 201–225.

Ashmore, P.E., and Parker, G. (1983). Confluence scour in coarse braided streams, *Water Resources Res.* **19,** 392–402.

Bailey, R.W. (1935). Epicycles of erosion in the valleys of the Colorado Plateau Province, *J. Geol.* **43,** 337–355.

Baker, V.R. (1978). Paleohydrology and sedimentology of Lake Missoula flooding in eastern Washington, Geol. Soc. Am., Spec. Paper 144, 79 pp.

Barnard, R.S. (1977). Morphology and morphometry of a channelized stream: The case history of Big Pine Creek Ditch, Benton County, Indiana, Purdue Univ. Water Resources Research Center Tech. Report No. 92, 86 pp.

Barr, D.I.H. (1968). Discussion of scale model of urban runoff from storm rainfall, *J. Hydraulics Div., Am. Soc. Civil Eng.* **94,** 2, 586–588.

Bathurst, J.C., Thorne, C.R., and Hey, R.D. (1979). Secondary flow and shear stress at river bends, *J. Hydraulics Div., Am. Soc. Civil Eng.* **105,** 1277–1295.

Beaty, C.B. (1970). Age and estimated rate of accumulation of an alluvial fan, White Mountains, California, U.S.A., *Am. J. Sci.* **268,** 50–70.

Beaty, C.B. (1974). Debris flows, alluvial fans and a revitalized catastrophism, *Zeit. Geomorph. Suppl. 21,* **2,** (IV), 39–51.

Beer, C.E., and Johnson, H.P. (1965). Factors related to gully growth in the deep loess area of Western Iowa, U.S. Dept. Agriculture, Misc. Pub. 970, pp. 37–43.

Begin, Z.B. (1979). Aspects of degradation of alluvial stream in response to base-level lowering, unpublished Ph.D. dissertation, Colorado State Univ., Fort Collins, Col. 239 pp.

Begin, Z.B., and Schumm, S.A. (1979). Instability of alluvial valley floors: A method for its assessment, *Am. Soc. Agric. Eng.* **22,** 347–350.

Begin, Z.B., Meyer, D.F., and Schumm, S.A. (1980a). Knickpoint migration in alluvial channels due to base level lowering, *Amer. Soc. Civil Eng. Jour. Waterways Port. Coastal and Ocean Div.* **106,** 369–388.

Begin, Z.B., Meyer, D.F., and Schumm, S.A. (1980b). Sediment production of alluvial channels in response to base level lowering, *Trans. Amer. Soc. Agric. Eng.* **23,** 1183–1188.

Begin, Z.B., Meyer, D.F., and Schumm (1981). Development of longitudinal profiles of alluvial channels in response to base level lowering: *Earth Surface Proc. and Landforms* **6,** 49–68.

Bendefy, L., Dohnalik, J., and Mike, K. 1967, Nouvelles methodes de l'etude gentique des cours de eau, *Int. Assoc. Scientific Hydrology* **75,** 64–72.

Bergstrom, F.W. (1980). Episodic behavior in badlands: Its effect on channel morphology and sediment routing, unpublished M.S. thesis, Colorado State Univ., Fort Collins, Col. 210 pp.

Bergstrom, F.W., and Schumm, S.A. (1981). Episodic behavior in badlands, Internal. Assoc. Hydrological Sci. Pub. 132, pp. 478–492.

Betson, R.P. (1964). What is watershed runoff?, *J. Geophysical Res.* **69,** 1541–1552.

Betson, R.P., and Marius, J.B. (1969). Source areas of storm runoff, *Water Resources Res.* **5,** 574–582.

Bettis, E.A., III, and Thompson, D.M. (1982). Interrelations of cultural and fluvial deposits in northwest Iowa, Assoc. Iowa Archaeologists Guidebook, April 1982, Iowa City 142 pp.

Beutner, E.C. (1968). Structure and tectonics of the southern Lemhi Range, Idaho, unpublished Ph.D. dissertation, Pennsylvania State Univ., University Park, 105 pp.

Beveridge, W.I.B. (1957). *The Art of Scientific Investigation*, 3rd ed., Random House, New York, 239 pp.

Black, P.E. (1970). Runoff from watershed models, *Water Resources Res.* **6,** 465–477.

Black, P.E. (1972). Hydrograph responses to geomorphic model watershed characteristics and precipitation variables, *J. Hydrology* **17,** 309–329.

Black, P.E., and Cronn, J.W., Jr. (1975). Hydrograph responses to watershed size and similitude relations, *J. Hydrology* **25,** 255–266.

Blank, H.R. (1970). Incised meanders in Mason County, Texas, *Geol. Soc. Am. Bull.* **81,** 3135–3140.

Blench, T. (1969). *Mobile-Bed Fluviology*, Univ. Alberta Press, Edmonton, Canada 300 pp.

Blissenbach, E. (1952). Relation of surface angle to particle size distribution on alluvial fans, *J. Sed. Petrology* **22**(1), 25–28.

Blissenbach, E. (1954). The geology of alluvial fans in semiarid regions, *Geol. Soc. Am. Bull.* **65,** 175–189.

Bluck, B.J. (1964). Sedimentation of an alluvial fan in southern Nevada, *J. Sed. Pet.* **34,** 395–400.

Bluck, B.J. (1976). Sedimentation in some Scottish rivers of low sinuosity, Trans. Royal Soc. Edinburgh **69,** 425–456.

Bones, J.G., and Ford, D.C. (1971). Simulating the development of river drainage networks, *Canadian Geogr.* **15,** 207–211.

Boothroyd, J.C. (1970). Recent braided-stream sedimentation, south central Alaska, *Am. Assoc. Petroleum Geol. Bull.* (Abstr.) **54,** 836.

Born, S.M., and Ritter, D.F. (1970). Modern terrace development near Pyramid Lake, Nevada, and its geologic implications, *Geol. Soc. Am. Bull.* **81,** 1233–1242.

Bowler, J.M., and Harford, L.B. (1966). Quaternary tectonics and the evolution of the Riverine Plain near Echuca, Victoria, *Geol. Soc. Austral.* **13,** 339–354.

Bradford, J.M., Farrell, D.A., Larson, W.E. (1973). Mathematical evaluation of factors affecting gully stability, *Soil Sci. Soc. Am. Proc.* **37,** 103–107.

Brakenridge, G.R. (1984). Alluvial stratigraphy and radiocarbon dating along the Duck River, Tennessee: Implications regarding floodplain origin, *Geol. Soc. Am. Bull* **95,** 9–25.

Braun, D.D. (1983). Lithologic control of bedrock meander dimensions in the Appalachian Valley and Ridge Province, *Earth Surface Proc. and Landforms* **8,** 223–237.

Bray, D.I. (1973). "Regime equations for Alberta gravel-bed rivers," in Nat'l Research Council Canada, *Fluvial Processes and Sedimentation,* Ottawa, pp. 440–452.

Bretz, J.H. (1924). The Dalles type of river channel, *J. Geol.* **24,** 129–149.

Brice, J.C. (1984). "Planform properties of meandering rivers," in *River Meandering*, Proc. Conf. Rivers '83, Am. Soc. Civil Engs., New York, pp. 1–15.

Bridge, J.S. (1977). Flow, bed topography, grain size and sedimentary structure in open channel bends: A three dimensional model, *Earth Surface Proc.* **2,** 401–416.

Bridge, J.S., and Jarvis, J. (1976). Flow and sedimentary processes in the meandering River South Esk, Glen Cova, Scotland, *Earth Surface Proc.* **1,** 303–337.

Brown, A.J. (1983). Channel changes in arid badlands, Borrego Springs, California, *Physical Geogr.* **4,** 82–102.

Brush, L.M., Jr. (1961). Drainage basins, channels, and flow characteristics of selected streams in Central Pennsylvania, U.S. Geol. Survey Prof. Paper 282-F, pp. 145–181.

Brush, L.M., Jr., and Wolman, M.G. (1960). Knickpoint behavior in noncohesive material: A laboratory study, *Bull. Geol. Soc. Am.* **71,** 59–74.

Bryan, K. (1928). Historic evidence on changes in the channel of the Rio Puerco, tributary of the Rio Grande in New Mexico, *J. Geol.* **36,** 265–282.

Bull, W.B. (1964a). Alluvial fans and near-surface subsidence in western Fresno County, California, U.S. Geol. Survey Prof. Paper 437-A, pp. A1–A71.

Bull, W.B. (1964b). Geomorphology of segmented alluvial fans in western Fresno County, California, U.S. Geol. Survey Prof. Paper 532-F, pp. 79–129.

Bull, W.B. (1968), Alluvial fans, *J. Geol. Education* **16,** 101–106.

Bull, W.B. (1972). Recognition of alluvial fan deposits in the stratigraphic record, Soc. Econ. Paleo Mineral. Spec. Pub. 18, pp. 63–83.

Bull, W.B., and Knuepfer, R.L.K. (1983). Stream adjustments to regional uplift and late Quaternary climate change, *Geol. Soc. Am.* (Abstr. with Programs 1983), p. 430.

Burnett, A.W. (1982). Alluvial stream response to neotectonics in the lower Mississippi valley, unpublished M.S. thesis, Colorado State Univ., Fort Collins, Col., 160 pp.

Burnett, A.W., and Schumm, S.A. (1983). Alluvial river response to neotectonic deformation in Louisiana and Mississippi, *Science* **222,** 49–50.

Carslaw, M.S., and Jaeger, J.C. (1959). *Conduction of Heat in Solids*, Clarendon Press, Oxford, U.K., 510 p.

Carlston, C.W. (1965). The relation of free meander geometry to stream discharge and its geomorphic implications, *Am. J. Sci.* **263,** 864–885.

Carryer, S.J. (1966). A note on the formation of alluvial fans, *New Zealand J. Geol. Geophys.* **9,** 91–94.

Carson, M.A. (1984). The meandering-braided river threshold, a reappraisal, *J. Hydrology* **73,** 315–334.

Carson, M.A., and Kirkby, M.J. (1972). *Hillslope Form and Process,* Cambridge Univ. Press, Cambridge, 475 pp.

Carter, C.S., and Chorley, R.J. (1961). Early slope development in an expanding stream system, *Geol. Mag.* **98,** 117–130.

Chang, H.H. (1982). Fluvial hydraulics of deltas and alluvial fans, *Jour. Hydraulics Div. Am. Soc. Civil Engs.*, **108**(11), 1282–1295.

Chang, H.H. (1984). Modeling of river channel changes, *J. Hydraulic Eng.* **110,** 157–172.

Chang, H.Y. (1967). Hydraulics of rivers and deltas, Ph.D. dissertation, Colorado State Univ., Fort Collins, Col., 176 pp.

Chang, H.Y., Simons, D.B., Woolhiser, D.A. (1971). Flume experiments on alternate bar formation, Am. Soc. Civil Engineering Proc. Waterways, Harbors, and Coastal Engineering Div. 97, WWI, pp. 155–165.

Chery, D.L., Jr. (1969). A discussion of "Potential of physical models for achieving better understanding and evaluation of watershed changes," in W.L. Moore and C.W. Morgan, Eds., *Effects of Watershed Changes on Streamflow*, Water Resources Symposium, Univ. of Texas Press, Austin and London, 26 pp.

Chorley, R.J. (1967). "Models in geomorphology," in R. J. Chorley and P. Haggett, Eds., *Models in Geography*, Methuen, London, pp. 59–96.

Chorley, R.J., and Kennedy, B.A. (1971). *Physical Geography: A Systems Approach*: Prentice-Hall, London, 37 pp.

Chow, V.T. (1959). *Open-Channel Hydraulics*, McGraw-Hill, New York, 680 pp.

Church, M. (1984). "On experimental method in geomorphology," in T.P. Burt and D. E. Walling, Eds., *Catchment Experiments in Fluvial Geomorphology*, Geo Books, Norwich, U.K., pp. 563–580.

Church, M. and Jones, D. (1982). "Channel bars in gravel-bed rivers," in R.D. Hey, J.C., Bathurst, and C.R. Thorne, Eds., *Gravel Bed Rivers*, Wiley, New York, pp. 291–338.

Colby, B.R. (1963). Fluvial sediments—a summary of source, transportation, deposition, and measurement of sediment discharge, U.S. Geol. Survey Bull. 1181-A, 47 pp.

Colby, B.R. (1964). Discharge of sands and mean-velocity relationships in sand-bed streams, U.S. Geol. Survey Prof. Paper, 47 pp.

Cole, W.S. (1930). The interpretation of intrenched meanders, *J. Geol.* **38,** 423–436.

Coleman, A. (1958). The terraces and antecedence of a part of the river Salzach, *Inst. British Geogrs. Trans.* **25,** 119–134.

Collins, E.W., Dix, O.R., Hobday, D.K. (1981). Oakwood Salt Dome, East Texas: Surface geology and drainage analysis, Univ. of Texas, Bur. of Economic Geol., Geol. Circular 81-6, 23 pp.

Cooke, R.U., and Reeves, R.W. (1976). *Arroyos and Environmental Change in The American South-West*: Clarendon Press, London, 213 pp.

Cotton, C.A. (1941). *Landscape,* Cambridge Univ. Press, Cambridge, U.K., 509 pp.

Dacey, M.F., and Krumbein, W.C. (1976). Three growth models for stream channel networks, *J. Geol.* **84,** 153–163.

Daniel, J.F. (1971). Channel movement of meandering Indiana streams, U.S. Geol. Survey Prof. Paper 732-A, 18 pp.

Daniels, R.B. (1960). Entrenchment of the Willow Drainage Ditch, Harrison County, Iowa, *Am. Jour. Sci.* **258,** 161–176.

Daniels, R.B., and Jordan, R.H. (1966). Physiographic history and the soils, entrenched stream systems and gullies, Harrison County, Iowa, USDA Tech. Bull. 1348.

Daubrée, A.G. (1879). *Etudes Synthetiques de Geologie Experimentale,* Dunod, Paris, 828 pp.

Davies, T.R., and Sutherland, A.J. (1980). Resistance to flow past deformable boundaries, *Earth Surface Proc.* **5,** 175–179.

Davies, T.R.H., and Tinker, C.C. (1984). Fundamental characteristics of stream meanders, *Geol. Soc. Am. Bull.* **95,** 505–512.

Davis, W.M. (1890). Structure and origin of glacial sand plains, *Geol. Soc. Am. Bull.* **1,** 195–202.

Davis, W.M. (1893). The Osage River and the Ozark Uplift, *Science* **22,** 276–279.

Davis, W.M. (1902). River terraces in New England, Harvard College, Museum of Comparative Zoology, Bull. 38, pp. 281–346.

Davis, W.M. (1909). *Geographical Essays*, ed. D.W. Johnson, Ginn and Co., Boston, 777 pp.

Davis, W.M. (1912). *Die erklärende Beschriebung der Landformen*, B.G. Teubner, Leipzig, 565 pp.

Davison, C. (1888a). Note on the movement of scree material, *Quarterly J. Geol. Soc. of London* **44,** 232–238.

Davison, C. (1888b). Second note on the movement of scree material, *Quarterly J. Geol. Soc. of London* **44,** 825–826.

DeBlieux, C. (1951). Photogeologic study in Kent County, Texas, *J. Oil and Gas* **50,** 86.

DeBlieux, C. (1962). Photogeology in Louisiana coastal marsh and swamp, *Gulf Coast Assoc. Geol. Soc. Trans.* **12,** 231–241.

Denny, C.S. (1965). Alluvial fans in the Death Valley Region, California and Nevada, U.S. Geol. Survey Prof. Paper 466, 62 pp.

Denny, C.S. (1967). Fans and pediments, *Amer. J. Sci.* **265,** 81–105.

Dickinson, W.T., Holland, M.E., and Smith, G.L. (1967). An experimental rainfall-runoff facility, Hydrology Paper No. 25, Colorado State Univ., Fort Collins, Col., 81 pp.

Dietrick, W.E., Smith, J.D., Dunne, T. (1979). Flow and sediment transport in a sand bedded meander, *J. Geol.* **87,** 305–315.

Dooge, J.C.I. (1973). Linear theory of hydrologic systems, U.S. Dept. of Agric., Agric. Res. Serv., Tech. Bull. 1468, 327 pp.

Dort, W. (1962). Multiple glaciation of southern Lemhi Mountains, Idaho, *Tebiwa* **6,** 2-17.

Dort, W. (1965). Glaciation in Idaho—a summary of present knowledge, *Tebiwa* **8,** 29–37.

Drew, F. (1873). Alluvial and lacustrian deposits and glacial records of the upper Indus basin, *Quarterly J. Geol. Soc. London.* **29,** 441–471.

Dunne, T. (1980). Formation and controls of channel networks, *Prog. Phys. Geogr.* **4,** 211–240.

Dunne, T., and Black, R.D. (1970a). An experimental investigation of runoff production in permeable soils, *Water Resources Res.* **6,** 478–490.

Dunne, T., and Black, R.D. (1970b). Partial area contributions to storm runoff in a small New England watershed, *Water Resources Res.* **6,** 1296–1311.

Dunne, T., and Leopold, L.B. (1978). *Water in Environmental Planning*, Freeman, San Francisco, 818 pp.

Dury, G.H. (1954). Contribution to a general theory of meandering valleys, *Am. J. Science* **252,** 193–224.

Dury, G.H. (1958). Tests of a general theory of misfit stream, *Inst. British Geogr. Trans.* **25,** 105–118.

Dury, G.H. (1964a). Principles of underfit streams, U.S. Geol. Survey Prof. Paper 452-A, 67 pp.

Dury, G.H. (1964b). Theoretical implications of underfit streams, U.S. Geol. Survey Prof. Paper 452-C, 42 pp.

Dury, G.H. (1966a). "The concept of grade," in G.H. Dury, Ed., *Essays in Geomorphology*, Heinemann, London, pp. 211–234.

Dury, G.H. (1966b). Incised valley meanders on the Lower Colorado River, *NSW Australian Geogr.* **10,** 17–25.

Dury, G.H. (1970). A resurvey of part of the Hawkesbury River, New South Wales after 100 years, *Australian Geogr. Studies* **8,** 121–132.

Dury, G.H. (1976). Discharge prediction, present and former, from channel dimensions, *J. Hydrology* **30,** 219–245.

Dury, G.H., Hales, J.R., and Robbie, M.B. (1963). Bankfull discharge and the magnitude frequency series, *Australian J. of Science* **26,** 123–124.

Dzurisin, D. (1975). Channel responses to artificial stream capture, Death Valley, California, *Geology* **3,** 309–312.

Eagleson, P.S. (1969). "Potential of physical models for achieving better understanding and evaluation of watershed changes," in W.L. Moore and W.C. Morgan, Eds., *Effects of Watershed Changes on Streamflow,* Water Resources Symposium, v. 2, Univ. of Texas Press, Austin and London, pp. 12–25.

Eckis, R. (1928). Alluvial fans of the Cucamonga District, southern California, *J. Geol* **36,** 3, 224–247.

Edgar, D.E. (1973). Geomorphic and hydraulic properties of laboratory rivers, Unpublished M.S. thesis, Colorado State Univ., Fort Collins, Col. 156 pp.

Edgar, D.E. (1984). "The role of geomorphic thresholds in determining alluvial channel morphology," in *River Meandering*, Proc. conf. Rivers '83, Am. Soc. Civil Engs., New York, pp. 44–54.

Einstein, H.A., and Shen, H.W. (1964). A study of meandering in straight alluvial channels, *J. Geophysical Res.* **69,** 5239–5247.

Ellis, D.C. (1912). A working erosion model for schools, U.S. Dept. Agriculture, Office of Experiment Stations, Circular 117, 11 pp.

Ellison, W.D. (1947). Soil erosion studies, part I, *Agri. Eng.* **28,** 145–146, 197–201, 245–248.

Engelund, F., and Skovgaard, O. (1973). On the origin of meandering and braiding in alluvial streams, *J. Fluid Mechanics* **57,** 289–302.

Eyles, R.J. (1968). Stream net ratios in west Malaysia, *Geol. Soc. Am. Bull.* **79,** 701–712.

Fahnestock, R.K. (1963). Morphology and hydrology of a glacial stream, White River, Mount Rainier, Washington, U.S. Geol. Survey Prof. Paper 422-A, 70 pp.

Ferguson, R.I. (1976). Disturbed periodic model for river meanders: *Earth Surface Proc.* **1,** 337–347.

Fink, J.H., Greeley, R., and Gault, D.E. (1981). Impact cratering experiments in Bingham materials and the morphology of craters on Mars and Ganymede, *Lunar Planetary Science Proc.* **12,** (part B), 1649–1666.

Flint, J.J. (1973). Experimental development of headward growth of channel networks, *Geol. Soc. Am. Bull.* **84,** 1087–1094.

Flueck, J.A. (1978). The role of statistics in weather modification experiments, *Atmosphere-Ocean* **16,** 60–80.

Folk, R.L. (1974). *Petrology of Sedimentary Rocks*, Hemphill, Austin, Tex., 182 pp.

Friedkin, J.F. (1945). A laboratory study of the meandering of alluvial rivers, U.S. Waterways Experiment Station, Vicksburg, Miss., 40 pp.

Funk, J.M. (1976). Climatic and tectonic effects on alluvial fan systems, Birch Creek Valley, east-central Idaho, unpublished Ph.D. dissertation, Univ. of Kansas, Lawrence, Kan., 246 pp.

Gage, M. (1970). The tempo of geomorphic change, *J. Geol.* **78,** 619–625.

Galloway, W.E., and Hobday, D.K. (1983). *Terrigenous Clastic Depositional Systems, Applications to Petroleum, Coal and Uranium Exploration,* Springer-Verlag, New York, 423 pp.

Gardiner, V., and Park, C.C. (1978). Drainage basin morphometry, *Prog. Phy. Geogr.* **2,** 1–35.

Gardner, T.W. (1973). A model study of river meander incision, unpublished M.S. thesis, Colorado State Univ., Fort Collins Col. 86 pp.

Gardner, T.W. (1975). The history of part of the Colorado River and its tributaries: An experimental study, Four Corners Geologic Society Guidebook, 8th Field Conference (Canyonlands), pp. 87–95.

Gardner, T.W. (1983). Experimental study of knickpoint and longitudinal profile evolution in cohesive, homogeneous material, *Geol. Soc., Am. Bull.* **94,** 664–672.

Gavrilovic, P. (1972). Experimente zur klimageomorphlogie, *Zeit. Geomorph.* **16,** 315–331.

Gessler, J. (1971). ''Aggradation and degradation,'' H.W. Shen, Ed., *River Mechanics*, Fort Collins, Col. v. 1, pp. 8-1–8-24.

Ghosh, S.N., and Roy, N. (1970). Boundary shear distribution in open channel flow, *J. Hydraulics Div.*, Am. Soc. Civil Engineers Proc., **96,** 967–994.

Gilbert, G.K. (1890). Lake Bonneville, U.S. Geol. Survey Monograph 1, 438 pp.

Gilbert, G.K. (1893). *The moon's face, a study of the origin of its features, Philos. Soc., Washington, Bull* **12,** 241–292.

Gilbert, G.K. (1917). Hydraulic mining debris in the Sierra Nevada, U.S. Geol. Survey Prof. Paper 105, 154 pp.

Glock, W.S. (1931). The development of drainage systems: A synoptic View, *Geogr. Rev.* **21,** 475–482.

Gole, C.V., and Chitale, S.V. (1966). Inland delta building activity of Kosi River, *J. Hyd. Div., Am. Soc. Civ. Eng. Proc.* **92,** 111–126.

Gorycki, M.A. (1973). Hydraulic drag—a meander initiating mechanism, *Geol. Soc. Am. Bull.* **84,** 175–186.

Graf, W.H. (1971). *Hydraulics of Sediment Transport*, McGraw-Hill, New York, 513 pp.

Graf, W.L. (1983). Variability of sediment removal in a semi-arid watershed, *Water Resources Res.* **19**(3), 643–652.

Grant, P.J. (1977). Recorded channel changes of the Upper Waipawa River, Rauhine Range, New Zealand, Water and Soil Tech. Publ. No. 6, 1977, pp. 1–18.

Greeley, R., Iverson, J.D., Pollack, J.B., Vdovich, N. and White, B. (1974). Wind tunnel studies of Martian aeolian processes, *Proc. Royal Soc. London*, **342,** 331–360.

Gregory, K.J. (1976). Changing drainage basins, *Geogr. J.* **142,** 237–247.

Gregory, K.J. (1977). *River Channel Changes*, Wiley-Interscience, New York, 448 pp.

Griffiths, G.A. (1979). Recent sedimentation history of the Waimakariri River, New Zealand, *J. Hydrology* (New Zealand) **18**(1), 6–28.

Guilcher, A. (1958). *Coastal and Submarine Morphology*, Wiley, New York, 274 pp.

Gustaufson, T.C. (1974). Sedimentation on gravel outwash fans, Malaspina Glacier forland, Alaska, *Sed. Petrology* **44,** 374–389.

Hack, J.T. (1957). Studies of longitudinal stream profiles in Virginia and Maryland, U.S. Geol. Survey Prof. Paper 294-B, pp. 45–97.

Hack, J.T. (1960). Interpretation of erosional topography in humid temperate regions, *Am. J. Sci.* **258-A,** 80–97.

Hack, J.T. (1965). Post glacial drainage evolution and geometry in the Ontonagon Area, Michigan, U.S. Geol. Survey Prof. Paper 504B, 40 pp.

Harvey, A.M. (1984a). Aggradation and dissection sequences on Spanish alluvial fans: Influence on morphological development, *Catena* **11,** 289–304.

Harvey, A.M. (1984b) "Debris flows and fluvial deposits in Spanish Quaternary alluvial fans: Implications for fan morphology, in E.H. Koster and R.J. Steel, Eds., *Sedimentology of Gravels and Conglomerates*, Canadian Soc. Petroleum Geol. Memoirs v. 10, pp. 123–132.

Harvey, M.D. (1980). Steepland channel response to episodic erosion, unpublished Ph.D. dissertation, Colorado State Univ., Fort Collins, Col., 253 pp.

Hawkes, H.E. (1976). The downstream dilution of stream sediment anomalies, *J. Geochem. Explor.* **6,** 345–358.

Hayward, J.A. (1978). Hydrology and stream sediments in a mountain catchment, unpublished Ph.D. dissertation, Univ. Canterbury, New Zealand, vol. 2, 82 pp.

Heede, B.H. (1974). Stages of development of gullies in western United States of America, *Zeit. Geomorph.* **18,** 260–271.

Henderson, F.M. (1961). Stability of alluvial channels, *J. Hydraulics Div., Am. Soc. Civ. Eng. Proc.* **87,** 109–138.

Henderson, F.M. (1966). *Open Channel Flow*, Macmillan New York, 522 pp.

Herail, B. (1984). Les comes de degection: Formes et sediments, *Soc. Nat. Elf. Aquitaine (production) BCREDP* **8,** 135–150.

Hester, B.W. (1970). Geology and evaluation of placer gold deposits in the Klondike area, *Inst. Mining Metall., Trans.* (Sec. B) **79,** B60–B67.

Hey, R.D. (1978). Determinate hydraulic geometry of river channels, *J. Hyd. Div., Am. Soc. Civ. Eng. Proc.* **104,** 869–885.

Hey, R.D. (1979). Dynamic process-response model of river channel development, *Earth Surface Proc.* **4,** 59–72.

Hickin, E.J. (1969). A newly identified process of point bar formation in natural streams, *Am. J. Sci.* **267,** 999–1010.

Hickin, E.J. (1972). Pseudomeanders and point dunes—a flume study, *Am. J. Sci.* **272,** 762–799.

Hills, R.C. (1971). The influence of land management and soil characteristics on infiltration and the occurrence of overland flow, *J. Hydrology* **13,** 163–181.

Holland, M.E. (1969). Colorado State University experimental rainfall-runoff facility—design and testing of rainfall system, Eng. Res. Center Pub. 69-70-MEH21, Colorado State Univ., Fort Collins, Col., 81 pp.

Holland, W.N., and Pickup, G. (1976). Flume study of knickpoint development in stratified sediments, *Geol. Soc. Am. Bull.* **87,** 76–82.

Holmes, A. (1965). *Principles of Physical Geology*, Ronald Press, New York, 1288 pp.

Hong, L.B., and Davies, T.R.H. (1979). A study of stream braiding: A summary, *Geol. Soc. Am. Bull.* **90,** 1094–1095.

Hooke, R.L. (1967). Processes on arid-region alluvial fans, *J. Geol.* **75,** 438–460.

Hooke, R.L. (1968a). Model geology: Prototype and laboratory streams: Discussion, *Geol. Soc. Amer. Bull.* **79,** 391–394.

Hooke, R.L. (1968b). Steady-state relationships on arid-region alluvial fans in closed basins, *Am. J. Sci.* **266,** 609–629.

Hooke, R.L. (1974). Processes on arid-region alluvial fans, California Institute of Technology Contribution 1393, pp. 438–460.

Hooke, R.L. (1975). Distribution of sediment transport and shear stress in a meander bend, *J. Geol.* **83,** 543–566.

Hooke, R.L., and Rohrer, W. L. (1979). Geometry of alluvial fans: Effect of discharge and sediment size, *Earth Surface Proc.* **4,** 147–166.

Hoppe, G., and Ekmar, S. (1964). A note on the alluvial fans of Ladtjovagge, Swedish Lapland, *Geog. Annal.* **46,** 338–342.

Horton, R.E. (1945). Erosional development of streams and their drainage basins: Hydrophysical approach to quantitative morphology, *Geol. Soc. Am. Bull* **56,** 275–370.

Howard, A.D. (1967). Drainage analysis in geologic interpretation, *Am. Assoc. Pet. Geol.* **51,** 2246–2259.

Howard, A.D. (1971a). Simulation of stream networks by headward growth and branching, *Geogr. Anal.* **3,** 29–50.

Howard, A.D. (1971b). Simulation model of stream capture, *Geol. Soc. Am. Bull.* **82,** 1355–1376.

Howard, A.D. (1971c). "Problems of interpretation of simulation models of geologic processes," in M.E. Morisawa, Ed., *Quantitative Geomorphology: Some Aspects and Applications*, Pub. in Geomorph., State Univ. of N.Y., Binghamton, pp. 61–82.

Howard, A.D., and Kerby, G. (1983). Channel changes in badlands, *Geol. Soc. Am. Bull.* **94,** 739–752.

Howe, E. (1901). Experiments illustrating intrusion and erosion, U.S. Geol. Survey, 21st Annual Report, Part 3, pp. 291–303.

Hoyer, B.E. (1980a). "The geology of the Cherokee sewer site," in D.E. Anderson and H.A. Semkin, Jr., Eds., *Holocene Ecology and Human Adaptations in Northwestern Iowa*, Academic, New York, pp. 21–66.

Hoyer, B.E. (1980b). *Geomorphic History of the Little Sioux River Valley*, Geological Survey of Iowa, Field Trip Guide Iowa City, Iowa (September 20 and 21, 1980), 94 pp.

Hubbard, G.D. (1907). Experimental physiography, *Am. Geogr. Soc. Bull.* **39,** 658–666.

Hubbert, M.K. (1944). Theory of scale models as applied to the study of geologic structures, *Geol. Soc. Am. Bull.* **48,** 1459–1520.

Hunt, C.B. (1969). Geologic history of the Colorado River, U.S. Geol. Survey Prof. Paper 669, pp. 59–130.

Hunt, C.B., and Mabey, D.R. (1966). The stratigraphy and structure, Death Valley, California, U.S. Geol. Survey Prof. Paper 494-A, 162 pp.

Inglis, C.C. (1949). The behaviour and control of rivers and canals, Central Waterpower Irrigation and Navigation Research Station, Poona, India Research Publication 13, 283 pp.

Ippen, A.T., Drinker, P.A., Jobin, W.R., and Shemdin, O.H. (1962). Stream dynamics and boundary shear distribution for curved trapezoidal channels, Mass. Inst. Tech. Rep. No. 47, 81 pp.

Ireland, H.A., Sharpe, C.F.S., and Eargle, D.H. (1939). Principles of gully erosion in the Piedmont of South Carolina, U.S. Dept. of Agriculture, Tech. Bulletin 633, 142 pp.

Jackson, M.L.W. (1981). Geomorphology and sedimentology of experimental fan deltas: unpublished M.S. thesis, Colorado State Univ., Fort Collins, Col., 91, pp.

Jaeggi, M.N.R. (1984). Formation and effects of alternate bars, *J. Hyd. Eng.* **110,** 142–155.

Jaggar, T.A., Jr. (1908). Experiments illustrating erosion and sedimentation, *Bull. Museum of Comparative Zoology Harvard College* **49** (Geological Series, v. 8), 285–305.

Janda, R.J., Nolan, K.M., Harden, D.R., and Colman, S.M. (1975). Watershed conditions in the drainage basin of Redwood Creek, Humboldt County, California, as of 1973, U.S. Geol. Survey Open File Report 75-568, Menlo Park, Calif. 257 pp.

Jarvis, R.S. (1977). Drainage network analysis, *Prog. Phy. Geogr.* **1,** 271–295.

Jefferson, W.S. (1902). Limiting width of meander belts, *Natl. Geogr. Mag.* **13,** 373–383.

Jin, D., and Schumm, S.A. (1986). "A new technique for modelling river morphology," in K.S. Richards, Ed., *Proc. First Internat. Geomorphology Conf.*, Wiley, Chichester, in press.

Johnson, A.M. (1970). *Physical Processes in Geology*, Freeman, San Francisco, 577 pp.

Johnstone, D., and Cross, W.P. (1949). *Elements of Applied Hydrology:* Ronald Press, New York, 276 pp.

Kashiwaya, K. (1983). A mathematical model for the temporal change of drainage density, *Japanese Geomorph. Union, Trans.* **4,** 25–31.

Kehew, A.E. (1982). Catastrophic flood hypothesis for the origin of the Souris spillway, Saskatchewan and North Dakota, *Geol. Soc. Am.* Bull. **93,** 1051–1058.

Keller, E.A., and Melhorn, W. (1973). Bedforms and fluvial processes in alluvial stream channels: Selected observations,'' in M. Morisawa, Ed., *Fluvial Geomorphology*, Publications in Geomorphology, State Univ. of N.Y., Binghamton, pp. 253–83.

Keller, E.A., and Melhorn, W.N. (1978). Rhythmic spacing and origin of pools and riffles, *Geol. Soc. Am. Bull.* **89,** 723–730.

Kelsey, H.M. (1977). Landsliding, channel changes, sediment yield and land use in the Van Duzen River Basin, north coastal California, 1941–1975, unpublished Ph.D. dissertation, Univ. of California, Santa Cruz, 237 pp.

Kelsey, H.M. (1980). A sediment budget and an analysis of geomorphic processes in the Van Duzen River basin, north coastal California, 1941–1975, *Geol. Soc. Am. Bull.* **91,** 1119–1216.

Kennedy, J.F. (1983). Reflections on rivers, research, and Rouse, *J. Hyd. Eng.* **109,** 1254–1271.

Kennedy, J.F., and Brooks, N.H. (1965). Laboratory study of an alluvial stream at constant discharge, U.S. Dept. Agriculture Misc. Pub. 970, pp. 320–330.

Khan, H.R. (1971). Laboratory study of alluvial river morphology, Ph.D. dissertation, Colorado State Univ., Fort Collins, Col., 189 pp.

King, P.B., and Schumm, S.A. (1980). *The Physical Geography (Geomorphology) of William Morris Davis*, Geobooks, Norwich, U.K., 217 pp.

Kinosita, R. (1961). Investigation of channel deformation in Ishikari River, Report of Bureau of Resources, Dept. of Science and Technology, Japan, pp. 1–174.

Kirkby, M.J. (1967). Measurement and theory of soil creep, *J. Geol.* **75,** 359–378.

Kirkby, M.J. (1976). Tests of the random network model and its application for basin hydrology, *Earth Surface Proc.* **1,** 197–212.

Kjerfue, B., Shao, C.C., and Stapor, F.W., Jr. (1979). Formation of deep scour holes at the junction of tidal creeks: An hypothesis, *Marine Geol.* **33,** 9–14.

Knighton, A.D. (1972). Changes in a braided reach, *Geol. Soc. Am. Bull.* **83,** 3813–3822.

Knighton, A.D. (1984). *Fluvial Forms and Processes.* Edward Arnold, London, 218 pp.

Knoll, K.M. (1973). Geology of alpine glacier stillstands, east-central Lemhi Range, Idaho, unpublished Ph.D. dissertation, Univ. of Kansas, Lawrence, 503 pp.

Kochel, R.C., and Johnson, R.A. (1984). Geomorphology and sedimentology of humid-temperate alluvial fans in central Virginia, USA, *Canadian Soc. Pet. Geol.* **54,** 109–122.

Kohnke, H., and Bertrand, A.R. (1959). *Soil Conservation*, McGraw-Hill, New York, 208 pp.

Konditerova, E.A., and Ivanov, I.V. (1969). Pattern of variation of the length of freely meandering rivers, *Soviet Hydrology* **4,** 356–364.

Kondrat'yev, N. Ye. (1968). Hydromorphological principles of computations of free meandering, *Soviet Hydrology* **4,** 309–335.

Krigstrom, A. (1962). Geomorphological studies of the sandur plains and their braided rivers in Iceland, *Geograf. Ann.* **44,** 328–346.

Lacey, G. (1930). Stable channels in alluvium, *Inst. Civ. Engs., Proc.*, **299,** 259–384.

Lane, E.W. (1955). Design of stable channels, *Am. Soc. Civ.* Engs **120,** 1–34.

Lane, E.W. (1957). A study of the shape of channels formed by natural streams flowing in erodible material, M.R.D. Sediment Series No. 9, U.S. Army Engineer Div., Missouri River, Corps of Engineers, Omaha, Neb.

Leggett, R.F., Brown, J.E., and Johnston, G.H. (1966). Alluvial fan formation near Aklauik, Northwest Territories, Canada, *Geol. Soc. Amer.* Bull. **77,** 15–30.

Leliavsky, S. (1955). *An Introduction to Fluvial Hydraulics*, Constable, London, 356 pp.

Leopold, L.B., and Langbein, W.B. (1962). The concept of entropy in land-scape evolution, U.S. Geol. Survey Prof. Paper 500-A, 20 pp.

Leopold, L.B., and Maddock, T., Jr. (1953). The hydraulic geometry of stream channels and some physiographic implications, U.S. Geol. Survey Prof. Paper 252, 56 pp.

Leopold, L.B., and Miller, J.P. (1956). Ephemeral streams—hydraulic factors and their relation to the drainage net, U.S. Geol. Survey Prof. Paper 282-A, pp. 39–84.

Leopold, L.B., and Wolman, M.G. (1957). River channel patterns: Braided, meandering and straight, U.S. Geol. Survey Prof. Paper 282-B, pp. 39–85

Leopold, L.B., Wolman, M.G., and Miller, J.P. (1964). *Fluvial Processes in Geomorphology*, Freeman, San Francisco, 522 pp.

Lewin, J., 1976, Initiation of bedforms and meanders in coarse-grained sediments, *Geol. Soc. Am. Bull.* **87,** 281–285.

Lewin, J., and Brindle, B.J. (1977). "Confined meanders," in K.J. Gregory, Ed., *River Channel Changes:* Wiley, Chichester, pp. 221–234.

Lewis, W.V. (1944). Stream trough experiments and terrace formation, *Geol. Mag.* **81,** 241–253.

Lidstone, C.D. (1981). The development and distribution of alluvial placer deposits, unpublished M.S. thesis, Colorado State Univ., Fort Collins, Col., 191 pp.

Linsley, R.K., Kohler, M.A., and Paulhus, J.L.H. (1949). *Applied Hydrology*, McGraw-Hill, New York, 689 pp.

Lisle, T.E. (1982). Effects of aggradation and degradation on riffle-pool morphology in natural gravel channels, northwestern California, *Water Resources Res* **18** 1643–1651.

Lubowe, J.K. (1964). Stream junction angles in the dendritic drainage pattern, *Am. J. Sci.* **262,** 325–339.

Lusby, G.C. (1970). Hydrologic and biotic effects of grazing versus nongrazing near Grand Junction, Colorado, U.S. Geol. Survey Prof. Paper 700-B, pp. B232–B236.

Lustig, L.K. (1965). Clastic sedimentation in Deep Springs Valley, California, U.S. Geol. Survey Prof. Paper 352 F, pp. 131–192.

Lustig, L.K. (1974). "Alluvial fans," *Encyclopedia Britannica*, 15th ed., pp. 611–617.

McDowell, P.F. (1983). Evidence of stream response to Holocene climate change in a small Wisconsin watershed, *Quaternary Res.* **19,** 100–116.

McGowen, J.H. (1971). Gum Hollow fan delta, Muces Bay, Texas: Bur. Econ. Geol., Univ. of Texas, Austin, Tex., Rept. No. 69, 91 pp.

McGowen, J.H. (1974). Alluvial fans and fan deltas: Processes and facies, *Geol. Soc. Am., Abs. Prog.* **5,** 116–117.

McGowen, J.H., and Groat, C.G. (1971). Van Horn Sandstone, West Texas: An alluvial fan model for mineral exploration, Bur. Econ. Geology, Austin, Univ. of Texas, Austin, Tx. Rept. No. 72, 57 pp.

McGowen, J.H., and Scott, A.J. (1974). Fan delta deposition: Processes, facies, and stratigraphic analogues, Abs., Ann. Meeting, Am. Assoc. Pet. Geol., pp. 60–61.

McLane, C.F., III (1978). Channel network growth: An experimental study, unpublished M.S. thesis, Colorado State Univ., Fort Collins, Col., 100 pp.

Macke, D.L. (1977). Stratigraphy and sedimentology of an experimental alluvial fan, unpublished M.S. thesis, Colorado State Univ., Fort Collins, Col., 105 pp.

Maddock, T., Jr. (1970). Indeterminate hydraulics of alluvial channels, *J. Hyd. Div., Am. Soc. Civ. Engs.* **96,** 2309–2323.

Makkaveev, N.E. (1961). *Experimental Geomorphology* (Russian translated into Chinese, 1966), Science Press, Peking.

Martinson, H.A. (1983). Channel changes of Powder River between Moorhead and Broadus, Montana, 1939 to 1978, U.S. Geol. Survey, Water Resources Inv. Rept. 83-4128, 62 pp.

Matthews, E.B. (1917). Submerged "deeps" in the Susquehanna River, *Geol. Soc. Am. Bull.* **28,** 335–346.

Maxwell, J.C. (1960). Quantitative geomorphology of the San Dimas Experimental Forest, California, Project NR 389-042, Dept. Geology, Tech. Report 19, Columbia Univ., New York, 95 pp.

Meade, R.H. (1982). Sources, sinks and storage of river sediment in the Atlantic drainage of the United States, *J. Geol.* **90,** 235–252.

Meade, R.H. (1985). Wavelike movement of bedload sediment East Fork River, Wyoming, *Environ. Geol.* **4,** 219–225.

Meade, R.H., Emmett, W.W., and Myrick, R.M. (1981). Movement and storage of bed material during 1979 in East Fork River, Wyoming, USA, Internat. Assoc. Hydrologic Sci., Pub. 132, pp. 225–235.

Melton, F.A. (1959). Aerial photographs and structural geology, *J. Geol.* **67,** 351–370.

Melton, M.A. (1958). Geometric properties of mature drainage systems and their representation in an E_4 phase space, *J. Geol.* **66,** 35–56.

Melton, MA. (1965). The geomorphic and paleoclimatic significance of alluvial deposits in southern Arizona, *J. Geol.* **73,** 1–38.

Meyer, D.F. (1986). Arroyo development, bedload transport and channel pattern: Experimental and field studies, unpublished Ph.D. dissertation, Colorado State Univ., Fort Collins, Col., 211 pp.

Meyer-Peter, E., and Muller, R. (1948). Formulas for bed load transport, Inter. Assoc. Hydraulic Res. 2nd Congress, Stockholm, pp. 39–65.

Mills, A.A. (1969). Fluidization phenomena and possible implications for the origin of lunar craters, *Nature* **224,** 863–866.

Milne, J.A. (1979). "The morphological relationships of bends in confined stream channels in upland Britain," in A.F. Pitty, ed., *Geographical Approaches to Fluvial Processes*, Geobooks, Norwich, U.K., pp. 241–260.

Minshull, R. (1975). *An Introduction to Models in Geography*, Longman, London, 162 pp.

Minter, W.E.L. (1976). Detrital gold, uranium and pyrite concentrations in the Precambrian Vaal Reef placer, Witwatersrand, South Africa, *Economic Geol.* **36,** 225–247.

Mizutani, T. (1985). Experimental study on the evolution of the longitudinal profile of an artificial slope composed of volcanic ash due to gullying, Tokyo Metropolitan Univ. Geogr. Reports No. 20, pp. 179–187.

Mollard, J.D. (1972). "Airphoto interpretation of fluvial features," in *Fluvial Processes and Sedimentation*, Nat'l Research Council, Canada, Ottawa, pp. 341–380.

Moore, R.C. (1926a). Origin of inclosed meanders on streams of the Colorado Plateau, *J. Geol.* **34,** 29–57.

Moore, R.C. (1926b). Significance of inclosed meanders in the physiographic history of the Colorado Plateau country, *J. Geol.* **34,** 97–130.

Morisawa, M.E. (1964). Development of drainage systems on an unpraised lake floor, *Am. J. Sci.* **262,** 340–354.

Morisawa M.E. (1985). *Rivers: Form and process*, Longman, London, 222 pp.

Mosley, M.P. (1972a). An experimental study of rill erosion, unpublished M.S. thesis, Colorado State Univ., Fort Collins, Col., 118 pp.

Mosley, M.P. (1972b). Gully systems in blanket peat, Bleaklow, North Derbyshire, *East Midland Geogr.* **5,** 235–244.

Mosley, M.P. (1973). Rainsplash and the convexity of badland divides, *Zeit. Geomorph. Suppl.* **18,** 10–25.

Mosley, M.P. (1974). Experimental study of rill erosion, *Am. Soc. Agri. Eng.* (Trans, 1984), 909–913.

Mosley, M.P. (1975a). Channel changes on the River Bollin, Cheshire, 1972–1973, *East Midlands Geogr.* **6,** 185–99.

Mosley, M.P. (1975b). Meander cutoffs on the River Bollin, Cheshire, in July 1973, *Revue de Geomorphologie Dynamique* **24,** 21–31.

Mosley, M.P. (1975c). An experimental study of channel confluences, unpublished Ph.D. dissertation, Colorado State Univ., Fort Collins, Col., 216 pp.

Mosley, M.P. (1976). An experimental study of channel confluences: *J. Geol.* **84,** 535–562.

Mosley, M.P. (1978). Erosion in the southeastern Ruahine Range: Its implications for downstream river control, *New Zealand J. Forestry* **23,** 21–48.

Mosley, M.P. (1981). Semi-determinate hydraulic geometry of river channels, South Island, New Zealand, *Earth Surface Proc. and Landforms* **6,** 127–137.

Mosley, M.P. (1982a). Analysis of the effect of changing discharge on channel morphology and instream uses in a braided river, Ohau River, New Zealand, *Water Resources Res.* **18,** 800–812.

Mosley, M.P. (1982b). Prediction of changes in the physical character of braided rivers in response to changing discharge, *J. Hydrology* (N.2) **22,** 18–67.

Mosley, M.P., and Parker, R.S. (1972). Allometric growth: A useful concept in geomorphology?, *Geol. Soc. Am. Bull.* **83,** 3669–3674.

Mosley, M.P., and Schumm, S.A. (1977). Stream junctions—a probable location for bedrock placers, *Economic Geol.* **72,** 691–697.

Mosley, M.P., and Zimpfer, G.L. (1978). Hardware models in geomorphology, *Progress in Physical Geogr.* **2,** 438–461.

Murata, T. (1966). A theoretical study of the forms of alluvial fans, *Geographical Rept., Tokoyo Metropolitan Univ.* **1,** 33–43.

Nash, J.E. (1957). The form of the instantaneous unit hydrograph, Intern. Assoc. Sci. Hydrology, Pub. 45, vol. 3, pp. 14–121.

Neef, E. (1966). Geomorphologische Moglichteiten für festellung junger Erdkrustenbewengungen, *Geologie* (Berlin) **15,** 97–101.

Noble, C.A., and Palmquist, R.C. (1968). Meander growth in artificially straightened streams, *Proc. Iowa Acad. Sci.* **75,** 234–242.

Nossin, J.J. (1971). Outline of the geomorphology of the Doon Valley, Northern U.P., India, *Zeit. Geomorph. Suppl.* **12,** 18–50.

O'Brien, M.J. (1984). An experimental study of drainage network modification, unpublished M.S. thesis, Colorado State Univ., Fort Collins, Col., 69 pp.

Ollier, C.D. (1981). *Tetonics and Landforms*, Longman, New York, 324 pp.

O'Loughlin, C.L. (1969). Stream bed investigations in a small mountain catchment, New Zealand, *J. Geol. Geophysics* **12,** 684–706.

Ouchi, S. (1983). Response of alluvial rivers to slow active tectonic movement, Ph.D. dissertation, Colorado State Univ., Fort Collins, Col., 205 pp.

Ouchi, S. (1985). Response of alluvial rivers to slow active tectonic movement, *Geol. Soc. Am. Bull.* **96,** 504–515.

Paine, A.D.M. (1985). Ergodic reasoning in geomorphology: A preliminary review, *Prog. Phys. Geogr.* **9,** 1–15.

Palmquist, R.C. (1975). Preferred position model and subsurface symmetry of valleys, *Geol. Soc. Am. Bull.* **86,** 1392–1398.

Parker, G. (1976). On the cause and characteristic scales of meandering and braiding in rivers, *J. Fluid Mech.* **76,** 457–480.

Parker, R.S. (1977). Experimental study of basin evolution and its hydrologic implications, unpublished Ph.D. dissertation, Colorado State Univ., Fort Collins, Col., 331 pp.

Parker, R.S., and Schumm, S.A. (1982). "Experimental study of drainage networks," in R., Bryan and A. Yair, Eds., *Badland Geomorphology and Piping*, Geobooks, Norwich, U.K., pp. 153–168.

Patton, P.C. (1973). Gully erosion in the semi-arid West, M.S. thesis, Colorado State Univ., Fort Collins, Col., 129 pp.

Patton, P.C., and Baker, V.R. (1976). Morphometry and floods in small drainage basins subject to diverse hydrogeomorphic controls, *Water Resources Res.* **12,** 941–952.

Patton, P.C., and Boison, P.J. (1986). Processes and rates of formation of Holocene alluvial terraces in Harris of Wash, Escalante River basin, south-central Utah, *Geol. Soc. Am. Bull.* **97,** 369–378.

Patton, P.C., and Schumm, S.A. (1975). Gully erosion, Northern Colorado: A threshold phenomenon, *Geology* **3,** 88–90.

Partheniades, E. (1965). Erosion and deposition of cohesive soils, *Am. Soc. Civ. Eng. Proc.* **1,** 105–138.

Pearce, A.J., and Watson, A. (1983). Medium term effects of two landsliding episodes on channel storage of sediment, *Earth Surface Proc. and Landforms* **8,** 29–39.

Peterson, D.W., Yeend, W.E., Oliver, H.W., and Mattick, R.E. (1968). Tertiary gold-bearing channel gravel in northern Nevada County, California, U.S. Geol. Survey Circular No. 566, 22 pp.

Petts, G.E. (1979). Complex response of river channel morphology subsequent to reservoir construction, *Prog. Physical Geogr.* **3,** 329–362.

Pickup, G. (1975). Downstream variations in morphology, flow conditions and sediment transport in an eroding channel: *Zeit. Geomorpho.* **19,** 443–459.

Pickup, G. (1977). ''Simulation modelings of river channel erosion'' in K.J. Gregory, Ed., *River Channel Changes,* Chichester, pp. 47 -60.

Pickup, G., and Warner, R.F. (1976). Effects of hydrologic regime on magnitude and frequency of dominant discharge, *J. Hydrology* **29,** 51–76.

Piekotowski, A.J. (1980). Formation of bowl-shaped craters, *Geochemica et Cosmochemica Acta* (Supplement 14) **3,** 2129–2144.

Piest, R.F., Bradford, J.M., and Spomer, R.G. (1975). Mechanisms of erosion and sediment movement from gullies, Proc. Sed. Yield Workshop, USDA Sedimentation Laboratory, Oxford, Miss. ARS-S-40, pp. 162–176.

Potter, P.E. (1978). Significance and origin of big rivers, *J. Geol.,* **86,** 13–33.

Pretorius, D.A. (1974). The nature of the Witwatersrand gold-uranium deposits, Univ. Witwatersrand, Econ. Geol. Res. Unit, Information Circular 87, 22 pp.

Pretorius, D.A. (1975). The depositional environment of the Witwatersrand goldfields: A chronological review of speculations and observations, Minerals Science Engineering, vol 7, pp. 18–47.

Quraishy, M.S. (1944). The origin of curves in rivers, *Current Science* **13,** 36–39.

Quraishy, M.S. (1973). The meandering of alluvial rivers, *Sind Univ. Res. J.* (Sci. Ser.) **7,** 95–152.

Rachocki, A.H. (1981). *Alluvial Fans*, Wiley, New York, 161 pp.

Radulescu, G. (1962). ''Deciphering of tectonic movements in the Quaternary territory of Rumania by means of geomorphical methods,'' *Problems de Geographie*, pp. 9–19.

Raju, R., Kittur, G., Dhandapani, K.R., and Kondap, D.M. (1977). Effect of sediment load on stable sand canal dimensions, *J. Waterways, Port, Coastal, Ocean Div., Am. Soc. Civ. Engs* **103,** 241–249.

Ramberg, H. (1981). *Gravity, Deformation and the Earth's Crust*, 2nd ed., Academic, London, 452 pp.

Raudkivi, A.J. (1967). *Loose Boundary Hydraulics*, Pergamon, Oxford, U.K., 331 pp.

Reilinger, R.E., and Oliver, J.E. (1976). Modern uplift associated with a proposed magma body in the vicinity of Socorro, New Mexico, *Geology* **4,** 583–586.

Reineck, H.-E., and Singh, I. (1975). *Depositional Sedimentary Environments*, Springer-Verlag, New York, 439 pp.

Reynolds, O. (1883). An experimental investigation of the circumstances which determine whether the motion of water shall be direct or sinuous, and of the laws of resistance in parallel channels, *Philosophical Trans. Royal Soc.* **174,** 935–982.

Rich, J.L. (1914). Certain types of stream valleys and their meaning, *J. Geol.* **22,** 469–497.

Rich, J.L. (1935). Origin and evolution of rock fans and pediments, *Geol. Soc. Am. Bull.* **46,** 999–1024.

Richards, K.S. (1978). Channel geometry in the riffle-pool sequence, *Geografiska Annaler* **60,** 23–27.

Richards, K.S. (1979). ''Channel adjustment to sediment pollution by the China Clay industry in Cornwall, England, in D.D. Rhodes and G.P. Williams, Eds., *Adjustments of the Fluvial System*, Kendall-Hunt, Dubuque, pp. 309–331.

Richards, K. (1982). *Rivers: Form and Process in Alluvial Channels*, Methuen, London, 358 pp.

Ritter, D.F. (1982). Complex terrace development in the Nenana valley near Healy, Alaska, *Geol. Soc. Am. Bull.* **93,** 346–356.

Roberts, M., and Klingman, P.C. (1970). The influence of landform and precipitation parameters on flood hydrographs, *J. Hydrology* **11,** 393–411.

Romey, W.D. (1982). Earth in my oatmeal, *EOS* **63,** 162.

Rouse, H. (1950). *Engineering Hydraulics*, Wiley, New York, 772 pp.

Rountree, K.M. (1982). Sediment yields from a laboratory catchment and their relationship to rilling and surface armouring, *Earth Surface Proc. and Landforms* **7,** 153–170.

Rozelle, J. (1978). Lead and molybdenum dispersion in an arid evironment, Sonora, Mexico, unpublished M.S. thesis, Colorado School of Mines, Golden, Col., 196 pp.

Rubey, W.W. (1952). Geology and mineral resources of the Hardin and Brussels quadrangles, Illinois, U.S. Geol. Survey Prof. Paper 218, 179 pp.

Ruhe, R.V. (1950). Graphic analysis of drift topographies, *Am. J. Sci.* **248,** 435–443.

Ruhe, R.V. (1952). Topographic discontinuities of the Des Moines lobe, *Am. J. Sci.* **250,** 46–56.

Ruppel, E.T. (1964). Strike-slip faulting and broken basin ranges in east-central Idaho and adjacent Montana, U.S. Geol. Survey Prof. Paper 501-C, pp. 14–18.

Ruppel, E.T. (1967). Late Cenozoic drainage reversal, east-central Idaho and its relation to possible undiscovered placer deposits, *Economic Geol.* **62,** 648–663.

Russ, D.P. (1982). Style and significance of surface deformation in the vicinity of New Madrid, Missouri, U.S. Geol. Survey Prof. Paper 1236, pp. 95–114.

Ryder, J.M. (1971a). The stratigraphy and morphology of paraglacial alluvial fans in south-central British Columbia, *Can. J. Earth Sci.* **8,** 279–298.

Ryder, J.M. (1971b). Some aspects of the morphometry of paraglacial alluvial fans in south-central British Columbia, *Can. J. Earth Sci.* **8,** 1252–1264.

Saito, K. (1981). Classification of alluvial fans in Tohoku District based on cluster analysis, *Geogr. Rev. Japan* **53,** 721–729 (in Japanese).

Saito, K. (1982). Classification of alluvial fans in Japan by topographical and geological data of drainage basins, *Geogr. Rev. Japan* **55,** 334–349 (in Japanese).

Sauders, I. and Young, A. (1983). Rates of surface processes on slopes, slope retreat and denudation, *Earth Surface Proc. and Landforms* **8,** 473–501.

Scheidegger, A.E. (1970). *Theoretical Geomorphology*, Springer-Verlag, New York, 435 pp.

Schumm, S.A. (1956). Evolution of drainage systems and slopes in badlands at Perth Amboy, New Jersey, *Geol. Soc. Am. Bull.* **67,** 597–646.

Schumm, S.A. (1960). The shape of alluvial channels in relation to sediment type, U.S. Geol. Survey Prof. Paper 352-B, pp. 17–30.

Schumm, S.A. (1963a). Sinuosity of alluvial rivers on the Great Plains, *Geol. Soc. Am. Bull.* **74,** 1089–1100.

Schumm, S.A. (1963b). Disparity between modern rates of denudation and orogeny, U.S. Geol. Survey Prof. Paper 454-H, 13p.

Schumm, S.A. (1968). River adjustment to altered hydrologic regimen, Murrumbidgee River and paleochannels, Australia, U.S. Geol. Survey Prof. Paper 598, 65 pp.

Schumm, S.A. (1969). River metamorphosis, *J. Hyd. Div., Proc. American Soc. Civil Eng.* **1,** 255–273.

Schumm, S.A. (1970). Experimental studies on the formation of lunar surface features by fluidization, *Geol. Soc. Am. Bull.* **81,** 2539–2552.

Schumm, S.A. (1973). "Geomorphic thresholds and the complex response of drainage systems," in M. Morisawa, Ed. *Fluvial Geomorphology*, Publications in Geomorphology, State Univ. of N.Y., Binghamton, pp. 299–310.

Schumm, S.A. (1974). Structural origin of large Martian channels, *Icarus* **22,** 371–389.

Schumm, S.A. (1976). "Episodic erosion: A modification of the geomorphic cycle," in W.N. Melhorn and R.C. Flemal, Eds., *Theories of Landform Development*, Publications in Geomorphology, State Univ. of N.Y., Binghamton, pp. 69–85. Reprinted 1982 by Allen and Unwin, London.

Schumm, S.A. (1977). *The Fluvial System*, Wiley, New York, 338 pp.

Schumm, S.A. (1981). Evolution and response of the fluvial system: Sedimentologic implications, Soc. Economic Paleontologists and Mineralogists Spec. Pub. 31, pp. 19–29.

Schumm, S.A. (1984). "River morphology and behavior: Problems of extrapolation," in C.M. Elliott, Ed., *River Meandering*, Am. Soc. Civ. Engs., New York, pp. 16–29.

Schumm, S.A. (1985). Explanation and extrapolation in geomorphology: Seven reasons for geologic uncertainty, *Trans. Japanese Geomorph. Union* **6,** 1–18.

Schumm, S.A. (1986). Alluvial river response to active tectonics: in Active Tectonics, Studies in Geophysics, National Academy Press, Washington, D.C., pp. 80–94.

Schumm, S.A., and Hadley, R.F. (1957). Arroyos and the semiarid cycle of erosion: *Am. J. Sci.,* **255** 161–174.

Schumm, S.A., Harvey, M.D., and Watson, C.C. (1984). *Incised Channels: Initiation, Evolution, Dynamics, and Control,* Water Res. Publ., Littleton, Col., 200 pp.

Schumm, S.A., and Kahn, H.R. (1971). Experimental study of channel patterns, *Nature* **233,** 407–409.

Schumm, S.A., and Khan, H.R. (1972). Experimental study of channel patterns, *Geol. Soc. Am. Bull.* **83,** 1755–1700.

Schumm, S.A., and Lichty, R.W. (1965). Time, space and causality in geomorphology, *Am. J. Sci.* **263,** 110–119.

Schumm, S.A., and Parker, R.S. (1973). Implications of complex response of drainage systems for Quaternary alluvial stratigraphy, *Nature* **243,** 99–100.

Schumm, S.A., Khan, H.R., Winkley, B.R., and Robbins, L.G. (1972). Variability of river patterns, *Nature* **237,** 75–76.

Schumm, S.A., Watson, C.C., and Burnett, A.W. (1982). Investigation of neotectonic activity within the Lower Mississippi Valley Division, Phase I: U.S. Army Corps of Engineers, Lower Mississippi Valley Division (Vicksburg), Potamology Program, Report 2, 158 pp.

Schwab, G.O., Frevert, R.K., Edminster, T.W., and Barnes, K.K. (1966). *Soil and Water Conservation Engineering*, Wiley, New York, 683 pp.

Seed, H.B., and wilson, S.D. (1967). The Turnagain Heights landslide, Anchorage, Alaska, *J. Soil Mechanics and Foundations Div., Am. Soc. Civ. Eng.* **93,** 325–353.

Seginer, I. (1966). Gully development and sediment yield, *J. Hydrology* **4,** 236–253.

Seginer, I. (1969). Random walk and random roughness models of drainage networks, *Water Resources Res.* **5,** 591–607.

Sestini, G. (1973). "Sedimentology of a paleoplacer: The gold bearing Tarkwaian of Ghana," in *Ores in Sediments*, Int. Union Geol. Sci. (Publ.), Ser. A., pp. 275–305.

Shahjahan, M. (1970). Factors controlling the geometry of fluvial meanders, *Internat. Assoc. Sci. Hydrology Bull* **15,** 13–23.

Sharp, A.L., Bond, J.J., Neuberger, J.W.E., Kuhlman, A.R., and Lewis, J.K. (1964). Runoff as affected by intensity of grazing on rangeland, *J. Soil and Water Conservation* **19,** 103–105.

Shepard, F.P. (1955). Delta-front valleys bordering the Mississippi distributaries, *Geol. Soc. Am. Bull.* **66,** 1489–1498.

Shepherd, R.G. (1972a). Incised river meanders: Evolution in simulated bedrock, *Science* **178,** 409–411.

Shepherd, R.G. (1972b). A model study of river incision, unpublished M.S. thesis, Colorado State Univ., Fort Collins, Col., 135 pp.

Shepherd, R.G., and Schumm, S.A. (1974). Experimental study of river incision, *Geol. Soc. Am. Bull.* **85,** 257–268.

Sherman, L.K. (1932). The relation of hydrographs of runoff to size and character of drainage basins, *Trans. Am. Geophys. Un.* **13,** 332–339.

Shreve, R.L. (1966). Statistical law of stream numbers, *J. Geol.* **74,** 17–37.

Shreve, R.L. (1969). Stream lengths and basin areas in topologically random channel networks, *J. Geol.* **77,** 397–414.

Shreve, R.L. (1975). The probabilistic-topologic approach to drainage-basin geomorphology, *Geology* **3,** 527–529.

Simons, D.B., and Albertson, M.I. (1960). Uniform water conveyance channels in alluvial materials: *J. Hydraulics Div., Am. Soc. Civ. Eng.* **86,** 33–71.

Simons, D.B., and Richardson, E.V. (1966). Resistance to flow in alluvial channels, U.S. Geol. Survey Prof. Paper 422-J, 61 pp.

Slaymaker, O., Dunne, T., and Rapp, A. (1980). Geomorphic experiments on hillslopes: Preface, *Zeit. Geomorph. Suppl.* **35, v–vii.**

Smart, J.S. (1968). Statistical properties of stream lengths, *Water Resources Res.* **4,** 1001–1014.

Smart, J.S. (1969). Topological properties of channel networks, *Geol. Soc. Am. Bull.* **80,** 1757–1774.

Smart, J.S. (1972). Channel networks, *Advances in Hydroscience* **8,** 305–346.

Smart, J.S., and Moruzzi, V.L. (1971a). Computer simulation of Clinch Mountain drainage networks, *J. Geol.* **79,** 572–584.

Smart, J.S., and Moruzzi, V.L. (1971b). Random-walk model of stream network development, *J. Research Development, IBM* **15,** 197–203.

Smart, J.S., Surkan, A.J., and Considine, J.P. (1967). Digital simulation of channel networks, Internat. Assoc. Sci. Hydrology Pub. 75, pp. 87–98.

Smith, D.D., and Wischmeier, W.H. (1962). Rainfall erosion, *Advances in Agronomy* **14,** 109–148.

Smith, N.D., and Smith, D.G. (1984). Williams River: An outstanding example of channel widening and braiding caused by bed-load addition, *Geology* **12,** 78–82.

Sneh, A. (1979). Late Pleistocene fan-deltas along the Dead Sea Rift, *J. Sed. Pet.* **45,** 541–552.

Soni, J.P. (1981). Laboratory study of aggradation in alluvial channels, *J. Hydrology* **49,** 87–106.

Sparling, D.R. (1967). Anomalous drainage pattern and crustal tilting in Ottowa County, Ohio, *Ohio J. Sci.* **67,** 378–381.

Stebbings, J. (1963). The shapes of self-formed model alluvial channels, *Proc. Inst. Civ. Eng.* **25,** 485–510.

Stevens, G.R. (1974). *Rugged Landscape, the Geology of Central New Zealand,* A.H. and A.W. Reed, Wellington, 286 pp.

Stevens, P.S. (1974). *Patterns in Nature*: Little, Brown, Boston, 240 pp.

Strahler, A.N. (1946). Elongate intrenched meanders of Conodoguinet Creek, Pennsylvania, *Am. J. Sci.* **40,** 359–362.

Strahler, A.N. (1958). Dimensional analysis applied to fluvially eroded landforms, *Geol. Soc. Am. Bull.* **69,** 279–300.

Sugitani, T. (1984). Balance and terrace formation in a tilting laboratory flume, *Geogr. Rev. Japan* (in press).

Sunamura, T. (1982). A wave tank experiment on the erosional mechanism at a cliff base, *Earth Surface Proc. and Landforms* **7,** 333–343.

Tanner, W.F. (1960). Helicoidal flow, a possible cause of meandering, *J. Geophys. Res.* **65,** 993–995.

Tarr, R.S., and von Englen, O.D. (1908). Representation of land forms in the physiography laboratory, *J. Geogr.* **7,** 73–85.

Tarr, W.A. (1924). Intrenched and incised meanders of some streams on the northern slope of the Ozark Plateau, *J. Geol.* **32,** 583–600.

Tator, B.A. (1958). The aerial photograph and applied geomorphology, *Photogram. Eng.* **14,** 549–561.

Thomson, J. (1879). On the flow of water round river bends, *Proc. Inst. Mechanical Engs.*, August 6.

Thorne, C.R., Murphy, J.B., and Little, W.C. (1981). Bank stability and bank material properties in the bluffline streams of northwest Mississippi, Appendix D, *Stream Channel Stability*, USDA, Sedimentation Laboratory, Oxford, Miss. 158 pp.

Thornton, E.G., and Romer, L.S. (1975). "Comparison of hydraulic and numerical tidal models," in *Symposium on Modelling Techniques,* Am. Soc. Civil Engs., New York, pp. 1311–1328.

Tiffany, J.B. (1935). Model experiment to determine the direction energy of a river, U.S. Waterways Expt. St. Tech. Memo. 61–1, 17 pp.

Tiffany, J.B., and Nelson, G.A. (1939). Studies of meandering of model streams, *Trans. Am. Geophysical Union* (Part IV), 644–649.

Till, R. (1974). *Statistical Methods for the Earth Scientist:* An *Introduction*, Wiley, New York, 154 pp.

Tinkler, K.J. (1971). Active valley meanders in south central Texas and their wider implications, *Geol. Soc. Am. Bull.* **81,** 1873–1899.

Tinkler, K.J. (1972). The superimposition hypothesis for incised meanders—a general rejection and specific test, *Area* **4,** 86–91.

Trimble, S.W. (1975). Denudation studies: Can we assume stream steady state? *Science* **188,** 1207–1208.

Twidale, C.R. (1966). Late Cainozoic activity of the Selwyn Upwarp, northwest Queensland, *J. Geol. Soc. So. Australia* **13,** 491–494.

Twidale, C.R. (1971). *Structural Landforms*: MIT Press, Cambridge, Mass., 247 pp.

U.S. Army Corps of Engineers (1970). Hydraulic design of flood control channels, Engineer Design Manual EM-1110-2-1601, Washington D.C., 195 pp.

Vanoni, V.A. (1946). Transport of suspended sediment by water, *Trans. Am. Soc. Civ. Eng.* **111,** 67–102.

Waananen, A.O. (1969). ''Urban effects on water yield,'' in W.L., Moore and C.W. Morgan, Ed. *Effects of Watershed Changes on Streamflow,* Univ. Texas Press, Austin, pp. 169–182.

Wallace, R.E. (1967). Notes on stream channels offset by the San Andreas Fault, southern coast ranges California, *Stanford Univ. Pub. (Geol. Sci.)* **11,** 6–20.

Walters, W.H., Jr. (1975). Regime changes of the lower Mississippi River, unpublished M.S. thesis, Colorado State Univ., Fort Collins, Col., 129 pp.

Ward, R.C. (1971). Small watershed experiments, Univ. of Hull, Occasional Papers Geogr., No. 18, 254 pp.

Wasson, R.J. (1974). Intersection point deposition on alluvial fans: An Australian example, *Geogr. Annaler* **56,** (Ser. A), 83–92.

Wasson, R.J. (1977). Last glacial alluvial fan sedimentation in the Lower Derwent Valley, Tasmania, *Sedimentology* **24,** 781–799.

Watson, C.C., Schumm, S.A., and Harvey, M.D. (1983). ''Neotectonic effects on river patterns,'' ASCE, in C. Elliott, Ed., *Proc. Conference on River Meandering*, New Orleans pp. 55–66.

Watson, R.A. (1969). Explanation and prediction in geology, *J. Geology* **77,** 488–494.

Weaver, W.E. (1984). Experimental study of alluvial fans, unpublished Ph.D. dissertation, Colorado State Univ., Fort Collins, Col., 423 pp.

Wegener, A.L. (1921). *Die Entstehung der Mondkrater*, Druck and Verlag von Friedrich Vieweg and John, Braunschweig, 49 pp. (translated in 1975, *The Moon* **14,** 211–236).

Wells, S.G., Dohrenwend, J.C., McFadden, L.D., Turrin, B.D., Mahrer, K.D. (1985). Late Cenozoic landscape evolution on lava flow surfaces of the Cima volcanic field, Mojave Desert, California, *Geol. Soc. Am. Bull.* **96,** 1518–1529.

Werrity, A. (1972). ''The topology of stream networks,'' in R.J. Chorley, Ed., *Spatial Analysis in Geomorphology*, Methuen, London, pp. 167–196.

Wertz, J.B. (1949). Logarithmic pattern in river placer deposits, *Economic Geol.* **44,** 193–209.

Wescott, W.A. (1979). The Yallahs delta, southeastern Jamaica: A model for island-arc collision coast fan delta deposition, Ph.D. dissertation, Colorado State Univ., Fort Collins, Col., 84 pp.

Wescott, W.A., and Ethridge, F. C. (1980). Fan-delta sedimentology and tectonic setting—Yallahs fan delta, southeast Jamaica, *Bull. Am. Assoc. Pet. Geol.* **64,** 374–399.

Wildman, N.A. (1981). Episodic removal of hydraulic-mining debris, Yuba and Bear River basins, California, unpublished M.S. thesis, Colorado State Univ., Fort Collins, Col., 107 pp.

Williams, G.P. (1978). Bank-full discharge of rivers, *Water Resources Res.* **14,** 6, 1141–1154.

Williams, G.P., and Wolman, M.G. (1984). Downstream effects of dams on alluvial rivers, U.S. Geol. Survey Prof. Paper 1286, 83 pp.

Williams, V.S. (1977). Neotectonic implications of the alluvial record in the Sapta Kosi drainage basin, Nepalese Himalayas, unpublished Ph.D. dissertation, Dept. Geol. Sciences, Univ. of Washington, Seattle, 80 pp.

Winder, C.G. (1965). Alluvial cone construction by alpine mudflow in a humid temperate region, *Can. J. Earth Sci.* **2,** 270–277.

Winslow, A. (1893). The Osage River and its meanders, *Science* **22,** 31–32.

Wischmeier, W.H., and Smith, D.D. (1965). Predicting rainfall erosion losses from cropland, U.S. Dept. Agri. Agricultural Handbook, 48 pp.

Wohletz, K.H., and McQueen, R.G. (1984). "Experimental studies of hydromagmatic volcanism," in *Explosive Volcanism: Inception, Evolution and Hazards*, Studies in Geophysics, National Academy Press, Washington, D.C., pp. 158–169.

Woldenberg, M.J. (1985). *Models in Geomorphology*: Allen and Unwin, London, 434 pp.

Wolman, M.G. (1955). The natural channel of Brandywine Creek, Pennsylvania, U.S. Geol. Survey Prof. Paper 271, 56 pp.

Wolman, M.G., and Brush, L.M., Jr. (1961). Factors controlling the size and shape of stream channels in coarse noncohesive sands, U.S. Geol. Survey Prof. Paper 282-G, pp. 183–210.

Wolman, M.G., and Gerson, R. (1978). Relative scales of time and effectiveness of climate in watershed geomorphology, *Earth Surface Proc.* **3,** 189–208.

Wolman, M.G., and Miller, J.P. (1960). Magnitude and frequency of forces in geomorphic processes. *J. Geol.* **68,** 54–74.

Womack, W.R. (1975). Erosional history of Douglas Creek, northwestern Colorado, unpublished M.S. thesis, Colorado State Univ., Fort Collins, Col., 76 pp.

Womack, W.R., and Schumm, S.A. (1977). Terraces of Douglas Creek, Northwestern Colorado: An example of episodic erosion, *Geology* **5,** 72–76.

Woodyer, K.D. (1968). Bankfull frequency in rivers, *J. Hydrology* **6,** 114–142.

Woolhiser, D.A. (1973). Hydrologic and watershed modeling—state of the art, *Trans. Am. Soc. Agr. Engs.* **16,** 553–559.

Woolsey, T.S., McCallum, M.E., and Schumm, S.A. (1975). Modeling of diatreme emplacement by fluidization, *Physics and Chemistry of the Earth* **9,** 29–42.

Würm, A. (1935). Morphologische Analyse und Experiment Schichtstufenlandschaft, *Zeit. Geomorph.* **9,** 1–24.

Würm, A. (1936). Morphologische Analyse und Experiment Hangentwicklung, Ernebnung, Piedmonttreppen, *Zeit. Geomorph.* **9,** 58–87.

Yang, C.T., Song, C.C.S., and Woldenberg, M.J. (1981). Hydraulic geometry and minimum rate of energy expenditure, *Water Resources Res.* **17,** 1014–1018.

Yen, C.-L (1970). Bed topography effect on flow in a meander, *J. Hydraulics Div., Am. Soc. Civ. Eng.* **1,** 57–73.

Young, R.W., and Nanson, G.C. (1982). Terrace formation in the Illawarra region of New South Wales, *Australian Geogr.* **15,** 212–219.

Yoxall, W.H. (1969). The relationship between falling baselevel and lateral erosion in experimental streams, *Geol. Soc. Am. Bull.* **80,** 1379–1384.

Ziemer, R.R. (1979). Storm flow response to road building and partial cutting in small streams of northern California, *Water Resources Res.* **15,** 907–917.

Zimpfer, G.L. (1975). Development of laboratory river channels, unpublished M.S. thesis, Colorado State Univ., Fort Collins, Col., 111 pp.

Zimpfer, G.L. (1982). Hydrology and geomorphology of an experimental drainage basin, unpublished Ph.D. dissertation, Colorado State Univ., Fort Collins, Col., 185 pp.

Zuchiewicz, W. (1979). A possibility of application of the theoretical longitudinal river's profile analysis to investigations of young tectonic movement, *Annal. Soc. Geol. Pologne* **49,** 327–342.

Author Index

(Only senior authors are cited.)

Subject Index